Commander Martin Marks OBE served in the Royal Navy as a mechanical engineer officer for 36 years including during the Falklands conflict as MEO of HMS *Fife*. He was the Admin Director of the Royal Naval Hospital, Haslar and was awarded an OBE for 'Services to Military Medicine'. He was also involved in the creation of The Diving Museum, Gosport. He co-authored *Barracks, Forts and Ramparts: Regeneration Challenges for Portsmouth Harbour's Defence Heritage* (Tricorn Books, 2020) with Celia Clark.

Stephen McLaughlin retired in 2017 after working for 35 years as a librarian at the San Francisco Public Library. In addition to contributing regularly to *Warship*, he is the author of *Russian and Soviet Battleships* (US Naval Institute Press, 2003, reissued 2021) and is co-editor of an annotated version of the controversial *Naval Staff Appreciation of Jutland* (Seaforth Publishing, 2016). He is currently working on a two-part study of Russian cruisers.

Kathrin Milanovich has contributed a series of articles to *Warship* on the warships of the Imperial Japanese Navy during the period 1890–1945. Recent contributions have included the cruisers of the *Takasago* class and the escort destroyers of the *Matsu* class.

Dirk Nottelmann is a marine engineer by profession, and is currently working for the German shipping administration. He has contributed to *Warship*, *Warship International* and various German magazines, as well as being the author of *Die Brandenburg-Klasse* (Mittler, 2002) and co-author of *Halbmond und Kaiseradler* (Mittler, 1999). His book *The Kaiser's Cruisers, 1871–1918* (with Aidan Dodson) was published by Seaforth in late 2021.

Ian Sturton is a regular contributor of articles and illustrations to naval publications, including *Warship*, *Warship International* and *Jane's Fighting Ships*. He edited *Conway's Battleships: The Definitive Visual Reference to the World's All-Big-Gun Ships* (Conway Maritime Press 1987, updated 2008).

Conrad Waters is the author of numerous articles on modern naval matte
Naval Review (Seaforth Publishing). He also edited Seaforth's *Navies in th*
shortlisted for the 2017 Mountbatten Award. His history of the Royal Navy
'Town' class cruisers was published in 2019 and a sequel on the *Fiji* class
2024. He is currently preparing a history of the *Leander* and *Arethusa* clas
with Peter Cannon.

Tom Wismann is a Marine Engineer by education and former Lieutenant-Commander in the Danish Maritime Home Guard. He has written and published a series of books on ships of the Royal Danish Navy and modern fortifications. He has for many years been a member of the board of The Danish Naval Historical Society and editor of the society's journal.

WARSHIP 2025

WARSHIP 2025

Editor: **John Jordan**

Assistant Editor: **Stephen Dent**

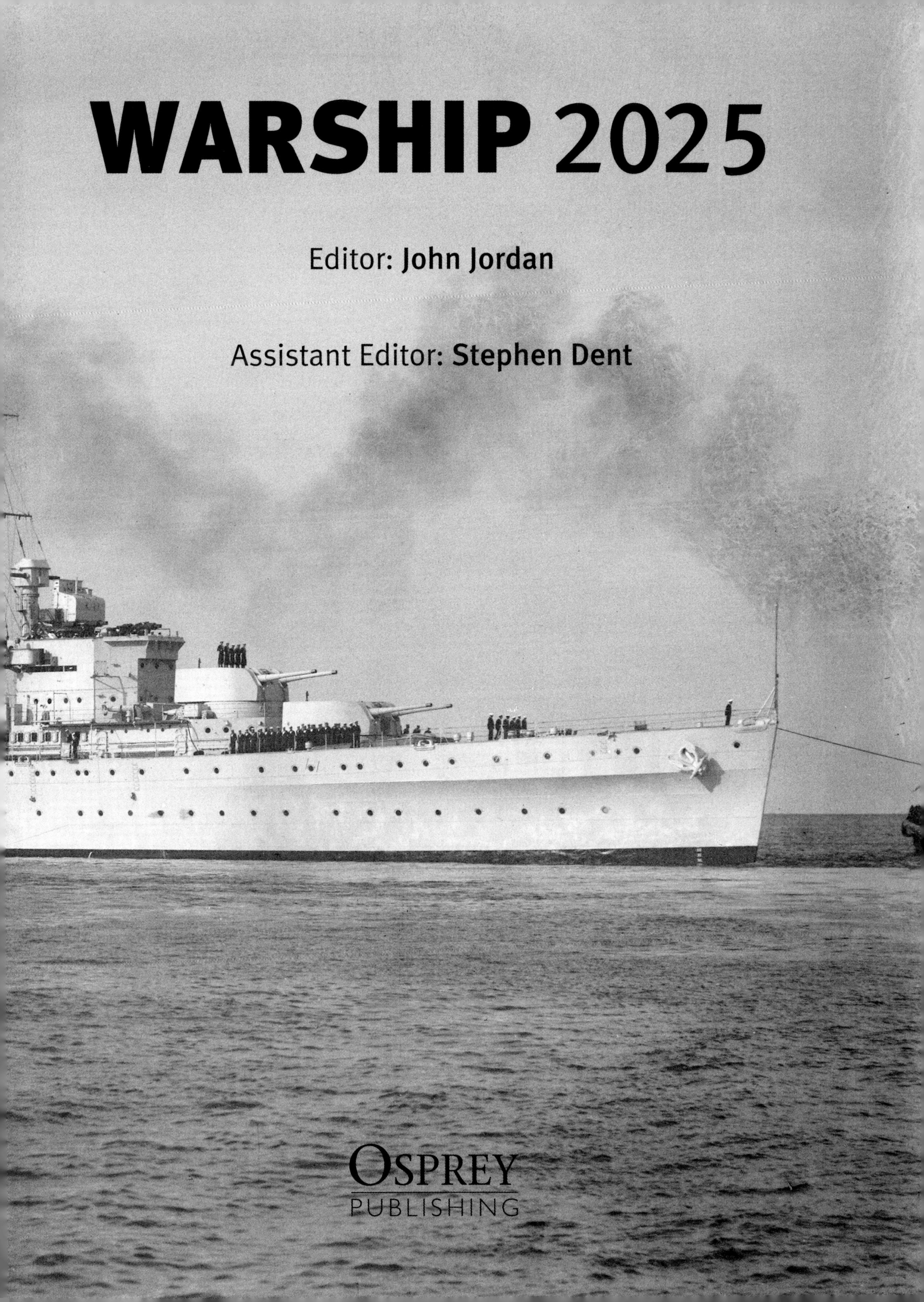

OSPREY
PUBLISHING

Title pages: HMAS *Hobart* arrives at Melbourne on 4 April 1939. The 6in main armament of these ships is the subject of a feature article by Peter Cannon in this year's annual. (Allan C Green collection, State Library of Victoria)

OSPREY PUBLISHING
Bloomsbury Publishing Plc
Kemp House, Chawley Park, Cumnor Hill, Oxford OX2 9PH, UK
29 Earlsfort Terrace, Dublin 2, Ireland
1385 Broadway, 5th Floor, New York, NY 10018, USA
E-mail: info@ospreypublishing.com
www.ospreypublishing.com

OSPREY is a trademark of Osprey Publishing Ltd

First published in Great Britain in 2025

A catalogue record for this book is available from the British Library

ISBN: HB 9781472868541; eBook 9781472868534; ePDF 9781472868565; XML 9781472868558

25 26 27 28 29 10 9 8 7 6 5 4 3 2 1

Cover design by Stewart Larking
Page layout by Stephen Dent
Printed and bound in UAE by Oriental Press

MIX
Paper | Supporting responsible forestry
FSC® C004800

Osprey Publishing supports the Woodland Trust, the UK's leading woodland conservation charity.

To find out more about our authors and books visit www.ospreypublishing.com. Here you will find extracts, author interviews, details of forthcoming events and the option to sign up for our newsletter.

CONTENTS

EDITORIAL

In the feature section of this year's *Warship* we are publishing the first instalments of no fewer than three new multi-part series. Hans Lengerer leads off with Part I of a two-part series on the IJN battleship reconstructions of the 1930s. Prevented by the Washington 1922 and London 1930 Treaties from laying down new ships, and restricted to three-fifths of the US Navy's battleship tonnage, the Imperial Japanese Navy resolved to completely rebuild its ageing dreadnoughts to modern standards in a bid to create one-on-one superiority over their American counterparts. The reconstructions were radical, labour-intensive and costly, and focused on improvements in gunnery, protection and speed; batteries of modern high-angle guns and close-range AA guns were incorporated into the rebuild, together with catapults and handling cranes for aircraft capable of reconnaissance and the spotting of gunnery. In this first part, Hans outlines the general principles that governed these reconstructions and looks in detail at the modifications to the battlecruisers of the *Kongō* class. The article makes extensive use of Japanese primary and secondary sources.

The second new series is from Dirk Nottelmann, who has agreed to follow up his ground-breaking series of articles on the development of the early German light cruisers with a similar study of the destroyers. Part I covers the experimental craft up to the mid-1880s. Three (possibly four) further instalments will take us up to the end of the Great War. The series will feature a large quantity of previously unpublished material.

In parallel with this is the Editor's new series on the French pre-First World War destroyers. Part I focuses on the300-tonne destroyer, 55 of which were completed for the French Navy between 1899 and 1910 with a further 20 built for export. Although the impetus for the design came from the entry into service of the first British 'torpedo boat destroyers' (TBD) during the mid-1890s, the *300-tonnes* was purely French in conception and was the brainchild of Jacques-Augustin Normand, the preeminent designer of torpedo boats for the French Navy. Parts II and III will cover the boats of the 450-tonne and 800-tonne types respectively. The series draws on the seminal French-language study of these ships by the late Henri Le Masson and is illustrated by the author's drawings, which are based on the official plans.

Peter Cannon has returned to the annual after an absence of five years with a detailed study of the Royal Navy's 6in Mark XXI twin turret. The Mark XXI was derived from the Mark XVIII turret mounted in the battleships *Nelson* and *Rodney*. Lessons were learned, with a particular emphasis on the simplification of the ammunition supply and loading arrangements to ensure a high practical rate of fire. The Mark XXI proved particularly successful; it was fitted in the five 'light' cruisers of the *Leander* class, the three Modified *Leander*s and the four *Arethusa*s, and became the benchmark for the modern British medium-calibre weapon systems that fought the Second World War.

This year's annual features two articles on submarines. Przemysław Budzbon focuses on the Soviet 'S'-class submarine, designed by the Germans and built with German assistance, and generally regarded as the best of the early Soviet submarine types. Przemek tells the story of the fraught negotiations between the Soviets and the Germans that laid the basis for the class, and details its further development prior to and during the Second World War. Less well known are the Danish coastal submarines of the 'H' class, which are the subject of a feature by Tom Wismann. Completed shortly before the Second World War, they saw no action due to Denmark's decision not to resist Nazi Germany's takeover in April 1940 and were scuttled in 1943, but they constituted an interesting and well-developed design that met the Danish Navy's special requirements. Tom's article is illustrated by a striking collection of photographs and plans from the Marinens Bibliotek archive.

There is an enduring fascination among naval enthusiasts for 'the ones that got away': the projects for major, often innovative types of warship that never fully materialised despite the drawing up of detailed plans. In the summer of 1911, the Imperial Russian Navy invited six Russian and seventeen of the most prominent foreign shipbuilders to participate in a competition for the design of a powerful battlecruiser. Stephen McLaughlin describes the process that eventually led to the four ships of the *Izmail* class, which were laid down but never completed. This article has been a collaborative project with the Editor, who has redrawn many of the plans submitted to the Imperial Russian Navy, which included vessels powered by multiple diesel engines.

In a similar vein, Michele Cosentino looks at the 'paper' projects for 'battle monitors' and battlecruisers proposed, often by former naval architects and private shipbuilders, for the *Regia Marina* of the interwar period. The 'battle monitor', which was intended to engage its opponents with its hull partially submerged, was a striking, if wildly impractical proposal, but some of the later 'armoured cruisers', armed with 10in or multiple 8in guns and intended to hunt down and sink Treaty cruisers, were impressive, well-developed designs.

Kathrin Milanovich continues her well-regarded series on IJN warships with a study of the protected cruisers of the *Chikuma* class laid down in 1910. Developed from the earlier *Tone*, *Chikuma* and her two sisters had a uniform main battery of eight 6in guns and were among the first Japanese ships to be powered by turbines: a competitive evaluation pitted *Chikuma* and *Hirato*, which were fitted with two-shaft American Curtis impulse turbines built under licence by Kawasaki, against

HMCS *Border Cities*, one of the Canadian-built 'emergency escorts' that feature in a ground-breaking study by Nigel Whybrow due to be published in next year's annual. The ship is shown in August 1944 shortly after her completion, either on trials or on transit from the builders at Port Arthur on Lake Superior. *Border Cities* and her sisters constituted a unique RCN variant on the *Algerine*-class. (Library and Archives Canada/Department of National Defence)

the third ship, *Yahagi*, which had Parsons reaction turbines built by Mitsubishi on four shafts.

Our final feature this year is Part II of the article by Aidan Dodson and Dirk Nottelmann on the German Flak ships of the Second World War, which was held over from last year due to lack of space. This part covers the conversion of the former Dutch coast defence ships *Hertog Hendrik* and *Jacob van Heemskerck*, and the old protected cruiser *Gelderland*, alongside the proposed conversion of the former Danish coast defence ship *Peder Skram*. All were to be rebuilt and rearmed with a modern anti-aircraft battery of 6–8 10.5cm HA guns and four 20cm *Flakvierling*, with 6-metre rangefinders for fire control and the FuMG 39 TD Würzburg air search radar for long-range detection.

This year's Warship Notes section leads with a piece by Cdr Martin Marks (RN Rtd) on the conversion following the Great War of the 'semi-dreadnought' battleship *Agamemnon* to serve as a radio-controlled target ship. The note was originally published as part of a newsletter for the preserved Steam Pinnace 199, for which Martin has been the project leader, and has been produced with the cooperation of the National Museum of the Royal Navy. Ian Sturton has followed up last year's Warship Note on the French 19th century exploration of the Upper Niger with a detailed account of the expedition of Jean-Baptiste Marchand of 1896–98 using the gunboat *Faidherbe*. The Editor has supplied a third Note with new information about the protection system of the French fast battleship *Dunkerque* based on official plans of the ship that have only recently been made available. And regular contributor Conrad Waters, whose authoritative book on the RN cruisers of the *Fiji* class was published in the summer of 2024, closes the section with some useful advice for prospective authors of naval history books.

Warship Gallery, which this year has been compiled by Assistant Editor Stephen Dent, features photographs taken by Royal Marine Herbert Sparrow during his time serving in China during 1927 and, subsequently, the eastern Mediterranean. These came to the attention of the Assistant Editor through an acquaintance, suggesting that there may be other untapped sources of such images that might be of interest to our readers.

The feature section of next year's annual is already in preparation. We are aiming to publish the second parts of the series begun this year, on the IJN's battleship reconstructions (by Hans Lengerer), and on the early German and French destroyers (Dirk Nottelmann and the Editor respectively). There will also be two major features from contributors new to *Warship*: Robert Brown will provide an analysis of RN carrier battle damage during the Second World War, while Nigel Whybrow will look at the emergency escort programmes developed by Britain, Canada and the USA in response to the successes of the German U-boats in 1941–42. Przemysław Budzbon embarks on a two-part study of the Imperial Russian destroyer *Novik*, which attained a world record speed of 37 knots in 1913 and inspired a series of similar ships laid down during the Great War. There will also be features on the IJN seaplane carriers of the 1930s, with a special emphasis on *Kamoi* and *Nisshin*, by Kathrin Milanovich, on the abortive Italian aircraft carrier *Sparviero* (Enrico Cernuschi), on RN cadet training ships (Aidan Dodson), and on the early and little-known actions of the Russo-Japanese War (Stephen McLaughlin).

John Jordan
1 April 2025

THE RECONSTRUCTION OF THE BATTLE FLEET OF THE IMPERIAL JAPANESE NAVY BETWEEN THE WARS: PART I

The ten Japanese dreadnoughts that survived the Washington Treaty were extensively rebuilt during the interwar period. In this, the first of a pair of articles, **Hans Lengerer** looks at the rationale for the reconstructions and describes the modernisation of the battlecruisers of the *Kongō* class.

The interwar period was shaped by the two Naval Arms Limitation Treaties, which prohibited the building of new capital ships and placed strict limitations on the reconstruction of existing units. These constraints flew in the face of established practice, which dictated that major warships be constantly updated to take into account progress in technology, particularly with regard to weaponry and propulsion, in order to adapt their offensive and defensive capabilities to the most recent advances and thereby prolong active service life.

The Washington Treaty determined capital ship tonnage allocations for the five major naval powers according to the following ratio: Britain and the USA 5; Japan 3; France and Italy 1.75. Japan was permitted to retain ten capital ships (four battlecruisers of the *Kongō* class and the six battleships of the *Fusō*, *Ise* and *Nagato* classes), one of which (*Mutsu*) had been rushed to completion, for a total of 315,000 tons. Fourteen incomplete or projected capital ships of the Eight-Eight Fleet had to be abandoned. The older 'semi-' and 'pre-dreadnought' ships completed before the *Kongō* class were all to be scrapped or relegated to training duties. This gave the IJN ten capital ships versus eighteen for the US Navy – a tonnage ratio of approximately 60 per cent – and compelled Japan to subject its strategy for a possible war against the USA to a fundamental review.

Under Washington Treaty rules, the service life of a capital ship was to be 20 years, and a replacement ship could be completed only after this period had passed. In addition, a so-called 'naval holiday' was agreed, meaning that no new capital ship could be commissioned before 1932. This was a lengthy period that would preclude the incorporation of new technology into the battle fleets of the major powers, leaving only obsolescent vessels in service by the end of the decade.

Many of the capital ships to be retained had been designed around 1910. Ten years later, not only had naval technology advanced, but the advent of aircraft and the submarine had radically changed the conditions under which future naval warfare would be conducted. The capital ships designed before the First World War had little protection against the new threats from the air and from below the surface. Moreover, the lessons drawn from the only major engagement between the British and German battle fleets, the Battle of Jutland, had effectively drawn a line between ships designed prior to 1916 and those designed in its aftermath. However, under the terms of the treaty, only Britain was permitted the construction of new capital ships (*Nelson* and *Rodney*), while the US Navy was allowed to complete two battleships armed with 16in guns in exchange for the completion of the Japanese *Mutsu*, which was similarly armed. The current construction programmes of France and Italy were effectively terminated, and their surviving capital ships, all of which had been completed before Jutland, would become quickly outdated compared to their more modern British and American counterparts.

Despite the constraints on new construction, both Britain and the USA were anxious to protect their older ships against the new dangers posed by submarines and aircraft, so the treaty included provision for a limited reconstruction within a limited increase of displacement. In Part 3 Section I ('Rules for Replacement') the contracting powers agreed that no retained capital ship should be reconstructed except for the purpose of providing means of defence against air and submarine attacks, and that for this purpose existing tonnage could be equipped with bulge or blister or anti-air deck protection, provided the increase in displacement did not exceed 3,000 long tons (3,048 metric tons) for each ship. No alteration in side armour, in calibre, number of and general type of mounting of main armament was permitted. However, there were some exceptions for France and Italy and one for Britain – a major reconstruction of the battlecruiser *Renown* was underway which involved fitting both vertical and horizontal protection.

This photo of *Kongō* was taken between April 1925 and July 1926; she became a gunnery training ship on 1 December 1925. No 1 main gun turret is trained to port while the other three turrets are trained on the axis. The original tripod foremast has been fitted with a control top and additional command platforms, and there is a prominent cowling on the first funnel to keep the bridge and foremast clear of smoke. (Lars Ahlberg collection)

The IJN began the modernisation of its capital ships with the oldest units, the battlecruisers of the *Kongō* class. Work on *Haruna* began in 1924, followed by her three sisters – reconstruction of the fourth ship, *Hiei*, had to be paused then modified following the London Conference of 1930. These four ships were successively modernised in accordance with the terms of the treaty: the horizontal and underwater protection was reinforced, but the propulsion system had to be left as it was. The speed of the ships declined accordingly, while protection was strengthened almost to battleship standard. Following reconstruction the three ships of *Kongō* class were reclassified as 'battleships' on 1 June 1931, and from this time use of the 'battlecruiser' classification was discontinued.

The London Treaty of April 1930 extended the 'battleship holiday' until 1936, thereby precluding new construction. It also stipulated that the number of capital ships remaining in service be reduced to the level envisaged by the Washington Treaty of 1922; excess vessels would either be scrapped or used for other permitted purposes. For Japan this meant that the former ten ships had to become nine, and *Hiei*, which had only recently begun her reconstruction, was to be redesignated a training ship; this would involve stripping one part of her armour and armament and reducing her service speed.

Both arms limitation treaties were due to expire at the end of 1936 and there was no prospect for their renewal. It was informally agreed at London that modernisation of the older battleships remaining in service could be more extensive than originally envisaged at Washington, and the IJN promptly embarked on radical reconstructions of her older battleships, beginning with *Fusō* and *Yamashiro*. The reconstruction, which was executed in two phases – for *Yamashiro* these were contiguous – lasted from 1930 to 1935. The four ships of the *Ise* and *Nagato* classes were then taken in hand, and the three *Kongō*s that remained in active service underwent a second reconstruction.

This second series of modernisations involved marked structural changes that extended to the general arrangement and the distribution of equipment and fittings. The appearance of all ten ships was radically altered. The IJN term used for this work was *kaizō* ('modernisation conversion') in order to distinguish it from the customary *kaisō* ('refit'). The extent to which the reconstruction increased the defensive capabilities of the older ships is open to question; the dated design of these units placed serious constraints on the effectiveness of 'add-ons' such as extra armour plating and bulges. However, the IJN invested massive resources, both financial and human, in the reconstructions, and the naval architects deployed considerable ingenuity in devising the required improvements, which were intended to bring Japan's battle fleet up to a uniform standard and enable it to defeat the US Navy in the envisaged 'decisive battle'.

Principal Elements of Modernisation

Before going into the specific details of reconstruction, it would be useful to give a brief overview of the principal works carried out for the modernisation of all Japanese capital ships. These were executed in either a single phase or two phases; even the two-phase modernisations were always recognised as part of a sequence.

The timing of the modifications was dependent on treaty provisions and informal agreements between the principal naval powers. The reinforcement of protection was a response to the advent of the submarine and aircraft in naval warfare; the increase in the angle of elevation of the main guns – initially a point of contention between Britain and the USA – was the result of the experience of long-range gunnery engagements during the First World War.

In general terms, for the IJN the principal elements of the first phase were as follows:

- The maximum angle of elevation of the main guns was improved from 20–25 degrees to 33 degrees to secure a range of about 30,000 metres. This distance was considered necessary for the execution of the outranging tactics judged by the IJN as indispensable in order to compensate for the inferior ratio in capital ship strength decreed by the Washington Treaty. In company with this the tripod foremast was reinforced and converted to a tower-like mast that could carry the many observation, fire control and spotting stations. Considerable stress was laid on the provision of the control equipment necessary for long-range engagement, using the main guns to secure early hits, and the function of the mast as the conning centre of the ship was extended.
- In response to the increase in the range of gunnery engagements, protection against plunging AP shell was improved by the reinforcement of the thickness of the protective deck over the ship's vitals. During the 1920s the effect of major-calibre AP shell fired from long range was considered more dangerous than that of bombs, which were relatively light, had limited penetrative power against armour, and carried only a relatively small explosive charge. The loss of three British battlecruisers at Jutland to hits by comparatively small-calibre (28/30cm) German shell served to reinforce this view; however, protection against large-calibre shell was also effective against bombs dropped by aircraft, and this argument was deployed by all three major naval powers to justify the modifications.
- The armour thickness of the face and the roof of the main gun turrets was increased to take account of the effect of large-calibre shell fired at long ranges. (During an engagement main gun turrets would be trained towards the enemy; turret side armour was considered sufficient to deflect enemy shell.)
- In response to the danger from the air, the anti-aircraft (AA) armament was strengthened.
- The submarine had proven its effectiveness during the First World War to an extent that had not previously been anticipated, so protection against torpedo hits was an urgent requirement. Bulges were added as anti-torpedo protection, backed up by reinforced longitudinal torpedo bulkheads over the vitals. Bulges not only provided underwater protection by creating a void space that could absorb much of the explosive force of a torpedo warhead, but provided buoyancy compensation for the additional weight of horizontal armour, thereby ensuring that the main belt continued to protect the waterline – it would otherwise have been partially or completely submerged. The reinforcement of the protection below the waterline was to counter not only the explosive effect of a torpedo warhead but also so-called 'underwater trajectory' shell – a phenomenon discovered accidentally in the course of trials with the incomplete battleship *Tosa*.[1] It was rumoured that the US Navy was developing a shell designed to strike below the main belt of a capital ship, and the IJN put considerable effort into developing this technology.
- Torpedo nets were removed due to their lack of effectiveness and vulnerability to damage.[2]
- The number of boilers was reduced and some were converted to burn oil (only *Haruna* retained her mixed-fired boilers). Coal bunkers were converted to heavy oil tanks, and additional tanks incorporated into the bulges to increase the radius of action. The arrangement of oil tanks in the bulges contributed also to the improvement of underwater protection. An additional benefit of the conversion from coal to oil firing was a reduction in the enormous volumes of smoke that led to early detection of a force beyond the horizon and hindered fire control during an engagement.
- Underwater torpedo tubes were fully or partially landed due to the limited opportunities for use once gunnery engagement ranges increased.
- Seaplanes were now carried for reconnaissance and spotting purposes, requiring the fitting of handling equipment including catapults, cranes and avgas tanks. The observation of fall of shot was considered a key element in fire control at longer ranges.
- The latest W/T (wireless telegraphy) equipment was fitted to reinforce ship-to-ship and ship-to-shore communication.
- The new equipment installed required an increase in electrical generation capacity.

The second phase of modernisation was carried out in preparation for the expiry of the naval arms limitation treaties and exploited to the full recent progress in propulsion technology. The re-arrangement of the machinery spaces, the replacement of the main turbines (except in the *Nagato* class) and other measures for an increase in speed were placed at the centre of these reconstructions. The *Kongō* class was prioritised to adapt them for their role in the newly-adopted night engagement that was planned to precede the decisive battle, which would take place during daylight hours. The speed of these former battlecruisers had been reduced as a consequence of the fitting of extra horizontal protection (and compensatory bulges) during their first modernisation, and the IJN wanted a maximum speed of 30 knots to fit them for the role of 'fast battleship'. The speed of the other classes of battleship was to be increased to a uniform 25 knots. The angle of elevation of the main guns was increased for a second time in order to implement out-ranging tactics, and fire control and associated systems were to be upgraded accordingly.

The principal works carried out during this second reconstruction were as follows:

- New oil-fired boilers replaced the original coal- and

Photographed on 2 December 1928 at Yokohama, *Kongō* retains the prominent cowling on the first funnel, and there is a clear view of the various platforms added to the foremast during the early 1920s. At the time this photo was taken *Kongō* was flagship of the Second Fleet and the flag of the C-in-C, Vice Admiral Otani Kōshirō, can be seen atop the foremast. The first reconstruction would begin nine months later, in September 1929. (*Sekai no Kansen*)

mixed-fired models (except in *Haruna*, which retained some of her original boilers with mixed firing). The new, larger boilers generated considerably more steam, and as far as practical each was located in its own watertight boiler room to improve underwater compartmentation and limit flooding.

- The original direct-drive steam turbines were replaced by a new, higher-powered Kampon type of Japanese design with single reduction gearing for higher power and improved fuel consumption. The new turbines effectively doubled the original output. This modification was subsequently extended to all the battleships except the *Nagato* class, which had been fitted with high-power geared turbines on completion.
- The stern was lengthened to secure a higher length/beam ratio, thereby reducing hull resistance at higher speeds and partially compensating for the increase in weight. The protection of the steering gear compartment was also reinforced and the rudder strengthened.
- The angle of elevation of the main guns was increased from 33 to 43 degrees, so that the maximum range of engagement was now in excess of 35,000 metres.
- The tower mast was rebuilt as a heavier 'pagoda'-type tower and was redesigned to accomodate the fire control directors for the main and secondary guns and their associated equipment, and also night battle facilities.
- The angle of elevation of the secondary 15cm guns was increased from 15 to 30 degrees, and two guns were landed, leaving a total of fourteen.
- An 'emergency flooding and counter-flooding system' was fitted in order to maintain/regain stability in the event of underwater damage – this was fitted retrospectively to *Haruna* and *Kirishima* after their reconstruction, but had been planned as part of it.
- Poison gas evacuation equipment was fitted in all the key compartments and spaces.
- The last of the submerged torpedo tubes were removed.

Reconstruction of the *Kongō* Class

The first ship to be taken in hand was the battlecruiser *Haruna*, which had experienced an explosion in main gun turret No 1 during an exercise on 12 September

1920. She underwent a four-year refit at Yokosuka NY which began in March 1924 and was completed in September 1928, by which time work had begun at Kure NY on her sister *Kirishima*. The remaining two ships were taken in hand in the autumn of 1929 (see Table 1), but although work on *Kongō* was duly completed at Yokosuka NY in March 1931, the reconstruction of *Hiei* (Kure NY) was suspended following the signing of the London Treaty of 1930, and she was rebuilt at Yokosuka as a gunnery training ship.

The reaffirmation of the 3:5 inferiority ratio in capital ships at the London Conference led to a further review of the IJN's strategy. The concept of the attrition of the US battlefleet during its advance accross the Pacific was now supplemented by the addition of a new, intermediate phase: the further reduction of the enemy's strength by a night engagement immediately prior to the 'decisive battle'. Light forces comprising destroyers and light/heavy cruisers were to be the main actors in this scenario, but in order to penetrate the outer defence of the ring formation of the US battlefleet, which it was thought would be constituted by heavy cruisers, it was envisaged that the *Kongō* class would be employed. After destroying this outer defence ring using their superior gun power, they would retire from the battlefield to fight in the decisive battle as part of the main force, leaving the light forces to penetrate to the centre of the US Navy formation and execute close-range torpedo attacks, causing confusion and serious damage to the battle line. In order to perform this role the *Kongōs* would need to keep pace with the light forces. Thus the primary goal of the second reconstruction was to restore the high speed of the ships that had been lost in the first reconstruction.

The second series of reconstructions was carried out in the same order as the first but with a more compressed timetable, beginning with *Haruna* in August 1933 and ending with *Kongō* in January 1937. *Hiei* would undergo a separate, more extensive reconstruction following the expiry of the Washington Treaty (see Table 2 and below).

First Reconstruction

Horizontal protection

The protective deck (middle deck) above the magazines and machinery spaces was reinforced to enable it to withstand the impact of 36cm (14in) AP shells fired from 20,000 to 25,000 metres (see Table 3).

Turrets and barbettes

As penetration of the magazines would undoubtedly have resulted in the loss of the ship, additional armour plating was fitted to the main gun turrets and the barbettes. The thickness of armour on the faces of the turrets was increased from 10in (254mm) to 11in (280mm), and the

A fine image of *Haruna* running her full speed trial at Tateyama on 25 May 1928 after her first reconstruction. Displacement was 33,825 tonnes, and she made 25.86 knots with 75,807shp. As the first ship of the class to undergo reconstruction, *Haruna* was in dockyard hands for more than four years (March 1924 to end July 1928). The work included the fitting of bulges, reinforcement of the deck protection, an increase in the elevation of the main guns, and further modifications to the foremast platforms. The engines were retained, but the original 36 boilers with mixed firing were replaced by 16 larger models, six of which were of the new oil-burning Kampon *Ro Gō* type; this made possible a reduction from the original three to two funnels. Bunker capacity was now 3,292 tonnes of furnace fuel oil and 2,661 tonnes of coal. Note the quantity of smoke emitting from the second funnel, which housed the uptakes for the boilers with mixed firing. (Fukui Shizuo collection)

thickness of the turret roofs was doubled from 3in (76mm) to 6in (152mm). There was similar reinforcement of the barbettes between decks.

Funnel uptakes and ventilation ducts

To prevent the passage of plunging shell through the large openings for the funnel uptakes and ventilation ducts in the protective deck, coaming armour was used to strengthen the horizontal protection as follows:

a) Funnel uptakes	6.5in (165mm)
b) Ventilation ducts to BR	7in (178mm)
c) Ventilation ducts to ER	4–5in (102–127mm)

Much of the armour plating used in this reconstruction had been intended for the capital ships of the 8-8 Fleet which had been cancelled under the terms of the Washington Treaty, the plating being rolled again to obtain the required thickness.

These measures resulted in a marked increase in the weight of protection, and the work also required skilful execution. However, immunity against 36cm AP shell was limited to a relatively small zone, and the defects inherent in the battlecruiser type remained. While horizontal protection was considerably strengthened, vertical protection, which was subject to treaty regulations, had to remain unchanged, so the balance between the two was lost.

Reinforcement of the underwater protection

Underwater protection was provided by (i) increases in the thickness of hull plating, (ii) the fitting of longitudinal torpedo bulkheads and (iii) the fitting of bulges capable of resisting the detonation of a 200kg (440lb) warhead.

The reinforcement of the hull plates was by either three (ammunition magazines) or four (machinery spaces) layers of 25.4mm HT steel fitted either at some distance from the hull side plating or directly over the plating, depending on the position. In addition, a longitudinal torpedo bulkhead with a maximum thickness of 2in (52mm) was fitted abreast the machinery spaces.

The bulges were fitted mainly to provide space for the expansion of the destructive explosive gases of a torpedo warhead, but also to restore buoyancy and stability in compensation for the 3,000-ton increase in weight. The upper part of the bulges was to be filled with watertight steel tubes 10in (254mm) in diameter similar to those planned for the contemporary British battleships *Nelson* and *Rodney*.[3] In theory the sealed tubes increased resistance to explosive forces and (provided they remained intact) preserved buoyancy; they were to be embarked immediately prior to the outbreak of war. A similar rationale dictated the subdivision of the bulges into watertight compartments (WTC) and heavy oil tanks outboard of the machinery spaces and WTC outboard of the powder magazines. The bulges were extended down to the double bottom in order to limit the effects of underwater diving shell. However, there can be little doubt that irrespective of bulges, double bottom, strengthening of hull plating and torpedo bulkhead, underwater protection was insufficient to withstand a plunging 36cm shell, as demonstrated by the *Tosa* experiments.

Table 1: **First Reconstruction**

	Haruna	*Kirishima*	*Kongō*	*Hiei*
1924	Mar	[1]	[1]	[1]
1925				
1926	Yokosuka	[2]	[2]	[2]
1927		Mar	[3]	[3]
1928	31 Jul	Kure		
1929			Sept	15 Oct
1930		31 Mar	Yokosuka	Kure › Yokosuka
1931		[4]	31 Mar	
1932				31 Dec
1933	[4]		[4]	

Notes:
[1] Modification to foremast (FC & S/L control)
[2] Increase from four to seven 8cm HA guns
[3] Conversion of foremast to tower (pagoda) type
[4] HA armament replaced by four twin 12.7cm

The first reconstruction of *Hiei* was halted on 24 April 1930. She was transferred to Yokosuka NY on 7 November where she was converted to a training ship (reclassified *Renshū Senkan* on 1 December 1932).

All except *Hiei* were re-classified as battleships on 1 June 1931.

Armament

The maximum elevation angle of the main guns was increased from 20 degrees (25 degrees in *Kongō*) to 33 degrees; theoretical range increased to 30,000 metres. Four of the eight submerged torpedo tubes were removed and the torpedo rooms converted to provide additional magazines and shell rooms for the main guns.

Table 2: **Second Reconstruction**

	Haruna	*Kirishima*	*Kongō*	*Hiei*
1933	1 Aug Kure			
1934	30 Sept	1 Jun		
1935		Sasebo	1 Jun	
1936		8 Jun	Yokosuka	26 Nov
1937		[1]	8 Jan	
1938				Kure
1939	[1]			
1940	[2]	[2]		31 Jan

Notes:
[1] Reinforcement of light AA by 10 x II 25mm MG
[2] Fitting of emergency damage control equipment

Table 3: **Reinforcement of Protection**

	Existing Thickness	Additional Thickness	Total
Boiler Rooms	19mm Nickel steel	76mm high-tensile steel	95mm
Engine Rooms	[as above]	82–89mm HT steel	101–108mm
Forward Magazines	[as above]	102mm NVNC [flat] 76mm HT [slopes]	121mm 95mm
Aft Magazines	[as above]	102–114mm NVNC	121–133mm
Lower Deck Aft	44mm	76mm	120mm
Ammunition Hoists [Turrets 1 & 2]	63mm KC	76mm KC	139mm
Below Superstructure Deck [Turret 3]	76mm KC	76mm KC	152mm
Below Upper Deck [Turret 4]	76mm KC	76mm KC	152mm

Notes:
HT High-Tensile
NVNC New Vickers Non-Cemented
KC Krupp Cemented

Ammunition stowage was thereby increased, probably from 90 to 120 shells per barrel.

The number of 8cm 40-cal 3rd Year Type HA guns was increased from four to seven; two of the additional guns were mounted on the forecastle deck amidships and the third abaft the mainmast. However, there was no dedicated fire control system, so the value of these weapons is questionable.

Modifications to the tripod foremast

The tripod foremast was now partially enclosed and supported multiple platforms, resulting in a markedly different silhouette. The platforms allowed for the fitting of larger and more complex components for the fire control systems in order to match the increase in engagement ranges for the main and secondary guns. Many of the bridge functions were moved into the new tower,

Kongō: Protection

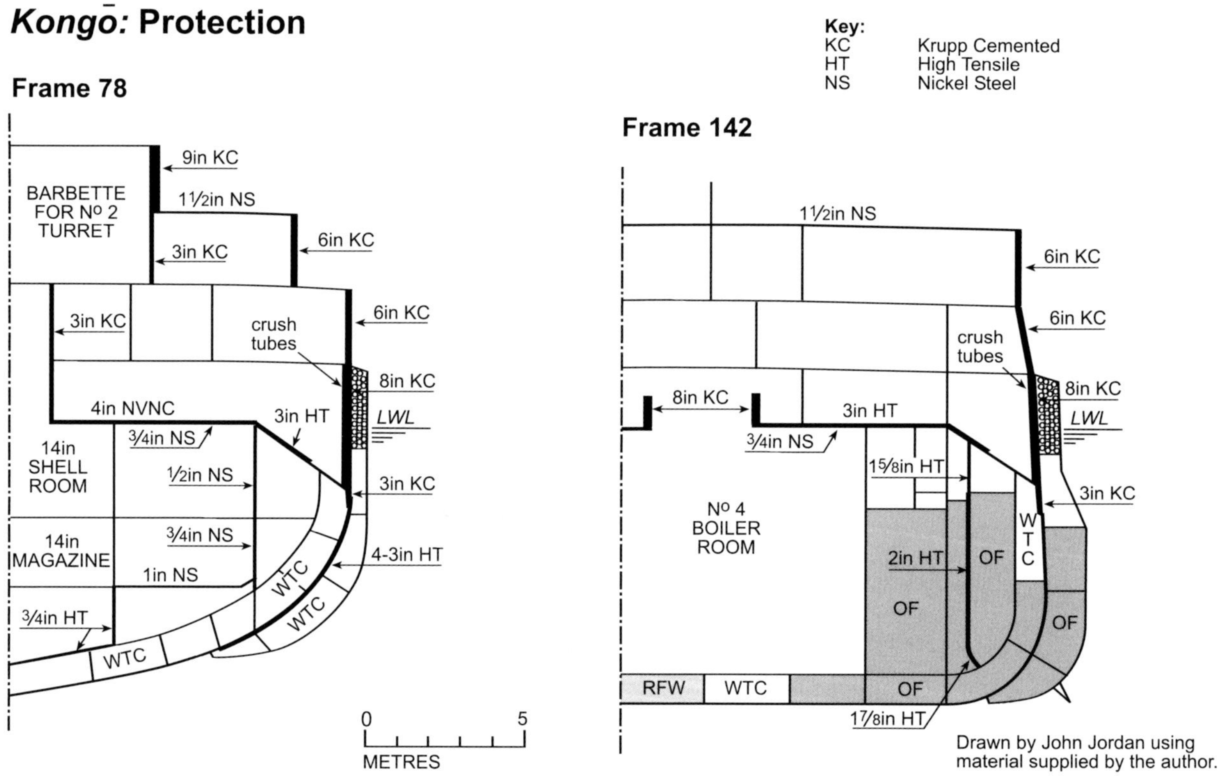

Drawn by John Jordan using material supplied by the author.

Kirishima at Kure Navy Yard on 10 March 1930 after completing her first reconstruction. She has been undocked only temporarily, and the hull has yet to be repainted; work on the ship was not completely finished until mid-May. Both funnels are fitted with prominent rain- and water-excluding caps. (Lars Ahlberg collection)

which housed an enlarged navigation bridge, signalling and lookout stations, chart and operations rooms, together with the main W/T office.

Machinery

The modifications made to the machinery plant in this first reconstruction were aimed at a reduction in smoke, an increase in endurance, and a saving in space and weight.

The 36 large-tube Yarrow (*I Gō* Kampon in the case of *Haruna*) mixed-fired boilers were replaced by sixteen *Rō Gō* Kampon small-tube units: ten with mixed firing and six burning heavy oil.[4] These boilers generated more steam than the older models but required less space, so the number of funnels could be reduced to two by removing the first, from which smoke and fumes often shrouded the optical stations mounted on the tripod foremast prior to reconstruction.

Coal capacity was reduced from 4,200 tons to 2,661 tons, while heavy oil stowage increased from 1,000 tons to 3,292 tons to give a theoretical range of 9,500nm at 14 knots – an increase of 1,500nm. Some of the original coal bunkers were reconstructed as heavy oil tanks, and some watertight compartments likewise converted, with caulking to make them oil-tight (OT). Some of the coal bunkers above the lower deck were converted into accommodation and stores.

Aircraft facilities

The space between main gun turrets Nos 3 and 4 on the superstructure deck was used for the installation of a catapult on the centreline, with rails and turntables on both sides for the accommodation of three/four Type 14 3-seater floatplanes for reconnaissance and spotting. They were recovered by a derrick fitted on the upper deck to starboard abaft No 3 turret. An avgas tank was installed close to the seaplane deck.

Second Reconstruction

New engines for higher speed

The first reconstruction was limited to changes in the type and number of boilers and the related auxiliary machinery; the original turbines were retained, and the additional weight of the horizontal protection, together with the fitting of compensatory bulges, meant a reduction in speed. However, the new strategy adopted post-1930 required fast battleships, ideally with a maximum speed of 30 knots. This implied a complete revision of the propulsion system.

Engines of the same type as mounted in the new cruisers of the *Mogami* class, complete with reduction gearing, were fitted. The only major difference was that the condenser was suspended beneath the LP turbine – in the *Mogami* class it was installed alongside the turbine.

Photographed in Dock No 5 of Yokosuka NY on 21 October 1929, *Kongō* is undergoing preparatory work for the fitting of the bulges; part of the crew remained on board, as indicated by the washing hung out to dry along the forecastle. The reconstruction period is given officially as from September 1929 to March 1931, but some sources claim that work began as early as December 1928. Until the completion of Dock No 4 at Kure NY in 1931, this dock was the largest in Japan; it was sufficiently wide to accommodate the *Nagato* class after reconstruction, and also *Akagi* and the fleet carriers of the *Shōkaku* class. (Lars Ahlberg collection)

The cruise turbines were connected to the inboard shafts and the exhaust channelled into the HP turbines on the wing shafts.

Haruna was the first ship to be taken in hand, but due to budgetary constraints not all the boilers could be changed. Six of the oil-burning boilers fitted in the first reconstruction, together with two intended for the battleship *Tosa* (expended as a target following the Washington Treaty) were improved by attaching air preheaters, and three new boilers were also installed for a total of eleven. The boilers were rated at 20kg/cm^2 and used saturated steam. Total horsepower was now 136,000shp.

The tight budget also prevented the change of the steering gear to the electric-driven hydraulic type, and *Haruna* retained the older-type steam-powered system. However, this proved to have insufficient power due to the increase of the speed, and maximum rudder angle was restricted from 35 to 30 degrees.

Kirishima and *Kongō* received the same engines as *Haruna* but their original boilers were replaced by eight new oil-fired models of the same type fitted in the cruisers *Suzuya* and *Kumano*, which used superheated steam. Theoretical engine power was thereby increased, but it was decided to restrict nominal horsepower to the 136,000shp of *Haruna*. These two ships had the IJN's ideal arrangement of four turbine sets in four engine rooms and eight boilers in eight boiler rooms, whereas *Haruna* had an unbalanced layout using eleven boilers, and pipework was complex compared to her sisters.

Armament changes

The elevation of the main 36cm guns was further increased from 33 to 43 degrees, as in the other IJN battleships, by deepening the gun wells and lowering the revolving structure bodily towards the ship's bottom. The guns were run out using compressed air. Maximum theoretical range was now in excess of 35,000 metres.

In order to extend the range of the secondary guns from 14,200 metres to 19,500 metres, the maximum angle of elevation was increased from 15 to 30 degrees by raising the ring supports in the casemates. The guns in the foremost casemate on both sides and all torpedo tubes were landed in order to provide partial compensation for the increase in weight.

This photo shows *Kongō* in dock at Yokosuka NY on 24 December 1928. She was the third ship of the class to be taken in hand for reconstruction, and the photo shows her during the preparatory work. The first funnel has already been landed and can be seen on the dockside, to the right of the crane. Details of No 2 main gun turret and the casemates for the secondary 15cm guns are clearly visible. Note the housing for the rangefinder atop the armoured conning tower. (Hans Lengerer collection)

The water-cooled slow-firing 40mm Vickers machine guns were replaced by the newly adopted Hotchkiss 25mm MG in twin mountings.

Conversion of the foremast to the characteristic pagoda-style tower mast

The conversion of the original tripod foremast to a pagoda-style tower mast had been executed during the first conversion, but further modifications were made to improve long-range gunnery and to complete the provisions for night battle.

These changes required an increase in the number of fighting, command and observation stations, the mounting of new equipment and the upgrading of existing items. Considerable effort was made to secure these improvements, and detailed plans were drawn up for comment by the various sections. There were a number of suggestions for changes, and the final arrangements resulted in a marked increase in size and complexity, with the foremast fitted with a plethora of sensitive instruments.[5] There were detail differences between the first three ships taken in hand, and *Hiei*'s foremast was markedly different because she was used as a model for the battleships of the *Yamato* class. The following enumeration (from top to bottom) refers to the levels for the first three ships modernised:

- Type 94 10-metre duplex rangefinder (RF) with lightning conductor and radio direction finder (RDF) antenna atop
- Main and HA fire control station with Type 94 main gun director and a number of pairs of binoculars (with and without target angle transmitters)
- Secondary gun control and upper lookout station with secondary gun directors, spotting sights and binoculars
- Target speed/course metering and control station with Type 13 and Type 92 *Sokutekiban* and 12cm telescopes
- Battle bridge
- Lower lookout station with sided 1.5-metre RF, 60cm signalling projectors and 25mm MG
- Navigation bridge
- Upper bridge with Nos 1 and 2 secondary gun auxiliary control stations and 12cm telescopes; (in front) platform with Type 94 4.5-metre RF and 25mm MG

- Lower bridge with the conning tower at the forward end, 25mm MG to the sides and HA fire control station

There were the customary wireless and telephone rooms, a signal distribution centre, the captain's sea cabin, rooms for orderlies and WCs.

The auxiliary fire control systems for the main and secondary guns were fitted in the upper part of a tall multi-level deckhouse abaft the mainmast. The computers for the LA and HA guns were located below the armoured deck.

The low angle fire control system for the 36cm main and 14cm secondary guns comprised a Type 94 director with a Type 92 computer, the HA fire control system the Type 91 director with a Type 89 computer. In accordance with the Fire Control Instruction of 1932 the installation provided for fire on forward and after bearings.

Finally, six Type 96 110cm searchlight projectors were sited around the first funnel in order to improve the night fighting capability. The searchlight command stations (*shōshashikisho*) were installed in the tower mast.

Damage Control

The fitting of an emergency flooding and counter-flooding system (termed a 'discharging' system by the IJN) was a later modification. It comprised compartments on both sides of the hull which could be flooded rapidly (ten compartments on either side each with a 250-tonne capacity) or slowly (thirteen compartments on either side each with a 750-tonne capacity). Using the rapid-flooding system a list up to 5.5 degrees could be corrected and the ship brought back to an even keel; the slow-flooding system could correct a list of up to 7.7 degrees. The installation of the necessary pipework and pumps was no easy undertaking.

Reconstruction of *Hiei*

One month before the expiry of the Washington Treaty on 31 December 1936, work was begun on the fourth

Kongō at the fitting out (*shōkai*) quay at Yokosuka NY on 20 December 1931, when the reconstruction was 86 per cent complete. The two modified funnels are now in place, suggesting that work on the boiler rooms has almost been completed. Like her sister *Kirishima*, *Kongō* received eight of the new Kampon *Ro Gō* oil-burning boilers. The 15cm secondary guns are mounted in their casemates; the guns could be dismounted and stowed inside the hull. (Fukui Shizuo collection)

Kongō photographed while running her first official full power trial on 20 July 1931 after her first reconstruction. The configuration of the funnels and the foremast distinguished her from *Haruna*. (*Sekai no Kansen*)

ship, *Hiei*, previously a gunnery training vessel, to bring her up to the same standard as her sisters. However, the reconstruction differed from that of her sisters because she was used to trial some of the features planned for the 'super-battleships' of the *Yamato* class:

- The arrangements of the tower mast were revised to the new specifications with the conventional bridge used for the night battle and an additional 'day battle' bridge in the upper position accessed by double ladders (the *Yamato* class would have two sets of double ladders). A new-model Type 98 director was mounted above and in the same vertical axis as the 10-metre rangefinder atop the tower. The cylindrical communications tube for the main director was angled aft to form a fourth 'leg' to the original tripod foremast, thereby minimising vibration. The new Type 94 HA fire control system replaced the earlier Type 91 on both sides of the lower part of the tower, and quadruple 13mm MG were mounted at the sides of the machine gun platform.
- The conning tower was replaced by a new model.
- One of the hydraulic pumps for the main guns was a turbo-pump purchased from Brown Boveri. It proved successful, and the *Yamato* class would mount four larger improved domestically-built models.
- A turbo-cooler of 80,000kcal was fitted experimentally for cooling the magazines (the *Yamato* class would be equipped with four 150,000kcal coolers for magazines and living spaces).
- Several components of the emergency flooding and counter-flooding system were designed as 'fast functioning types' for the *Yamato* class. The system would be further developed for the *Yamato*s, which had remotely-controlled water inlet flaps.

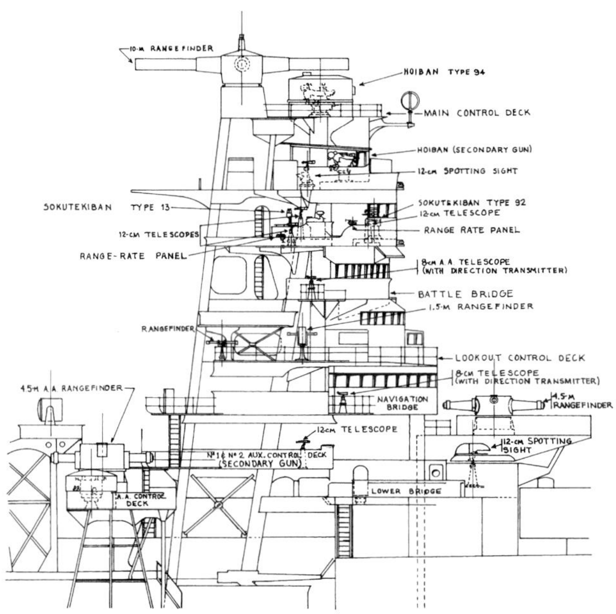

The forward tower of *Kongō* following her reconstruction. See text for details. (US Naval Technical Mission to Japan Report)

The photo shows *Haruna* running trials off Sukumo Bay during July/August 1934 after her second reconstruction. The new mission of the *Kongō* class, namely the destruction of the protective ring of cruisers and destroyers around the US main fleet as it advanced towards Japan, required a complete renewal of the propulsion machinery and the lengthening of the stern. There were now eleven Kampon *Ro Gō* type oil-burning boilers and four sets of Kampon geared turbines to generate the 136,000shp necessary for a top speed of 30 knots. Displacement increased by about 6,000 tonnes. Note the revision of the foremast platforms and the higher second funnel. (Lars Ahlberg collection)

- The extension of the stern was as in *Hiei*'s sister ships, but the torpedo bulges were enlarged in order to reduce draught in the full load condition and to ensure that the protective deck remained above the waterline.

Because these ships had been designed as battlecruisers, the thickness of the belt armour (8in/203mm) was considered insufficient. Reinforcement by inclining the armour belt was considered and tests carried out. However, this would have resulted in a slight increase in weight, and the technical problems involved would have extended the time necessary for reconstruction and increased the demands on dockyard labour, so the proposal was not implemented.

Principal Sources:

Books:

Kaigun zōsen gijutsu gaiyō (Outline of Naval Shipbuilding Technique).

Shōwa zōsen-shi (History of Shipbuilding in the Shōwa Era), Vol I.

Fukui Shizuo, *Nippon no gunkan* (Japanese Naval War Vessels).

Kirishima photographed from a US Navy warship at Amoy on 21 October 1938, following her second reconstruction. There are a number of detail differences from *Haruna*, in particular the shape and height of the second funnel. Note the floatplane atop the catapult, with a second stowed on rails abaft turret No 3. (NHHC, NH-50903)

Hiei as a gunnery training ship, seen here off Yokosuka in 1932. Under the terms of the London Treaty of 1930 she lost No 4 main gun turret, had her side belt removed, and speed was reduced to 18 knots. In consequence the first funnel disappeared, the second was replaced by a slim 'pipe', and freeboard increased; she now displaced only 19,500 tonnes. (Kure Maritime Museum)

Journals:
Gakken, Sekai no kansen, Maru, Maru Special, Maru Graphic.

Endnotes:

1 See the author's article in *Warship 2020*.
2 Action damage to the hinged booms sometimes resulted in the nets snagging propellers.
3 The results of the research undertaken by then-Rear Admiral (Eng) Hiraga Yuzuru in Britain and the USA in 1924 were a significant influence on the conversion of the Japanese capital ships. Due to his friendly relations with leading British naval architects, Hiraga obtained confidential data about the newly-built battleships of the *Nelson* class, and in particular about their horizontal and vertical protection systems. However, the IJN did not solely rely upon information from abroad but undertook its own research during trials with the incomplete battleship *Tosa*. Most of the decisions were made after Hiraga's return from abroad and the evaluation of results of the trials.
4 Steam pressure was 19.3 bar, steam temperature 200°C. During trials on 25 May 1928 *Haruna* obtained 25.86 knots with 75,807shp on a displacement of 33,825 tons
5 In order to minimise vibration supporting struts for the main rangefinder assembly were fitted at the rear of the foremast.

Hiei running her full speed trials off Sukumo Bay on 5 December 1939 following her reconstruction as a high-speed battleship at Kure NY. On a displacement of 36,332 tonnes she obtained 29.9 knots with 137,970shp. Note the tower replacing the original foremast, which trialled the arrangements for the *Yamato* class, and the absence of rain-excluding caps on the funnels. (Lars Ahlberg collection)

THE 'BLACK GANG'

THE DEVELOPMENT OF TORPEDO BOATS IN THE IMPERIAL GERMAN NAVY (Part I)

Dirk Nottelmann tells the story of the development of the German *Torpedoboote*, beginning with the spar torpedo-carrying boats of 1870 and ending in 1918 when, in addition to their 'classical' tasks, it was envisaged that they would replace the small cruiser as the primary scouting unit of the Imperial German Navy (IGN), thereby abandoning their original role that subordinated them to the battlefleet.

In the years between 1870 and 1918, the torpedo arm in the Imperial German Navy rose from obscure beginnings to unexpected heights of recognition and expectation which it would ultimately fail to fulfil in the First World War. The present study is necessarily focused primarily on the technical development of the boats and their weaponry; operational history is excluded, not least because of the large number of boats commissioned during the period in question. It is envisaged that this series of articles will eventually be expanded into a separate publication that will cover all aspects of the boats' history.

The term 'torpedo carrying boats' in the IGN originated during the early period, when the word 'torpedo' was synonymous with either an immobile mine or an explosive device rigged to a spar; these early torpedoes were not independently powered. The first experiences with moored torpedo-mines in German waters extended back to the first Danish Schleswig-Holstein War of 1848, when a young Prussian artillery lieutenant, Werner Siemens,[1] supported by his brother-in-law, professor August Himly[2] of Kiel university, prepared and laid a row of comparatively crude explosive containers in the waters outside the inner port of Kiel. One of these defensive 'mines' was accidently detonated before being laid, shattering a large number of windows in the surrounding area. This did not go unnoticed by the Danes, who subsequently avoided coming too close to these 'hellish' devices. However, it would be stretching the definition to describe these craft as 'torpedo boats'!

Laying mines off Pillau in the Baltic during the Franco-North German War. This was a complex and slow operation, as demonstrated by this engraving in the *Illustrierte Zeitung* of December 1870. (Author's collection)

Two lines of development can be clearly made out, beginning some two decades later. The first involves studies and experiments undertaken by the Austrian Navy's Captain Johann Luppis, at a time when Austria was still the leading power in the German Federation. Luppis invented and promoted his 'coast-protector' from 1860: a small boat driven by clockwork and controlled by wires containing an explosive charge. He would later seek technical advice from, and eventually collaborate with Robert Whitehead,[3] then technical director of the Stabilimento Tecnico Fiumano (STF), in the development of the first 'automobile torpedo'. Successful testing of this new weapon from 1866 led to the first negotiations between the newly-established navy of the North German Federation (NGF) and STF in June 1869 over the possible acquisition of 'Whitehead torpedoes', as they would henceforth be generally known.

These initial negotiations were led by *Korvettenkapitän* (Commander) Graf von Monts,[4] and the subsequent reports sent to Berlin were evidently not encouraging, as the NGF refrained from acquisition of the new torpedoes, and decided instead to pursue the second option available when political tensions with France came to a head, leading to a declaration of war in July 1870. Reports of successful attacks using improvised spar torpedoes during the US Civil War seem to have made a big impression on the NGF. Due to the marked inferiority of the nearly non-existent North German naval forces when faced with the second largest navy in the world, it was clear from the outset that the stance of the Navy would be strictly defensive; it was charged with protecting the many estuaries and the extended coastline

The paddlewheel dispatch vessel *Falke*, acquired on 25 August 1870 during the war to be employed as a scout and a carrier of Harvey torpedoes. (Author's collection)

from surprise attacks or the much-feared landing operations. Against a blockading fleet these weapons seemed to offer some prospect of success, especially in coastal waters. Therefore, shortly after the commencement of hostilities, a new military unit designated the Voluntary Naval Defence Forces was established to conduct mine and torpedo work. There would be two independent flotillas: one comprising seven pulling cutters, each equipped with a single spar torpedo, the other consisting of seven[5] steam-powered tugboats collected from along the River Elbe, six of which would be equipped with a spar torpedo, while the seventh would tow a Harvey torpedo.[6] Also equipped with the last-named special device was the recently-acquired paddlewheel dispatch vessel *Falke*, which had originally been built speculatively in Scotland as a blockade-runner for the Confederacy.

Despite the lack of any success by these flotillas over the few months when France demonstrated its naval presence in the North Sea and the Baltic, the NGF administration nevertheless persisted on this new path, and would order purpose-built steam-driven boats to perform the same function during the closing stages of the war (or possibly shortly after the signing of the Treaty of Frankfurt on 10 May 1871 – the exact date remains unclear).

Spar Torpedo Boats

Little is known about the orders for these six boats (in two batches of three), even less about their origin and their basic planning. Unusually, both groups would be allocated the numbers I–III, distinguished only by the shipyard's name as prefix. The first batch, ordered slightly in advance of the second, seems to have been ordered from the Devrient shipyard,[7] located in the vicinity of Danzig. These boats would be significantly larger than those of the second batch, displacing 34 metric tons compared with 20 tonnes. A contemporary newspaper gives the following description of the Devrient boats:

> Special attention has recently been given by the Naval Ministry to the concept of converting torpedoes, those formidable defensive weapons, into weapons of attack. (...) These new boats, the smallest screw-driven and steam-powered vessels in our navy, are intended to place torpedoes below the enemy's ships and then make them explode. The boats are built entirely from iron and have a distinctly slim shape, with a length of 65 feet and a breadth of 10 feet. Their diminutive size, in combination with an inconspicuous grey paint-scheme, will allow them to approach the enemy without detection at dawn and in darkness while their crew, under cover of the turtleback steel deck, will be perfectly safe under small-arms fire. The last point is particularly important, as the boat must operate in close proximity to the enemy's vessels; consequently, no crew member will ever be exposed during an attack. (...)[8]

In order to enhance the intended stealthy qualities of the boat, the designers introduced a distinctly novel feature which was far ahead of its time: an oil-fired boiler which, it was hoped, would avoid the emission of dark smoke during the approach. These efforts were in vain: during 18 months of trials the engineers were unable to achieve stable, smokeless combustion due to the crude design of the oil burners. In anything other than calm seas, the oil, supplied by a pump connected to the main engine, would drop in uncontrolled fashion from the firing grooves, which were simply constructed using tubes with the upper halves cut off. The result was incomplete combustion, generating clouds of thick smoke; moreover, the spilled oil tended to drain into the boiler room bilges. Thus, after the conclusion of the trials, all the boats were reconstructed to burn coal. In general terms, their performance was less than impressive: maximum speed did not exceed 8 knots, which effectively precluded their employment against ships underway and restricted them to attacks on moored targets.

The armament of these boats remains something of an enigma, despite the well-known fact that it initially consisted of a single spar torpedo with an explosive charge of 17kg black powder. Equally well-documented is that the spar could be worked entirely from the inside of the boat using a special winch. Only recently, after

Engraving of the new Devrient torpedo boats, which had only recently been completed, in the *Illustrierte Zeitung* of November 1871. For a long time, this illustration would remain the only visual depiction of the boats. It served to confuse matters in that the operation of the spar was difficult to determine from the engraving. (Author's collection)

combining the few illustrations of the boats that have survived with the knowledge of techniques later employed on boats built for foreign navies, has it been possible to assess its construction and operation with any certainty: it has now been determined that there was a second spar, attached to the tip of the first to form a triangle, with both legs running along the forefoot of the boat on either side. Any arrangement with less than two fixed guiding points would have caused the single spar to veer away from the intended direction when deployed – a hoisting/lowering line from the end of the bowsprit not counting as a fixing point.

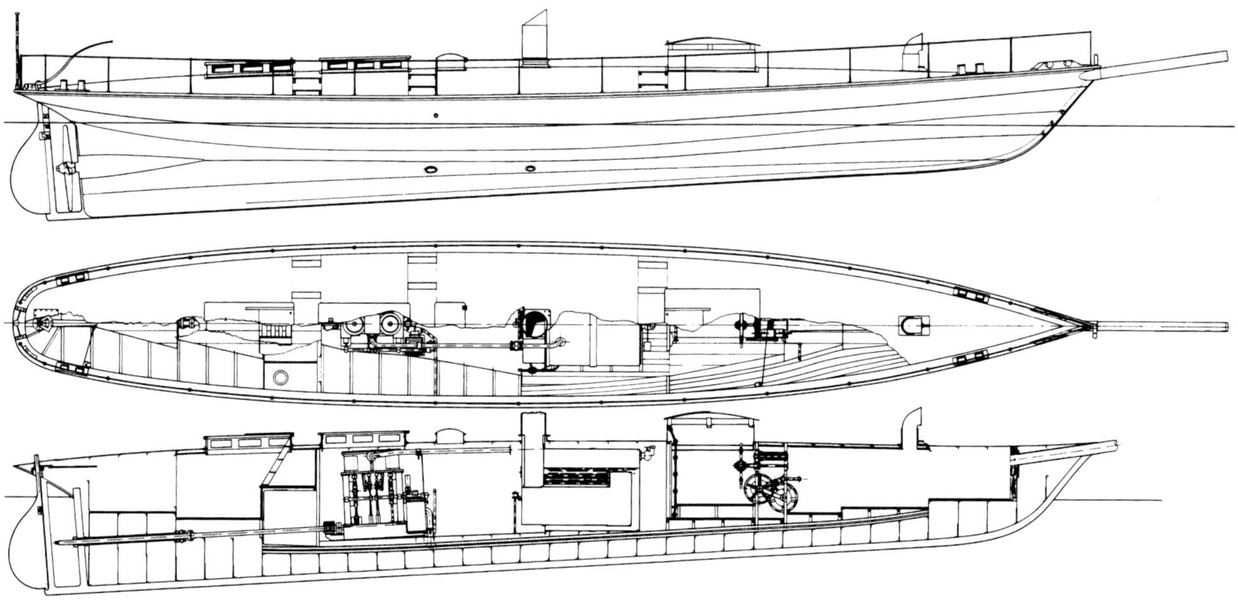

Drawings based on the original plans of the Devrient boats, highlighting their fine lines, which nevertheless did not result in a speed of more than 8 knots. The device for the spar torpedo is not depicted. (Drawn by the author)

An early launch tube similar to those fitted atop the stem when the boats were modified in 1881. (Author's collection)

Even less is known about how the conversion, around 1881, to a single torpedo tube was executed except for the vague description of the 381mm tube being 'fixed at the bow'. This would certainly have involved removal of the bowsprit, as this would otherwise have obstructed ahead fire from the fixed, axial tube.

Given all these shortcomings, it is unsurprising that the active careers of these boats were brief. On 15 December 1875 the spar torpedo devices had already been removed and the boats were redesignated *Minenleger* (Minelayers) *4–6*, serving in the local harbour defence of Wilhelmshaven and being capable of laying one mine apiece.[9] The above-mentioned conversion to mount a fixed torpedo tube in 1881 would result in a further redesignation, as *Torpedoboot I–III*; the boats remained in service as part of the harbour defence forces until about 1885, when they were stricken and sold for scrap.[10]

The history of the second trio, ordered from the C Waltjen & Co shipyard[11] at Bremen in 1871, is even less well-documented. Designated 'Waltjen torpedo boats' *I–III*, they feature neither in contemporary newspapers nor in other official publications; not a single illustration of them is known to have survived. Their armament is said to have comprised a spar torpedo, which was somehow attached to the bow below the waterline. As with the Devrient boats, around 1875 the spar torpedo devices were removed and they were redesignated as *Minenleger 1–3* in the local harbour defence of Wilhelmshaven. Later, they received an abbreviated designation: *W 1-3*. The first, *W 1*, would be prepared for a prolonged lifespan of another three years in 1901, before being handed over in Summer 1902 to the mine depot at Friedrichsort (Kiel) as *F 1*. After that, its track is lost – in contrast to its sisters. *W 2/F 2* had likewise been transferred to Friedrichsort in the summer of 1902, and would continue its service as a minelaying training vessel until January 1906, to be finally discarded in April. *W 3/F 3* had already been transferred in the summer of 1901 to Friedrichsort, for the same purpose, and shared the fate of its sister.

Shortly after the introduction of these novel torpedo boats, the administration of the Navy underwent a major reorganisation. On 30 November 1871, Kaiser Wilhelm I decided on the creation of a separate 'Imperial Admiralty', which would now be subordinated to him directly in military matters – instead of being a branch of the War Ministry – and to the Chancellor in administrative matters. The new structure was to become effective on 1 January 1872. It was to be headed by *Generalmajor* Albrecht von Stosch (1818–1896), a much-respected organiser, albeit with little expertise in naval affairs. Von Stosch, promoted to the dual rank of *General* and *Admiral* in 1875, would serve in this capacity until 1883; his initial deputy as Director of Admiralty was Rear Admiral Ludwig Henk (1820–1894).

Parallel to the establishment of this new institution, the German parliament approached Chancellor Bismarck, requesting a review of the extent of progress with the Fleet Foundation Plan of 1867 now that it had reached the half-way mark. Bismarck responded in a memorandum of 8 March 1872, which was published two months later. However, the Diet was not entirely satisfied with his response, principally because the 1867 Plan was grossly under-funded and under normal circumstances (*ie* without the irregular expenditure incurred by the war with France) would have already been over budget. Consequently, on 28 May 1872, the parliament demanded that the original plan be revised; this would eventually result in von Stosch's Fleet Foundation Plan[12] of April 1873. The updated plan included the projected acquisition of 28 'torpedo vessels', of an unspecified type save for a division into 'large' and 'small' boats. The existing six Devrient and Waltjen boats were included in the 'small' boat category.

The tug *Notus* around the turn of the century; currently the only known image of one of the first three former *Torpedodampfer*. (Author's collection)

The only known depiction of *Ulan* in her original configuration, published in the *Illustrierte Zeitung* of December 1876. There is no visible indication of the 'hellish device' beneath the bows. (Author's collection)

There was a subsequent order for three additional experimental spar torpedo boats from the Vulcan shipyard, located close to Stettin. These were of completely different design and size; two of them were a type of paddlewheel steam tug, the third was driven by a 'hydromotor' – a waterjet engine similar to those of the British gunboat *Waterwitch*.[13] Initially known as *Torpedodampfer* (torpedo steamer) *I–III*, they would later be named *Notus*,[14] *Zephyr*[15] and *Rival*.[16] The former two were converted to tugs in 1875; the latter was converted to paddlewheel propulsion, the waterjet engine having proved as unsuccessful as that of her British counterpart, before being employed as *Minenleger IV* in the local harbour defence of Wilhelmshaven. Unfortunately, only one image of these three vessels seems to have survived.

Marginally more interesting is the story of the subsequent *Torpedodampfer IV*, later christened *Ulan*,[17] ordered from the Stettiner Maschinenbau-Anstalt und Schiffsbauwerft AG (formerly Möller & Hollberg) in 1875. She would be the last German torpedo vessel to be armed with a spar torpedo. Intended for employment on the high seas rather than in coastal waters, she displaced 440 metric tons fully loaded. Initially, the spar had to be attached to the keel by divers before the boat went into action – it was a 'one-shot' weapon only. Later on, a mechanical device would be employed to angle the spar upwards, thereby easing the arming process.

The attack profile was also unusual: having closed the intended target, the crew was to secure the steering for a straight run, apply full steam ahead, and disembark into an 'unsinkable' float, which was normally towed astern on a long line. After detonation of the charge on impact, the men were to re-embark on their mothership (hopefully finding it still afloat!) and return home. For better survivability, not only had a collision bulkhead been installed, but the plating of the fore part of the vessel had been doubled, the space between the two shells being filled with cork as a shock absorber. Furthermore, the construction of the main engine had been specially modified to withstand longitudinal forces resulting from the impact and the detonation of the charge.[18] Fortunately, this concept never needed to be tested in real life! At the end of the 1870s *Ulan*, like earlier torpedo craft armed with the spar torpedo, would be converted to launch mobile 381mm Whitehead torpedoes from a single tube mounted in the bows below the waterline. She would subsequently serve as an experimental tender to the new torpedo trials ship *Blücher* and her successors until sold in 1919. Following her conversion, her profile was much-changed, the exhaust gases of the four Scotch boilers being led into a single funnel – she had two as completed.

These early developments seem to confirm that after six years of experimentation in the IGN the spar torpedo was now considered to be a dead end, with few useful lessons beyond the futility of further experiments – a conclusion also reached by other navies at this time.

Orders from Abroad

For completeness, it is worth noting that the last spar torpedo boats built in German shipyards had their origins in two independent orders from Russia in 1877/78. Successes with spar torpedoes during the Russo-Ottoman War, together with fears of an intervention by the Royal Navy in the eastern Baltic in the course of the war, had encouraged the Russian Navy to pursue this path and to place orders for a total of 111 boats with domestic and certain foreign shipyards. Most, but not all were derived from designs by the British company of Alfred Yarrow.

One of the yards to react quickly to the Russian request for tenders was that of Ferdinand Schichau, Elbing, which proposed an experimental design by Schichau's son-in-law and chief engineer Carl Ziese.[19] Following its acceptance, this became the first of a long line of torpedo boats delivered from that yard to a wide

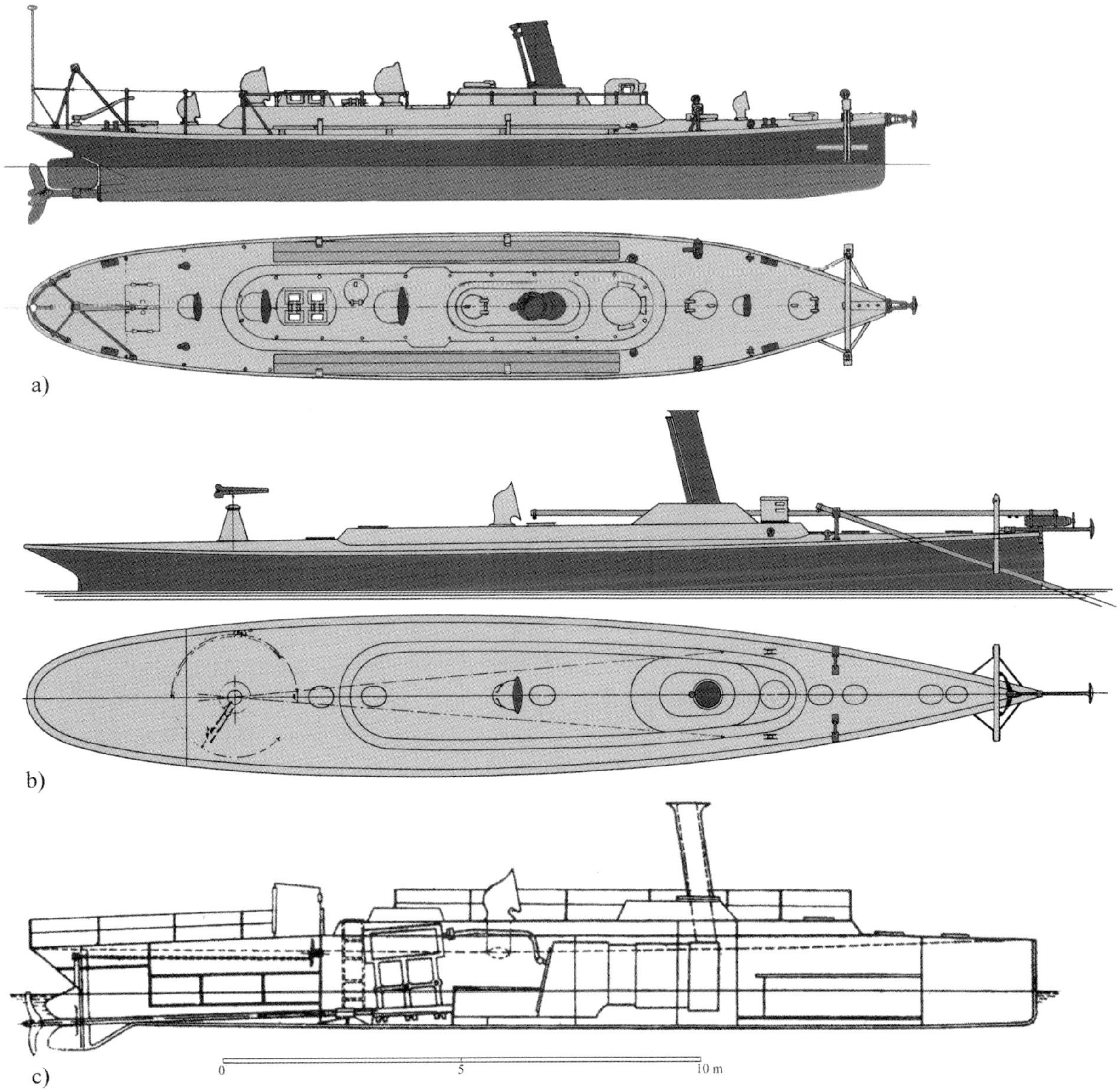

Three different images of the first Schichau boats built for Russia. The upper illustration is of *Karabin*, which paved the way for the subsequent order of ten slightly enlarged boats. The middle drawing is of *Bomba* with her spar torpedoes rigged and an additional 37mm Hotchkiss revolver cannon mounted. The lower drawing is of *Luk*, most probably after her conversion to a dispatch vessel, with the spars dismantled and the steering-position moved aft. (Author's drawing based on Russian sources)

variety of customers right up until 1945. The boat received the yard number 115 and was christened with the name *Karabin*;[20] she was delivered by rail to St Petersburg in the Autumn of 1877. Comparative trials on 28 November 1877 in the Kronshtadt Bight against a boat designed by Yarrow went well, particularly in terms of speed, resulting in a subsequent order for ten similar but slightly enlarged boats in February 1878, the so-called *Bomba* class (yard numbers 133–142). The first of these would sail for St Petersburg on 7 July 1878; it made the transit under its own power in order to demonstrate its capabilities, arriving at its destination within ten days; the remaining boats followed during the subsequent months, so that all ten would be delivered within six months from contract.[21] Apart from their successful delivery and their broadly successful subsequent trials, comparatively little is known about the careers of these ten boats and even less about their predecessor *Karabin*, which was possibly never in full commission with the Russian Navy. In general, experience would demonstrate that they were too small, even for the comparatively sheltered waters of the Gulf of Finland. Thus, most of the boats were converted to launches in the 1880s and discarded before the turn of the century, with the sole exception of *Kopje*, which was renamed *No 102* on 2 September 1885, again renamed *Kopje* in 1897, and ulti-

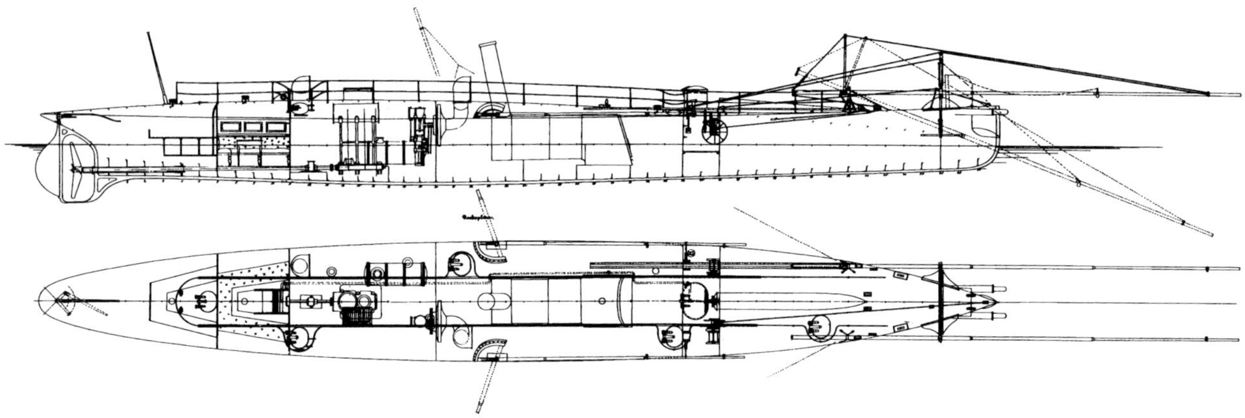

Sketch drawing of Vulcan's *Samopal*, with her mixed armament of automobile and spar torpedoes. (Author's collection)

mately renamed *Fontanka* on 21 April 1921; she survived until at least 1922, when her track was finally lost.

A second order would be placed with AG Vulcan, Stettin, for the delivery of two boats, to a general design by Yarrow but equipped with compound engines designed and built by Vulcan; besides spar torpedoes they were fitted with a bow tube for 381mm mobile torpedoes. Work commenced, after delivery of plans in February 1878, under the yard numbers 81 (*Raketa*) and 82 (*Samopal*).[22] Following several re-designations, both boats were ultimately discarded in 1908. The British magazine *The Engineer* claimed in its edition dated 1 February 1884 that the boats delivered from other than British yards were of such inferior quality that some of them had literally broken into pieces on delivery, the rest being subsequently offered up to auction, with any bid accepted. However, the Vulcan shipyard declined to respond to this provocation; it was in possession of a letter from an *aide-de-camp* to the Tsar stating that 'the workmanship is of such outstanding quality that I feel obliged to offer my special appreciation to you!'[23]

Introduction of the Whitehead Torpedo

After the termination of the war against France, a special commission for submarine weapons would be established in September 1871, led again by von Monts, which initially dealt with mines, spar and Harvey torpedoes only. Following the formulation in 1873 of the new goals for future naval development by von Stosch and a demonstration of his most recent 'fish torpedo' by Robert Whitehead at Wilhelmshaven at around the same time, the new naval administration finally felt convinced that it was safe to proceed in this new direction. It concluded a contract with Whitehead's newly-established company, erected from the shambles of the former STF at Fiume, for the subsequent production and delivery of 100 torpedoes of an improved type. It was agreed that the IGN would purchase its torpedoes exclusively from Whitehead until 1880 under a licence arrangement that gave the IGN freedom to construct the bodies and engines of the torpedoes elsewhere provided that the secret horizontal steering apparatus was obtained from Whitehead directly. The first batch of torpedoes, designated C/74, would arrive in Germany from 1876 onwards, followed by two further improved batches of C/76 and C/77 weapons; both these later models were equipped for the first time with contra-rotating propellers for improved straight-running capabilities.

Parallel to this first order, a carrier-vessel for running tests of this new, potentially lethal weapon was required. Initially, the German military attaché in London, *Oberstleutnant* von Schroetter,[24] was ordered to 'investigate reports concerning "small vessels, of striking velocity"'. Without the name being explicitly mentioned in the files, this was most probably Thornycroft's launch *Miranda*,[25] the model for the first designated torpedo launching craft, the Norwegian *Rap* (1872).[26] Schroetter managed to secure an invitation for a trials trip, and duly reported home: the boat had attained a maximum speed of 18.5 knots while he was on board, and was available at a price of £900. Von Stosch seems initially to have been quite enthusiastic about the possibilities of these small boats, as he was about the 'fish torpedo' in general, but his Deputy Director of Admiralty, Rear Admiral von Henk, had some reservations:

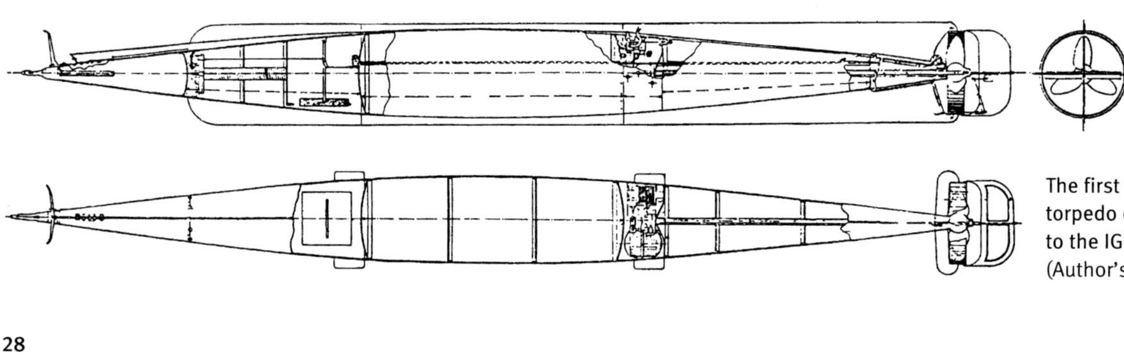

The first Whitehead torpedo delivered to the IGN. (Author's collection)

> In my view, it may be sufficient to have recognised the potential of such small speedy vessels, as a factor has doubtless been created to be reckoned with in the future, especially in view of [the failure of] related attempts on our own side.'[27]

While von Schroetter was actively investigating possible carrier vessels, the authorities at home had not been idle either, provisionally installing a 381mm 'torpedo-gun' for trials aboard the old wooden I. class gunboat *Basilisk* of 1862 vintage. She received the designation '*Torpedokanonenboot*' [torpedo gunboat] for her subsequent trials, which took place from 24 March to 16 May 1874. Trials in Wilhelmshaven roads were attended by von Stosch on at least one occasion.

While the acquisition of small boats from abroad had been postponed for the immediate future, von Stosch continued to look for an appropriate way of introducing the 'fish torpedo' and the acquisition of a suitable carrier vessel. He next turned his attention to Italy, and in particular to the proposed design of the so-called 'Mattei torpedo craft', which ultimately emerged as the Italian 'torpedo cruiser' *Pietro Micca* of 1875.[28] As early as 18 April 1874, von Stosch wrote a memo regarding the great interest his 'colleague', Admiral di Saint Bon,[29] had expressed in the Italian Diet for realising this particular project as soon as possible. Von Stosch had become curious not least in view of the projected high speed of 17 knots, which would enable the boat to easily 'drill its torpedoes into the enemy's flanks.'[30] And, like Saint Bon, von Stosch saw the comparatively low price of 500,000 francs as another key factor. He asked the German embassy in Rome to obtain further information, either via the Italian Government or, possibly, by approaching Mattei directly. In an initial reply dated 1 July 1874 von Stosch received information to the effect that the Italian Government was still undecided, so no further details could yet be reported. However, two weeks later von Stosch received word from Saint Bon that, in the interest of good relations between Italy and Germany, he was prepared to reveal some details of the project with the proviso that these should be kept strictly secret. Von Stosch was more than happy to conform to this request, and it was agreed that Senior Engineer Zarnack[31] would be withdrawn from his recent post in London to visit the engine shops at Genoa, where the machinery for the boat was under construction, and also the model on display in Rome, as the keel had not thus far been laid. Unfortunately, Zarnack's report of October 1874 has not survived, so its impact on German plans must remain speculative – if there was any at all. In the end, *Pietro Micca* seems to have been a relatively unsuccessful design, not up to expectations, and in stark contrast to the boat which may have been inspired by her (see below).

Von Stosch's next move would be to secure funding for the acquisition of a despatch vessel ('Aviso' in German terminology) in the 1875 Estimates, which ultimately emerged as *Zieten*.[32] This would remain the only such order by the IGN during von Stosch's tenure as Head of Admiralty and ran counter to his convictions, which favoured encouraging domestic shipbuilding activity. (The much later acquisition of model torpedo boats from companies such as Yarrow and Thornycroft – see below – was initiated under his successor Caprivi.) *Zieten*'s design history remains obscure apart from the well-known fact that she was offered by Thames Ironworks of Blackwall, London. It is unclear who was responsible for the design of such a unique vessel, which had no true predecessor, at least in the UK. However, the following observation from an unexpected source may shed some light on the matter. In 1884 the Royal Navy's former chief designer, Sir Edward Reed, stated:

> As to sea-going torpedo vessels, the German government took the lead in the production of this type of ship (...) an example of naval enterprise worth remembering to the credit of Germany.[33]

Although this eulogy alone does not constitute proof, the author is inclined to believe that Reed himself may have had more than a hand in her design. With the exception of her two 381mm torpedo launch tubes below the waterline, one in the stem, one in the stern (and the short-lived possibility of employing towed Harvey torpedoes), *Zieten* was initially unarmed, though she would shortly receive four 8.7cm guns for (minimal) self-defence in the event of her employment as a despatch or reconnaissance vessel. This outfit was augmented in 1877 by a deck-

Pietro Micca as completed, date unknown. (Author's collection)

mounted 'torpedo-gun' for trials only, designed by Whitehead in collaboration with the Torpedo Trials Command. *Zieten* would become one of the most long-lived ships of the IGN, serving initially (as planned) as a torpedo trials vessel and later in a variety of tasks, ranging from 'scout' to fishery protection vessel and, finally, picket duties during the war.

After her delivery – she would run her initial trials in

The earliest known image of *Zieten*. It shows her shortly after delivery in the first of her many configurations. (Author's collection)

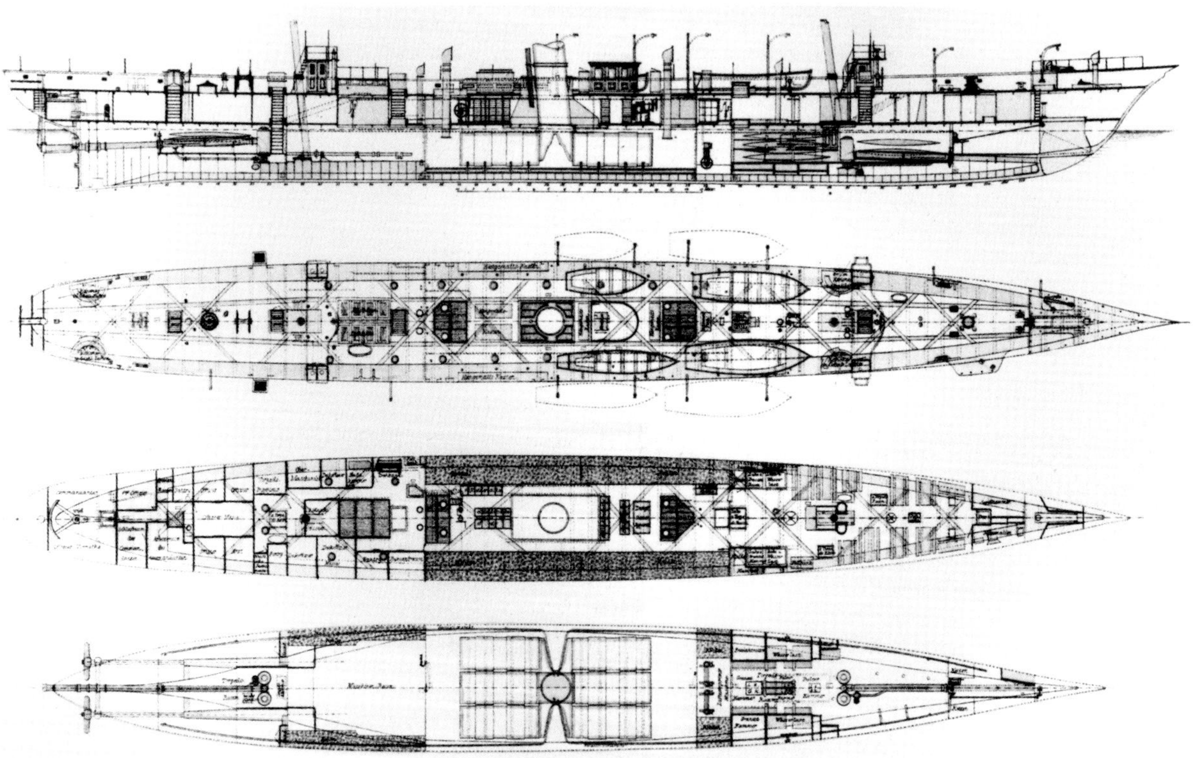

Set of drawings prepared by the Imperial Admiralty of *Zieten* in her second configuration, already mounting the four 8.7cm-guns on her upper deck. Note the bow and stern torpedo tubes. (Author's collection)

the UK in July 1876 before commissioning on 1 August, subsequently arriving at Wilhelmshaven on 11 August – efforts were concentrated primarily on making the torpedo itself an effective weapon. On 16 May 1877 von Stosch would choose and direct the promising young *Kapitänleutnant* Alfred Tirpitz, who had been officer in charge of warheads and detonators within the TVK since 1 January, to accompany the Head of the Torpedo Service, Cdr Eduard Heusner, on a trip to Fiume. The purpose of this visit was not only to learn directly from Whitehead all the key aspects relating to the torpedo's function, production, and operation, but also to be able to address all the technical problems that had emerged during the tests at home. The order also appointed Tirpitz to the post of torpedo officer on board *Zieten* on his return, and to train German personnel using his freshly-acquired knowledge. Von Stosch closed his order with the prospect for Tirpitz that, if he lived up to (Stosch's) expectations he would follow Heusner as soon as his superior was appointed to a seagoing command. As history tells us, Tirpitz more than fulfilled von Stosch's expectations, both before and after his return from Fiume at the end of June. He would remain at the TVK until 30 September 1879, before being appointed to the Admiralty as responsible officer for all torpedo-related matters, a post he held until 1886. During the summer campaigns of 1878–80 he was temporarily attached to *Zieten* as her commanding officer at the same time and subsequently, until 1884, as the CO of her successor as torpedo trials vessel, *Blücher*.

Tirpitz's exhausting mission to Fiume bore some fruit – not the least building up good personal terms with Whitehead – and he would return during the winter of 1877/78 for a second, even more difficult mission. His instructions were to convince Whitehead to take back, or hold back a large part of von Stosch's initial order of 100 torpedoes totalling 50 weapons. The trials commission had become increasingly dissatisfied with the workmanship and performance of the new weapon from Fiume, one of the key factors being that most of the torpedoes were found to be too heavy for the waters of the Baltic, which were less dense than those of the Adriatic. Happily for both parties, the torpedo was the weapon of the day, and Whitehead had no difficulty selling the 50 torpedoes elsewhere. The remaining half of the order was modified to some extent at the newly-founded Friedrichsort depot after arrival, to be made 'war capable', in the words of Tirpitz. These reconstructed 350mm C/77 (Fiume

Nothing was left to the imagination for the readers of the *Illustrierte Zeitung* of 22 January 1881, when these detailed illustrations of the devastating effects of the torpedo hit on *Barbarossa* on 28 July 1880 were published. (Author's collection)

Mk IIIC) torpedoes would receive the designation C/79, and would constitute the core weapon for the subsequent experiments and trials until the introduction of the locally-manufactured Schwartz-kopff torpedoes (of which more later).

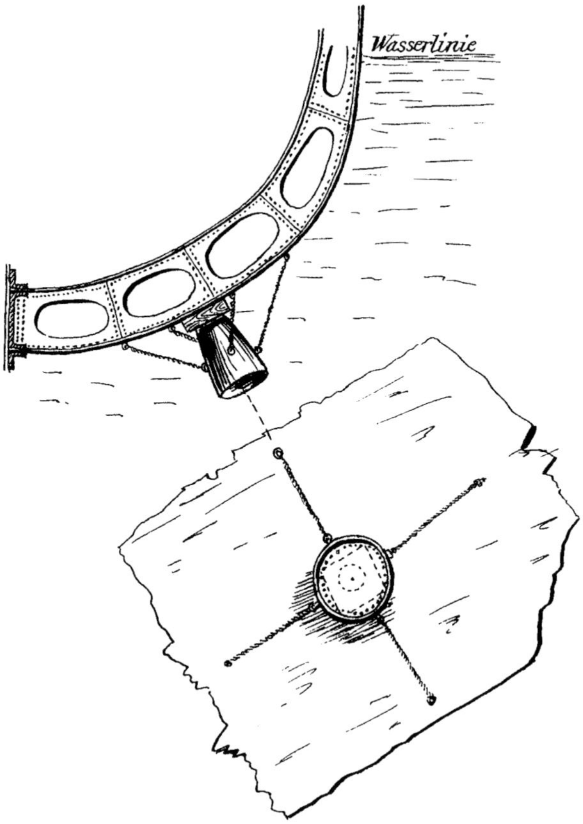

The attachment of the explosive charge beneath the ship's bottom. (Author's collection)

One major highlight of Tirpitz's work, recounted in a slightly off-hand manner in his later memoirs, occurred on 28 July 1880, when he was ordered by von Stosch to demonstrate the progress of the new weapon in a live trial against a ship in the presence of Crown Prince Friedrich Wilhelm (later the '99-day Emperor' Friedrich III) and his sons. The old hulk *Barbarossa* – once the *Britannia* of Cunard heritage, and one of the few survivors of the luckless federal fleet of 1848 – had been selected as the target. Tirpitz would later admit the risk he had taken in displaying a weapon which was still not ready for service and whose trial firings remained 'a true lottery, despite weeklong preparation, as to whether it finds its target or ends up careering wildly around.'[34] The scale of the task remains unclear: some sources state that *Zieten* had been moored some 400 metres away from *Barbarossa*, while others claim that she had first crossed the Kiel Fjord at full speed before delivering her shot at this distance. The latter course appears less likely given the high stakes on that day, with von Stosch's reputation still suffering from the *Grosser Kurfürst* affair[35] that had occurred two years previously. Another spectacular failure would not have made his position any easier; however, everything went surprisingly well, and the performance of the torpedo exceeded even the most optimistic expectations of the experts, despite apparently running and striking its target at surface level (see the accompanying illustrations). In the words of a contemporary correspondent:

> Immediately, the decidedly devastating effect of the torpedo on the fore part of "Barbarossa" became evident due to a high-rising plume of water. The ship, hit between bow and midships, took on a forward list and, filling with water, was obviously sinking, while the gunwales would remain visible, as the depth of water at this position was not very great. What had been a ship some moments before was now a wreck![36]

The remains of *Preussischer Adler*, lying on a mudflat outside Wilhelmshaven. The effects of the detonation below the fore part are clearly visible, while the crack in the hull aft seems to have been caused by sagging while the ship was lying on an uneven keel. (Author's collection)

Much less well-known is the preceding trial of 23 June 1880, which in a way prepared for the above, when the effect of detonating a torpedo warhead close to a ship's bottom was evaluated. The target chosen was the former Prussian iron postal steamer *Preussischer Adler* of 1846, taken into service in July 1862 as a despatch vessel and Royal Yacht but since 1878 hulked at Wilhelmshaven, the hull being completely worn out. The ship was moored in a creek within the Jade mudflats, leaving about 1.5 metres of water below the keel at high tide.[37] The explosive charge was the warhead of a standard C/79-torpedo, containing a charge of 19.5kg wet guncotton, secured to the bottom of the hull two metres below the waterline and electrically triggered. Unsurprisingly, the effect was as devastating as the trial at Kiel one month later. Following detonation, the fore part of the hulk lifted some 1.5 metres before rapidly sinking to the bottom, despite the adjacent bulkheads fore and aft remaining intact.[38] The charge having proved its effectiveness even against iron structures, the trial at Kiel could be prepared with confidence.

The sinking of *Barbarossa* would be *Zieten*'s ultimate achievement, as her successor in the role of torpedo trials ship, the new *Blücher*, was already at hand. Originally approved as 'covered corvette C' in the 1875 Estimates, work was begun in November 1875 and the keel laid in March 1876 alongside her sister-ship, already under construction and later to be named *Bismarck*, at the Norddeutsche Schiffbau AG at Kiel.[39] She was launched eighteen months later, on 20 September 1877, christened by the head of the Baltic Naval Station, Rear Admiral Reinhold Werner.[40] Between December 1878 and January 1879 she underwent her initial trials, after which she was laid up for the whole year. It is not clear precisely when the decision was made to convert her to a trials ship; a covered corvette was comparatively large for a pure torpedo training vessel, and had to be taken from the very limited number of available ships of this type for service abroad. However, on 10 August 1880 *Blücher* would commission for the first time in her new capacity. According to an early report by Tirpitz, then still in command of *Zieten*, she was far from being completely fitted out during these first two months of trials, which is unsurprising in view of the fledgling nature of the torpedo service:

> The trials with the Sb. torpedo-gun commenced in the final week of August, initially from a static ship. During these trials the working of the gun [*sic*] and the very convenient measures for taking torpedoes on board were demonstrated, as well as the comparability of the trial results when launching the torpedoes from a height of 0.8m, as compared with the height of 0.7m of the previous gun mounted on the minelayer. That is why it was decided to continue on the following day from a moving ship.
>
> Because of the limited time available during this summer's season, caused by the late completion of BLÜCHER's armament, I chose to investigate primarily whether an increase in launch speed would have any positive impact on the ability to launch on the broadside. (...) The trials commenced at speeds of 5 knots and 8 knots, at angles of 90° and 45° in the forward direction, and at the lowest and highest permitted launch pressures. The distance to the target varied between 300 metres and

A straight run into the target on the torpedo range in Wik Bight within the Kiel Fjord. A steam launch is in pursuit to recover the precious torpedo. (Author's collection)

> 350 metres. (…) However, it will most certainly be impossible to terminate these trials in the current year. In order to obtain valid results further thorough and systematic trials will be necessary and, because of the belated commencement of the BLÜCHER trials, I was not confident from the outset about what could be achieved this year![41]

To sum up, it would be fair to state that recent experiences still left room for major improvements before the new weapon could be considered a substantial addition to a ship's fighting power in anything other than ideal conditions (*ie* no wind, Sea State 0 and a target broadside on). In 1881 *Blücher*'s role was modified and she served not only as a trials platform but also as a training vessel for future officers who wished to qualify for the torpedo branch. In order to perform this function, she would spend the remaining years of her early career in the confined waters of Kiel Fjord. Her training and trials mission would subsequently be interrupted by sporadic deployments with newly-acquired torpedo boats and participation in the regular autumn manoeuvres, in which she generally simulated an ironclad due to a shortage of available vessels of this type. In 1886, when *Blücher* underwent a thorough overhaul and modernisation at Wilhelmshaven naval dockyard, the Navy thought about converting her back into a cruiser-frigate, but this came to naught. The wooden covered corvette *Elisabeth*, which had just returned from her last major cruise in March 1886, was selected as her possible replacement. She would replace *Blücher* as static training vessel at Kiel during her refit, but the idea was dropped once the latter was able to return to duty in the newly formed Torpedo Inspectorate. In subsequent years, *Blücher* would be equipped successively with all kinds of launching arrangements for torpedoes plus a few guns, mainly for saluting purposes. The best overview dates from around 1900 (see accompanying table).

Ulan had likewise returned on 20 July from her own major reconstruction, and was able to support *Blücher* in the various trial scenarios to come. During her early trials, it was quickly established that the installation of the submerged bow torpedo tube was much more successful than that on *Zieten*, thereby demonstrating the value of the conversion.

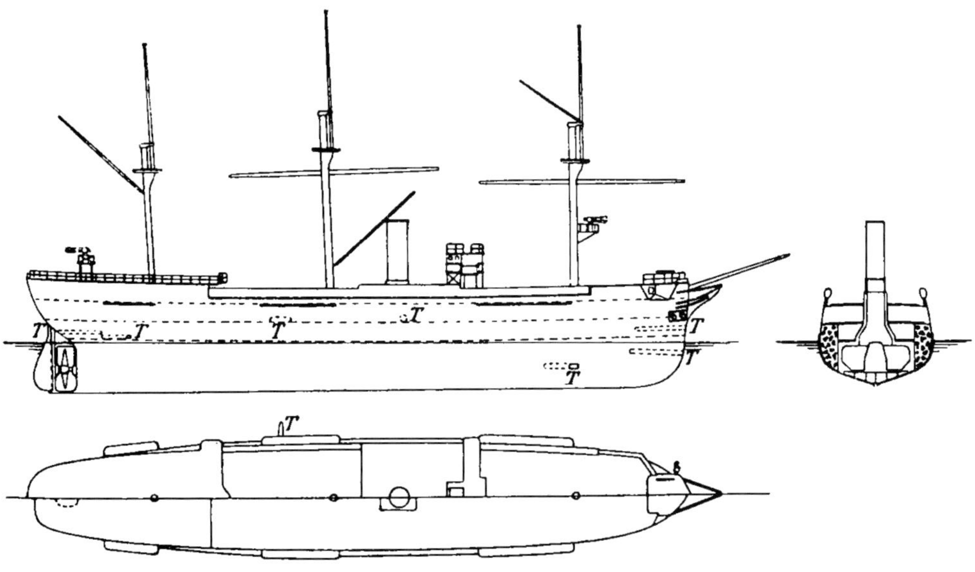

The official sketch drawing of *Blücher*'s torpedo tube armament around 1900 – see also the accompanying table. (Author's collection)

Blücher: Torpedo Arrangement c1900

Designation:	Type:	Location:	Orientation:
Bow/stern tubes:			
Bow TT	35cm (sub)	beside stem below tween deck	straight ahead
Bow TT	45cm (a/w)	port side on tween deck	trainable 35°P–6°S
Stern-TT	35cm(a/w)	tween deck	straight aft
Broadside tubes:			
Starboard	45cm (a/w)	12.5m aft of stem	30° forward of beam
Port	45cm (a/w)	battery deck, 34m aft of stem	trainable 45°fwd–45°aft
Starboard	35cm (a/w)	tween deck	trainable 41°fwd–39°aft
Port	35cm (a/w)	battery deck	trainable 70°fwd–70°aft

Note: In peacetime there were only two 8cm/25 boat guns mounted on the forecastle for saluting purposes. In wartime, when it was intended to employ the ship as a transport, the gun armament was to be augmented by a second pair of 8cm/25 boat guns astern on the poop deck, as well as thirteen 3.7cm Hotchkiss revolver cannon, distributed around the bulwarks.

The torpedo boat *Schütze* shortly after delivery, alongside the converted *Ulan*. Note the experimental mounting of a 37mm Hotchkiss revolver cannon atop the forward conning position of *Schütze*. Astern lies one of the German-style Rendel gunboats of the *Wespe* class. (Author's collection)

The First Boats

At this point, the turn of the 1870s into the 1880s, it should be pointed out that thus far the IGN had been primarily concerned with the task of developing the torpedo from an experimental into a functional weapon, but had left to one side the question of a carrier vessel. In Tirpitz's words: 'In our navy in particular, the question of a suitable torpedo boat had neither been the object of any serious scrutiny nor, until about 1881, had it even been considered.' He would tone down this somewhat dismissive statement immediately by taking

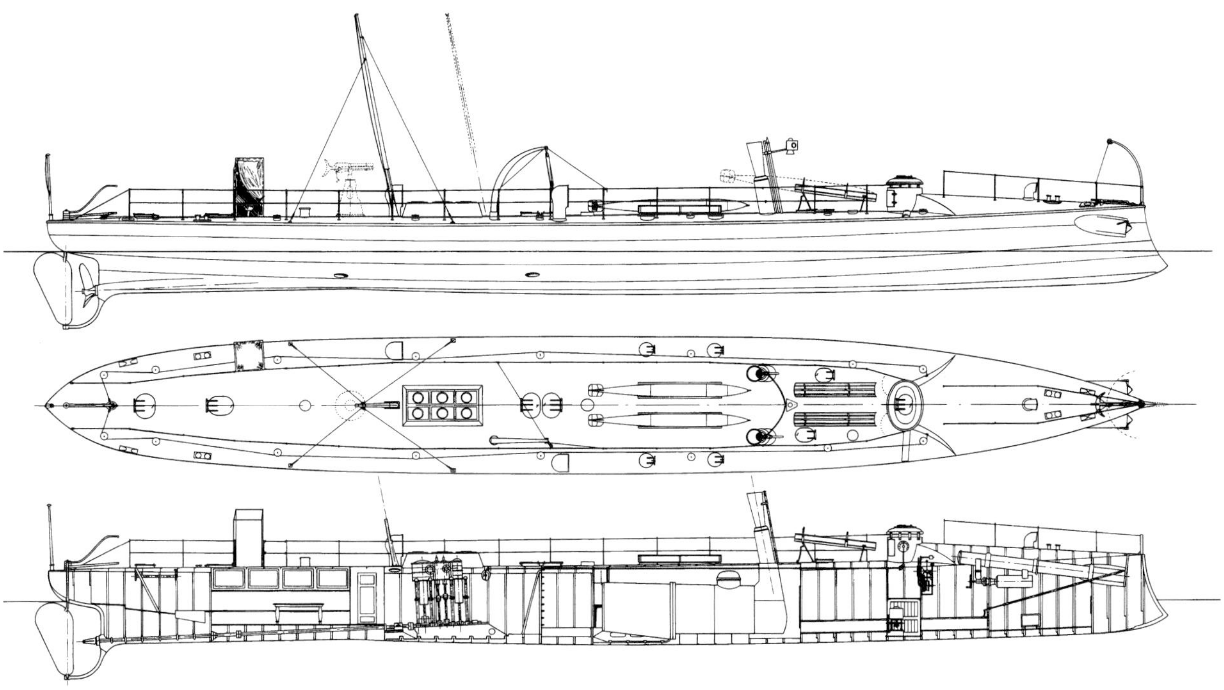

External profile and plan and longitudinal section of *Tapfer*. The broken lines indicate the revised position of the mast when the boat was rearmed with an additional 37mm Hotchkiss revolver cannon. (Drawn by the author)

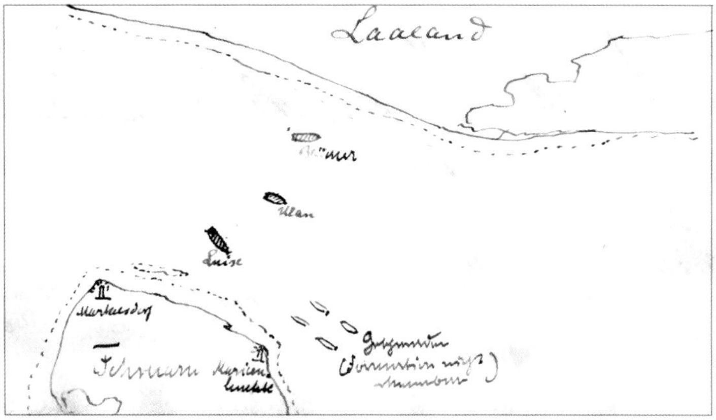

The location of the Battle of the Fehmarn Belt, between the Danish island of Laaland and the German island of Fehmarn. Admiral von Wickede had chosen to steer his ironclad squadron closer to the German side (track not depicted here), where the torpedo boats lay in ambush. (Author's collection)

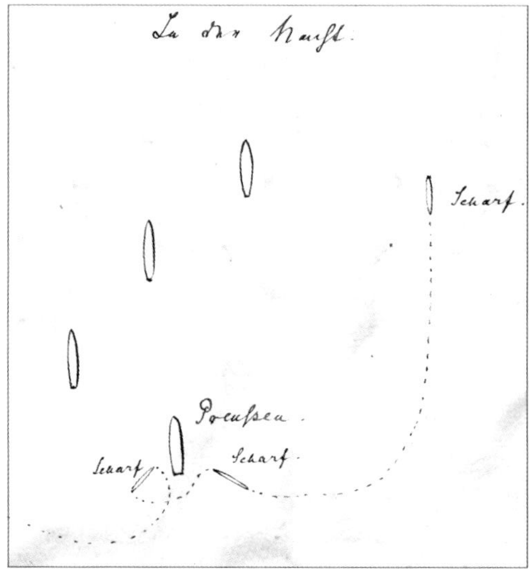

The track of the successful attack of *Scharf* on *Preussen*. Without anti-torpedo boat guns or searchlights the larger vessels were virtually defenceless against such attacks. (Author's collection)

his usual stance – one later adopted also by Caprivi – that the hasty introduction of a weapon still under development would have done more harm than good. This axiom would be valid throughout his entire career, and would lead to many quarrels with Wilhelm II and his more 'progressive' colleagues. This state of affairs continued for another two years; however, in the final year of von Stosch's tenure – he would leave office on 20 March 1883 – the situation changed dramatically and contrary to expectations.

In February 1882 the Russian General Skobelev[42] delivered a fiery speech in Paris which was taken in Germany as the first indication of an increasingly belligerent attitude on the part of Russia, despite the recent accession to the throne of Tsar Alexander III and the prolongation of the so-called 'Three Emperors Pact' (*Dreikaiserbund*) in June 1881. The IGN was quick to react, calling for an increase in the originally-planned acquisition of three small torpedo boats to seven in the 1882/83 Estimates – a development that had been under discussion for some time. (The arming of larger warships with 'torpedo-guns' is outside the focus of this article.) In the longer term, a total of ten large and twelve smaller boats were to be acquired. The Diet seems to have been equally concerned by Skobelev's speech, and the additional funding was readily made available, resulting in the immediate placing of an order with the AG Weser shipyard once the Estimates were approved in early April 1882. In the words of Tirpitz:

> That is how a heightening of tension with Russia prompted by Skobelev's speech in Paris forced us into an overly hasty order for seven boats of the 'Schütze' class,[43] with reduced times for delivery. The boats had been designed at the Admiralty as a close copy of the [French] Normand boats No. *60–74*,[44] to be constructed at the Weser yard.

The design was initiated by Chief Constructor Dietrich; Engineer (constructor) Gaede and Navy Privy Counsellor Uthemann[45] were responsible for the engineering aspects. The latter would be particularly innovative, exchanging, one year after delivery, the original 500ihp compound engine of *Scharf* for the first German triple expansion engine, with very satisfactory results: horsepower increased by *c*100ihp and maximum speed by one knot while the weight of the new engine remained the same. The deadline for delivery of all boats had been set for the end of July, excluding the torpedo tubes, which would be installed later at Wilhelmshaven dockyard. This deadline was comfortably met, the final unit being delivered on 10 July; each of the boats cost around 200,000 Marks. However, Tirpitz's concerns about excess haste were echoed in a private letter by Lt-Cdr Paul Jaeschke, one of the torpedo specialists working alongside him:

> Until recently, we had received only one boat here; the others are to follow in the coming days. Being truly 'firsts' from our industry, they show several shortcomings, mainly developing insufficient top speed. 'Schütze' is capable of just 15 knots at full power – this ... does not compensate for the light scantlings adopted for construction in order to secure a high maximum speed. The deck plating bends under your feet when you walk on it.[46]

The 1882/3 Estimates – still prepared under the direction of von Stosch – did show a certain stage of maturity regarding the torpedo question, as they provided not only funding for the first true torpedo boats, but for the acquisition of a total of 278[47] 3.7cm Hotchkiss revolver cannon, which had been approved as anti-torpedo boat guns for introduction and distribution across the fleet after thorough testing which ended in February 1881.[48] The boats of the *Schütze* class would later be retrofitted with one of these guns, having been virtually defenceless

A cutaway, full-scale model of an 'A'-class boat used for presentation at exhibitions. (Author's collection)

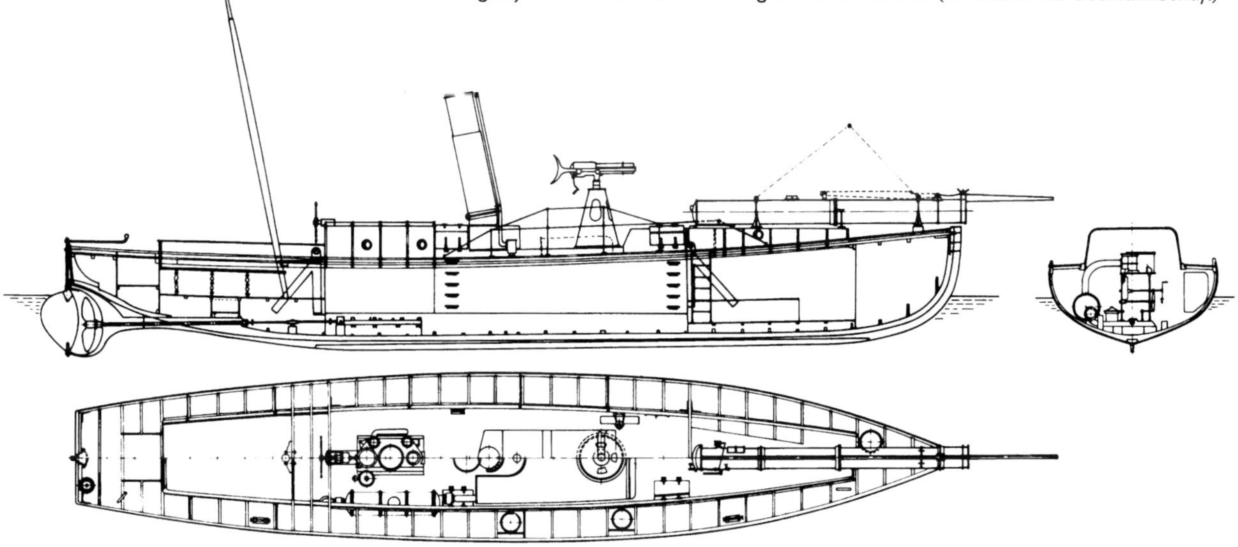

A slightly modified as-fitted drawing of the same boat. (*Handbuch der Seemannschaft*)

as originally completed. The revolver cannon was initially mounted atop the forward conning position, but was soon moved to a special pedestal abaft the engine room skylight. The size of the boats was determined by one very important consideration: they had to be sufficiently small to be handled by a single officer only, due to the chronic shortage of officers and men.

Even before this later conversion, on 10/11 September 1882, shortly after their delivery, the boats were to prove their worth during the annual manoeuvres, fighting the mock 'Battle of the Fehmarn Belt.' The concept behind the manoeuvre had been that an attacking force 'from the east,' consisting of four armoured frigates[49] plus a despatch vessel, would attempt to pass the Belt, which was around 8nm wide at its narrowest point, at an undefined hour and approaching on an undefined course, with the aim of mounting an attack on the defences of Kiel harbour some 40nm to the southwest. The defending forces comprised three torpedo-carrying vessels under the command of Tirpitz, simulated by *Blücher*, the flushdeck corvette *Luise*, *Ulan*, and four torpedo boats, *Schütze* (*V*), *Scharf* (*VII*), *Kühn* (*IX*), *Vorwärts* (*X*), under Jaeschke, who was temporarily embarked in *Schütze*. The attackers – unsurprisingly – chose night time to pass through the narrows, but made the mistake of hugging the Fehmarn coast, exactly where the torpedo boats were lying in wait. Following detection of the 'enemy' force, the presence of which was reported by *Ulan*, Jaeschke searched and closed up his squadron, which was credited with at least one lethal hit on *Preussen* by *Scharf*, the latter having closed to 250 metres undetected.[50] In his later report, Jaeschke would highlight the difficulty of keeping the boats in formation in darkness, having lost sight of at least one during the attack – a finding that would subsequently be repeated whenever these small boats operated together in numbers in anything but a calm sea and good visibility. However, even if the subsequent daylight action, in which the torpedo boats were

The unusual shape of the hull of this unit – at least for a boat of German design – suggests that it may be one of the mysterious 'White boats'; it is depicted during a trial run. (Author's collection)

repelled during their attack on the defences of Kiel, could have been considered a setback, von Stosch was sufficiently impressed by the results to support Tirpitz's demand for a further 18 torpedo boats[51] for trial purposes in the 1883/84 Estimates – the last for which he would be responsible.

Prior to the unplanned construction of the *Schütze* class, the Admiralty focused briefly on the acquisition of second class boats to be transported on larger units of the fleet and deployed close to a hostile anchorage or port for attack. The concept was not original, as virtually every navy had embarked on similar projects (*eg* the Royal Navy's HMS *Hecla*). In 1882 the Imperial Dockyard Danzig received an order to construct a prototype boat from wood for testing. Beyond its general specifics, cited below, little is known about this boat and its subsequent trials:

> The hull is constructed on the same principle as steam launches, with diagonal planking, and is protected against sinking by a number of air-tight boxes made from copper. The planking is of mahogany to reduce hull weight. Dimensions are 14.5m x 2.9m x 1.3m. The engine is on the compound principle, delivering 100ihp. Armament consists of Whitehead torpedoes and a Hotchkiss revolving cannon mounted in the bows. Torpedoes are launched by lowering two open metal tubes, mounted on either side of the hull, by a simple lever via a transmission. During transit the tubes remain in the raised position, to be lowered immediately before the attack. The torpedoes are forced out of their tubes using a compressed air cartridge.
>
> An extended trial run, held at Danzig, gave a speed of 11 knots. (...) The construction of the engine had been awarded to the firm of F Schichau by the Admiralty. (...) The design of the engine is a speciality of the F Schichau company from Elbing, which had already built its first compound engine for the IGN in 1873.[52]

The results must have been promising, as the boat later became attached to Tirpitz's trials command at Kiel, and the Type 'A' steam launch would subsequently be developed from it; it was slightly larger than the experimental boat and was equipped with a single removable 450mm torpedo tube in the bows.

In addition to this boat, the IGN apparently placed orders with the British shipyards of JS White, Cowes, for three second class boats made of wood, and J Thornycroft, Chiswick, for a single boat at a cost of £4,700. While nothing has come to light about the history of the three White boats other than the barest possible information on their existence, the latter boat may have been identical with boat *No IV – Th 2*, mentioned briefly above.[53] The idea must have originated with von Stosch in early 1882 – without having been executed immediately – only to be questioned by his successor von Caprivi 18 months later, having browsed the existing files after he had taken over, in a letter to an unnamed recipient. We will return to it in Part II of the article, which will cover the further developments under the latter's tenure. *(to be continued)*

Endnotes:

1 Following a career in the Army, Werner (von) Siemens (1816–1892) was co-founder of the renowned electro-technical works of Siemens & Halske (the Siemens AG of today).

2 August Friedrich Karl Himly (1811–1885), Professor of Chemistry at Kiel University, was married to one of Siemens' sisters.

3 Robert Whitehead (1823–1905) was a British engineer and is generally recognised as the inventor of the torpedo.

4 Alexander Graf von Monts (1832–1889) was a member of the Naval Ministry section responsible for coastal and naval artillery; he was later promoted to Vice Admiral, and was appointed Head of Admiralty/Supreme Command.

5 Sources disagree on numbers; one claims there were as many as 14 steam tugs in total.

6 The Harvey torpedo was a floating device towed at an angle of 45 degrees, the shape of the body combining with vanes to keep it on course. The attacking ship had to pass its target at a distance of about 40–60 metres at which point, through a variation in towing speed, the torpedo would sink below waterline level for greater effect, being detonated by a special lanyard. This was a very complex device with lots of possibilities for failure.

7 The shipyard was founded at Danzig in 1856 by master shipwright Felix Devrient (1825–1907) and a partner, when the former left the Prussian Navy following a disagreement with leading master shipwright Elbertzhagen. Devrient had previously been responsible for the construction of the paddle corvette *Danzig*, the first domestically-built steam warship in the Prussian Navy.

8 *Leipziger Illustrierte Zeitung*, November 1871.

9 The decision was made public by order of von Stosch in Naval Bulletin No 241: 'The six small steamers, up to now known as "Waltjen'sche resp. Devrient'sche torpedo-boats", will in future receive the designation (steam-) "minelayers", under consecutive numbers; thus, the former will retain their present numbers 1–3, while the numbers of the latter will be raised by three digits so that the former boat No. 1 will become "Minelayer No. 4".'

10 The records are not entirely consistent on their whereabouts, because there is reference to at least one document within the Tirpitz files to torpedo trials on board '*Minenleger V*', formerly (and subsequently) '*Torpedoboot Nr. II*,' which were carried out at Kiel in 1880, while '*Torpedoboot Nr. I*' would take part in the autumn manoeuvres of 1882 at Kiel (see below). Up to the present, the scarce information on

their fates suggests that all three boats were based at Wilhelmshaven during the first half of the 1880s.

11 The C Waltjen & Co shipyard had been founded in 1857 and would become the nucleus of the renowned AG Weser shipyard founded in 1872.

12 In reality, it had never been a plan, rather the presentation of a 'Memorandum on the Further Development of the German Navy, and the requirements regarding material and funding evolving from it.'

13 *Waterwitch* (1866) is commonly reported as having been almost unsteerable under engine power and therefore a complete failure.

14 *Notus* (2 August 1873) – 303mt; 37.2x6.7x2.5m; 1 spar torpedo; speed *c*10kts.

15 *Zephyr* (22 June 1874) – 129mt; 30.5x5.3x1.7m; 1 spar torpedo; speed *c*9kts.

16 *Rival* (2 September 1874) – 129mt; 30.5x5.3x1.7m; 1 spar torpedo; speed *c*6.5kts.

17 *Ulan* (3 April 1876) – 440mt; 36.0x8.0x4.1m; 1 spar torpedo; speed *c*12kts.

18 The simplicity of contemporary engines is evidenced by an entry in *Ulan*'s log to the effect an engine room mechanician was reprimanded on one occasion in the following manner: 'I punish ensign n.N. with three days in the brig, because he brought the M.E. to a halt by simply putting his hand into it.'

19 Carl Heinrich Ziese (1848–1919) was born and partly brought up in St Petersburg; this would later facilitate contacts with the Russian authorities.

20 *Karabin* (September 1877) – 11mt; 19.8x2.6x0.6m; 2 spar torpedoes.

21 *Bomba* (September 1878) – 20mt; 21.7x3.4x0.7m; 2 spar torpedoes. All boats exceeded their design speed of 16 knots, with *Bulava* attaining an impressive 17.56 knots.

22 *Raketa*, *Samopal* (June 1878) – 30mt; 25.6x3.3x0.63m; 2 spar torpedoes, 1 381mm TT; speed *c*15kts.

23 Höft M; *Der Vulcan*, Pt III, 157.

24 Gustav von Schrötter (1830–1919) would later rise to the rank of Lieutenant-General.

25 See in particular the recent work of L Brown: *Royal Navy Torpedo Vessels 1870–1914*, Seaforth Publishing (2023).

26 See Abelsen F: 'The Torpedo Boat *Rap*', in *Warship* No 38, April 1986.

27 Duppler J; *Der Juniorpartner: England und die Entwicklung der Deutschen Marine 1848–1890*, Mittler (1985), 61.

28 *Pietro Micca* (1 August 1876) – 527mt; 61.9x6x3.6m; 1 TT; speed (in service) 12.9kts.

29 Simone Pacoret di Saint Bon (1828–1892), was an Italian admiral and served as Navy Minister 1873–1876 and 1891–1892.

30 BA-MA RM1/16.

31 Only Zarnack's surname has survived, besides the information that his normal employment was with the Imperial Dockyard Danzig.

32 *Zieten* (9 March 1876) – 1,001mt; 69.5x8.6x4.6m; 2 381mm TT; speed *c*16kts. She was named after Hans Joachim von Zieten (1699–1786), a cavalry general under King Frederick the Great.

33 Reed, EJ: *Modern Ships of War* (New York 1888), 127.

34 Tirpitz: *Erinnerungen*, several editions.

35 The brand-new turret ironclad *Grosser Kurfürst* was in collision with the ironclad *König Wilhelm* off Folkestone on 31 May 1878. The former sank immediately with heavy loss of life. The incident discredited von Stosch and his 'system' significantly, and would ultimately become a major factor in his retirement in 1883.

36 *Leipziger Illustrierte Zeitung*, 22 January 1881.

37 For unknown reasons, the date and location of this trial has been mis-recorded in German naval history ever since: almost all the reports give the date as 26 June 1879, and the location as Kiel-Wik – close to the location of the subsequent test; they also state that the sinking had been accomplished by two torpedoes.

38 BA-MA N253/14, fol 36f.

39 The Norddeutsche Schiffbau AG was founded in 1867, went bankrupt in 1879 and, some years later, would become the Germania-Werft, which was taken over Krupp in 1896.

40 Reinhold (von) Werner (1825–1909) was one of the few officers who had started their career in the short-lived Federal Navy of 1848, before being enlisted in the Prussian Navy in 1852. He would end his career as Head of the Baltic Naval Station in 1878, having survived a serious quarrel with Bismarck in 1873. After his retirement he authored a number of books on the Navy.

41 BA-MA N253/14, fol 37ff.

42 Mikhail Dmitriyevich Skobelev (1843–1882), Governor of Minsk and leader of the Pan-Slavic Party in Russia. In February 1882, in front of a delegation of Serbian students, he encouraged each Slavic country to join Russia and France in their fight against Germany and Austria-Hungary with the historic words: 'The road to Constantinople leads through the Brandenburg Gate [Berlin].'

43 *Schütze* (11 May 1882), *Flink* (24 May 1882), *Scharf* (30 May 1882), *Tapfer* (6 June 1882), *Kühn* (12 June 1882), *Vorwärts* (19 June 1882), *Sicher* (26 June 1882) – 56mt; 32.6x3.9x2.1m; 2 x 350mm TT, (1 x 3.7 revolver cannon); speed *c*17 knots. In addition to being named, these boats received the numbers *V-XI* (later abandoned), thereby continuing the system which had begun with the three Devrient boats (*I–III*), and the small Thornycroft boat *Th 2* (*IV*) – see below.

44 *Nos 60-64* (*65-74*) – 46mt; 33.8x3.3x2m; 2 381mm TT; speed *c*20kts.

45 Dipl Eng Friedrich Uthemann (1851–????), rose to the rank of Navy Privy Counsellor for construction and Head of the Technical Office of the torpedo branch. Engineer Gaede, on the other hand, has left no traces in the history of the Navy, having quit the service early.

46 Paul Jaeschke: *Letters 1877–1889*, private publication, 238.

47 Because of the large sum, it was split between the Estimates of 1882/83 and 1883/84.

48 Von Stosch had initially been critical of this new weapon: 'Hotchkiss seems to me more of a toy than a weapon. However, if it proves its value, I will accept one Hotchkiss aboard. The true king of weapons remains the gun, which is thus the most dangerous opponent for men and ships.'

49 *Friedrich Carl* (flag), *Friedrich der Grosse*, *Preussen*, and *Kronprinz*.

50 Jaeschke himself speaks of only *Friedrich Carl* surviving the attack; see Paul Jaeschke, *Letters*, 243.

51 The timing of this request is odd, as Jaeschke had already mentioned an order for this precise number previously, in a private letter written on 4 September 1881; see Paul Jaeschke, *Letters*, 192.

52 *Mitteilungen auf dem Gebiete des Seewesens 1882*, 79.

53 The II. class boat had been derived from Thornycroft boats Nos *76-95*.

THE TWIN 6in Mark XXI TURRET

Introduced in the *Leander*-class light cruisers in 1933, the twin Mark XXI turret, mounting 6in Mark XXIII guns, became the benchmark for the modern British medium-calibre weapon systems that fought the Second World War. **Peter Cannon** examines the design, construction and equipment performance of one of the most successful and influential cruiser main armament systems of the mid-20th century.

Following the development of large and expensive 8in gun-armed cruisers, the Royal Navy's 1929 decision to revert to building smaller ships carrying 6in guns is attributable to two factors: the incompatibility of attaining sufficient numbers with available finances along with the the need to to build a ship better suited to performing a range of fleet tasks. While powered multiple gun mountings had long served at sea, the development of similar arrangements for the hitherto hand-worked 6in calibre constituted a significant step in British cruiser development. Multiple mountings made the most efficient use of limited centreline space and provided several advantages over the preceding single, less complex mountings. These requirements led to the twin Mark XXI turret, carrying the new Mark XXIII 6in gun, being developed for the light cruisers of the *Leander* class.

The *Leander*s were originally conceived as battlefleet cruisers, expected to both punch through as well as blunt attacks by enemy destroyer flotillas, but evolved into an all-purpose design of higher displacement than initially envisaged. HMS *Leander*, ordered as part of the Navy's 1929 construction programme, entered service in 1933 featuring a four-twin-turret arrangement: 'A' and 'B' grouped forward and 'X' and 'Y' aft. She was joined by *Achilles*, *Neptune* and *Orion* of the 1930 and *Ajax* and *Amphion* of the 1931 programmes. *Amphion*, as well as 1932's *Sydney* (RAN) and *Apollo*, were completed to the Modified *Leander* design. Specific refocusing on the battlefleet role produced the similar, but smaller *Arethusa* class with three Mark XXI turrets: 'A' and 'B' forward and 'Y' aft. *Arethusa*, *Galatea*, *Penelope* and *Aurora* were ordered at the rate of one per year over the 1931 to 1934 programmes. *Apollo* and *Amphion* were transferred to the RAN as *Hobart* and *Perth* in 1938 and 1939 respectively, while *Leander* and *Achilles* served in the New Zealand Division of the Royal Navy and in the RNZN after its establishment in 1941. The technical data in the present article refers to the four-turret ships unless specifically noted otherwise.

Development

The mountings designed for the *Leander* class were influenced by, but not derivatives of the experimental twin Mark XVII, carrying the older Mark XII gun, trialled in the cruiser *Enterprise* in 1926, as well as the Mark XVIII secondary armament of the battleships of the *Nelson* class (1927) with the new Mark XXII gun. To maintain high rates of fire and take advantage of increases in range facilitated by fire control systems of capital ship complexity, the primary design philosophy centred around streamlining ammunition supply and easing the handling burden on gun crews. Older designs relied entirely upon manual

Table 1. **Ships Equipped with the Twin 6in Mark XXI Turret System**

Ship	Class	Programme	Ordered	Completed
HMS/HMNZS *Leander*	*Leander*	1929	18.02.1930	24.03.1933
HMS/HMNZS *Achilles*	*Leander*	1930	16.02.1931	06.10.1933
HMS *Neptune*	*Leander*	1930	02.03.1931	23.02.1934
HMS *Orion*	*Leander*	1930	24.03.1931	18.01.1934
HMS *Arethusa*	*Arethusa*	1931	01.09.1932	23.05.1935
HMS *Ajax*	*Leander*	1931	01.10.1932	12.04.1935
HMS *Amphion* / HMAS *Perth*	Modified *Leander*	1931	01.12.1932	06.07.1936
HMAS *Sydney*	Modified *Leander*	1932	10.02.1933	24.09.1935
HMS *Apollo* / HMAS *Hobart*	Modified *Leander*	1932	01.03.1933	13.01.1936
HMS *Galatea*	*Arethusa*	1932	21.03.1933	14.08.1935
HMS *Penelope*	*Arethusa*	1933	05.02.1934	13.11.1936
HMS *Aurora*	*Arethusa*	1934	01.03.1935	12.11.1937

Forward main armament group of 'A' and 'B' turrets in HMAS *Perth*, *circa* October 1940. Sighting doors, each comprising two separate panels, are open, and the splinter shield for each gun is visible inside the gun ports. Note the rounded nature of the splinter-resistant gunshield of the Mark XXI twin turret. (Australian War Memorial, 005237/09)

handling, which was aided by simple loading trays in newer weapons. They could not reach high elevations as the requisite increased trunnion height rendered loading impractical. It was argued that powered ammunition hoists and manually operated but gravity-assisted loading systems to mitigate the heavy work of loading, built into protected, fully-enclosed mountings, would reduce fatigue, increase reliability and better maintain ammunition supply to achieve the high rates of fire of which 6in guns were capable. While *Nelson*'s mountings featured power ramming at a fixed elevation, those developed for *Leander* reverted to hand ramming to economise on machinery weight. This was essential to increase rates of fire while working within restrictive naval arms limitation treaties in what was a very tight design. A high angle of elevation reflected contemporary Admiralty interest in controlled 6in anti-aircraft fire.

The Mark XXI mountings were designed to Admiralty requirements by Vickers-Armstrongs Ltd and manufactured at their facilities in both Barrow-in-Furness and Elswick, while the Mark XXIII guns were supplied by either Elswick, the Royal Gun Factory in London's Woolwich Arsenal or William Beardmore & Co's Parkhead Forge works in Glasgow. The mounting was a 'short trunk' design, with a flashtight ammunition handling lobby breaking the hoisting arrangements between magazine stowage and the gunhouse. A short revolving trunk, suspended from the gunhouse into this lobby on the ship's fixed structure, offered a lighter mounting requiring less power to train and favoured cost-effective production. It also allowed a concentration of magazines and shell rooms that minimised the extent (and cost) of armour protection. Despite emerging considerably heavier than anticipated, this arrangement was used in all British cruisers until the *Belfast* sub-group of the larger *Southampton* class ordered in 1936.[1] During development, the decision was taken to increase the weight of British 6in projectiles from the relatively light 100lb to 112lb to improve ballistics and penetrative power. This change, approved in May 1931, came too late for *Leander* but was applied to all subsequent ships through relatively minor gun modifications.[2]

General Description

The twin Mark XXI system incorporated a revolving structure, comprising a fully enclosed gunhouse atop a turntable and working chamber, with dedicated ammuni-

HMAS *Sydney*'s forward main armament group of 'A' and 'B' turrets with seamen washing down the forecastle. Sighting doors are closed, 'B' turret's guns have non-ceremonial tampions mounted and 'A' turret's centreline Officer of the Turret's periscope is visible atop the turret roof. The photo was taken in Sydney Harbour on 5 February 1940. (Frederick Grimes, Mitchell Library, State Library of New South Wales and courtesy ACP Magazines Ltd, FL9619610)

tion hoists suspended into an ammunition handling compartment within the ship's fixed structure. The revolving structure facilitated mounting training and gun elevation while being supplied with 6in projectiles and separate uncased propellant cartridges from the fixed structure. Each gun rested in its own cradle capable of individual elevation and depression between +60° and -5° respectively, and could be loaded anywhere between +5° and +12.5°. While primarily power-operated, turrets could be kept in action in the event of a loss of mains power by hand-driven control and battery backup maintaining all systems including laying, training, ammunition supply, fire control and firing circuits.

The gunhouse was rotated and the guns elevated by hydraulic engines, while the revolving structure also carried hydraulic shell and cordite hoists, one for each gun. The mountings were designed to achieve eight rounds per gun per minute under favourable conditions (to be examined below), while the hoists lifted sufficient ammunition to supply two guns, in the event of a breakdown, under most combat conditions. While the relatively light 31lb cordite propellant cartridges were manually handled in the gunhouse, shell supply used a series of gravity-assisted trays delivering projectiles to the hand-worked breech of the gun. They were then rammed, likewise by hand with a cushion-headed wooden stave, into the chamber, followed by the propellant cartridge thrown in behind it.

Within the ship's fixed structure, cordite cartridges and shell were stowed in magazines and shell rooms in the hold, deep below the waterline. A single magazine supplied cordite cartridges for a pair of turrets with each containing a dedicated flashtight handling room for each turret. Cordite cartridges were withdrawn from their containers by hand and passed into the handling rooms via flashtight scuttles before being sent up to the ammunition lobby that interfaced with the revolving structure. 'Y' turret, which had its lobby directly above the handling room, utilised two flashtight 'hand-ups', while the remainder relied upon single electrically-powered hoists. Each turret had its own dedicated shell room, with projectiles also being delivered to the ammunition lobby by electric hoist. Within the lobbies, ammunition was transferred by hand to the revolving hoists suspended below the gunhouse to supply the guns.

The mountings were directed by the ship's primary fire control system, which comprised a single aloft Director Control Tower (DCT), mounted on the upper bridge, and the Transmitting Station (TS) with its electro-mechanical computer, the Admiralty Fire Control Table (AFCT). While turret crews received training and elevation values from the fire control system, they manually applied them to their equipment, hence fighting the turret themselves. The electrical data transmission systems of the early 20th century developed sufficient power only to drive remote visual indicators, such as the pointers in training and elevation receivers, to convey information, instructions and orders from the controlling director and supporting calculating instruments to the gun mountings. They were not remotely controlled, as with modern servo systems, where control signals are amplified into the power necessary to initiate and control their movement. This link in the chain was filled by highly skilled men converting received information into movement to train the turrets and elevate the guns to very high degrees of precision, the guns being remotely fired by the director layer in the DCT. Each turret was incorporated into an integrated system of fire control communications using powered telephones, voicepipes and mechanical indicators. In the event of breakdowns, each turret was fitted with a pedestal-mounted Local Director Sight (LDS) to fight the mounting independently of centralised control.

The 6in BL, Mark XXIII Gun

The 50-calibre Mark XXIII was an all-steel, built-up construction weapon comprising a tapered 'A' tube, shrunk-on jacket, breech ring and breech bush. A large chamber behind the bore facilitated the loose introduction of a silk-bagged propellant cartridge behind the shot seating. A Welin interrupted screw breech mechanism was hand-worked by a breech mechanism (BM) lever. The BM lever was mounted on the outboard side of the breech ring, to the left on left-hand guns and on the opposite side for right-hand guns, so that each mechanism opened outboard, away from the turret axis.

An obturation pad that compressed and helped seal the gun when fired was placed against the face of the breech block, while in front of the pad and securing it to the breech block was a round piece of steel in the shape of a mushroom called the vent axial – but often referred to as the 'mushroom head'. The underside of the head rested against the obturator pad, with the steel stalk of the mushroom going through a hole in both the pad and the breech block. A penetration hole forming a tube down the centre of the vent axial that allowed the cordite charge to be fired was known as the tube chamber. This enabled the insertion of single-use 1-inch vent tubes to fire the gun. The tube, a gunpowder magazine in a metal cylinder, was sealed by working an electric firing lock secured to the rear of the vent axial, the lock also serving to eject spent tubes when opened. Electrical firing circuits initiated the tube by heating a striker needle in the lock to fire flaming pellets into the gun chamber and ignite the cordite cartridge.

The design of the 6.91-ton gun achieved an efficient balance between penetrative power against both vertical and horizontal armour at practical fighting ranges, and longevity. While still a high-powered weapon, muzzle velocity (MV) in the Mk XXIII was reduced from earlier practice to a relatively modest 2,700ft/sec (new gun) for a maximum range of 24,800 yards, a performance exceeded by most foreign contemporaries. However, owing to inherent fire control limitations, maximum range surpassed those at which hits could reasonably be expected, while British projectiles retained sufficient striking velocity to do significant damage. The choice of MV influenced the attainment of an impressive barrel life

Mk XXI Mounting

Sectional Elevation at Left Hand Gun

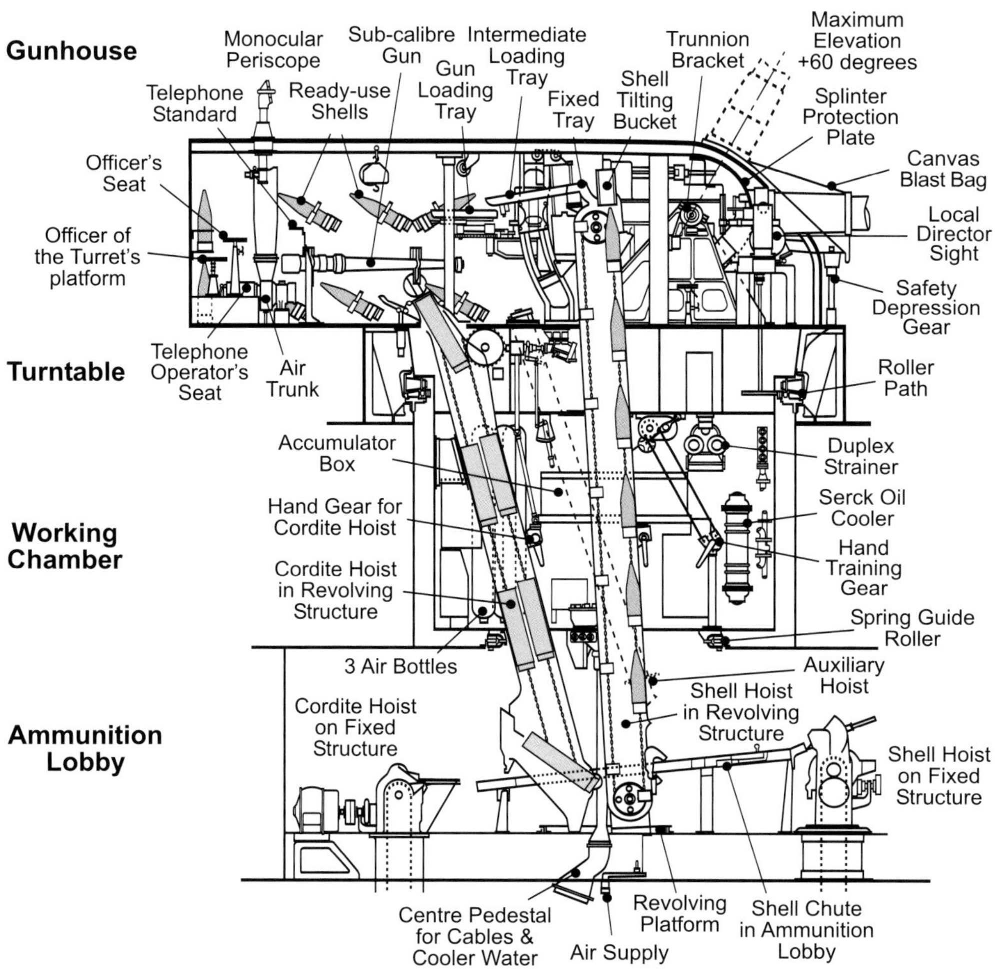

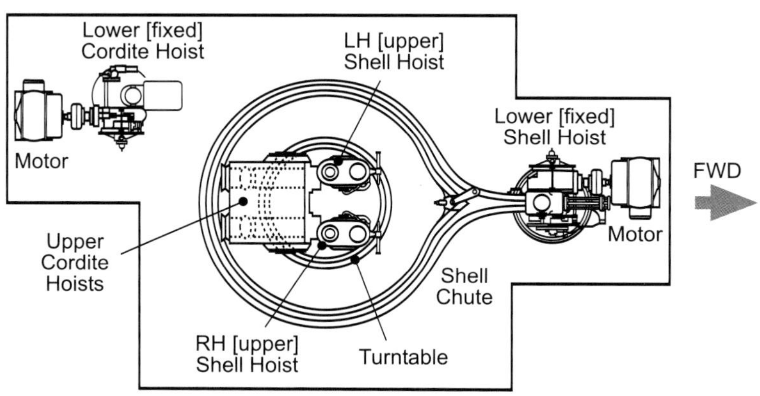

Ammunition Lobby

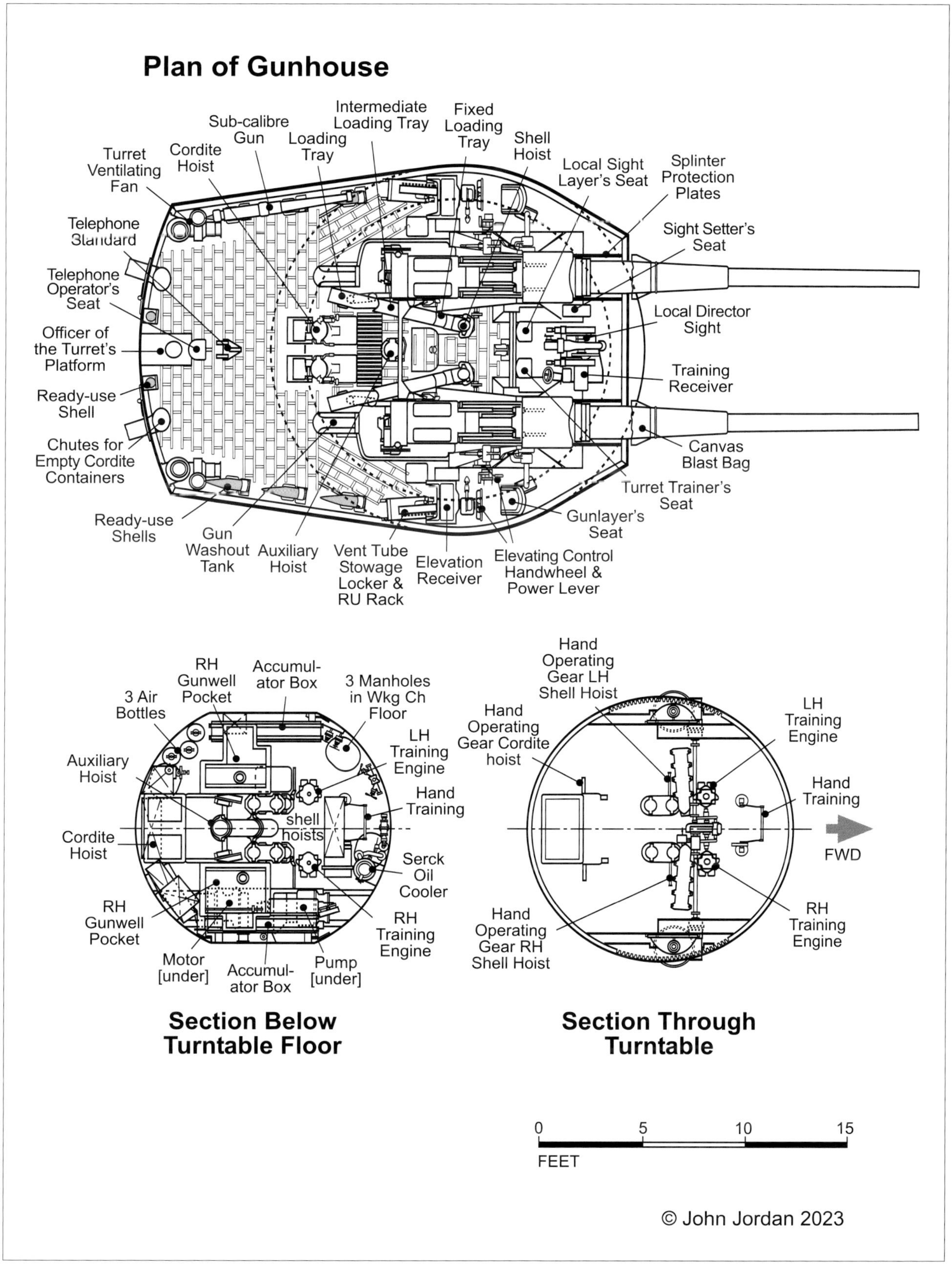

General arrangement of gunhouse, turntable compartment, working chamber and ammunition lobby, as drawn for HMAS *Sydney*'s 'A' turret. Drawn from Vickers-Armstrongs plan No N1060.

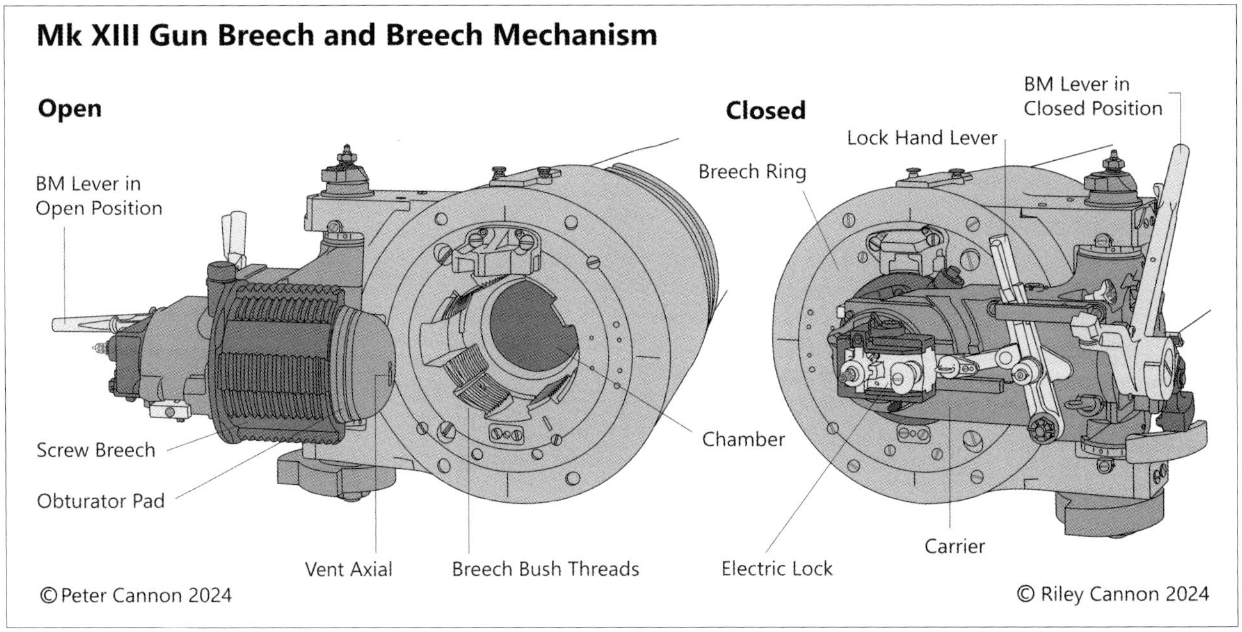

Mark XXIII gun breech mechanism, open and closed. Drawn from BR 224/45, Plate No 5.

HMS *Orion*'s original 'Y' turret right-hand Mark XXIII gun mounted in ex-HMS *Belfast*'s Mark XXIII 'Y' turret. The breech mechanism, with its interrupted screw thread, is swung out with the BM lever horizontal to the right. Atop the carrier arm is the air blast cam, pipework connections and air blast lever, the compressed air being led into the gun chamber through the copper piping attachment on the breech ring, to the left of the vent axial's 'mushroom head' and obturator pad. One side of the breech safety change-over switch is secured to the left side of the breech ring with the other side on the breech mechanism, just to the right of the screw thread. (Peter Cannon, London 7 August 2016)

of 1,020 effective full charges, allowing guns to retain their long-range fighting efficiency far longer than weapons that theoretically outperformed them. Combined with individual, widely-spaced gun cradles, these design decisions avoided numerous accuracy and gun wear difficulties that seriously compromised the effectiveness of comparable foreign weapons.[3]

Ammunition

Mark XXIII gun armed cruisers primarily used Common Pointed Ballistic Capped (CPBC) projectiles in the anti-shipping role. CPBC was a base percussion-fused, semi-armour piercing round fitted with delay/non-delay settings. It was complemented by nose-fused High Explosive (HE) ammunition fitted with direct action percussion fuzes for shore bombardment or either time-combustion or time-mechanical fuzes for anti-aircraft duties. Designed stowage allowed for 200 rounds per gun (rpg); all ships commissioning with outfits of 180rpg of CPBC, 20rpg of HE as well as 31rpg of inert practice ammunition. May 1940 saw the CPBC/HE ratio altered to 170/30 to reflect the increased use of the guns against aircraft while a higher proportion of HE was embarked for specific bombardment operations.

Propellant outfits comprised silk-bagged cartridges of Solventless Carbamite (SC) cordite, introduced in 1927, packaged in cardboard containers, although stocks of the obsolescent Modified (MD) cordite were commonly used in prewar practice firings. The standard service charge for the Mark XXIII gun, SC 150, was a 'full flash' powder and, notwithstanding the obvious advantages of concealment, it was the blinding effect on control personnel at night that proved its greatest disadvantage. As early British flashless powders were heavier than SC, matching the ballistic qualities of service ammunition resulted in charges too large for existing hoists, a problem not resolved until 1945. In the interim, beginning in December 1942, ships were issued with limited quantities of flashless propellant issued in two parts, which complicated supply to the guns but proved effective in action.[4]

Revolving Structure

The gunhouse sat atop a turntable compartment supported by the ship's structure with a working chamber suspended beneath. The gunhouse floor was 30lb (¾in) non-cemented (NC) armour with a fully enclosed gunshield of 40lb (1in) NC. Intended for ships of strictly

Right-hand Mark XXIII gun in ex-HMS *Belfast*'s 'B' turret with breech mechanism swung out. The front of the carrier arm shows the electric lock in the open position mounted in its box slide, with the lock operating lever standing vertical adjacent to the horizontal BM lever. The expendable lead is fitted to the front of the lock and is connected to the breech mechanism side of the breech safety change-over switch. Behind and to the right is the gunlayer's position with elevation receiver, handwheel and power lever. All of this equipment is identical to that used in the Mark XXI turret. (Peter Cannon, London 15 August 2016)

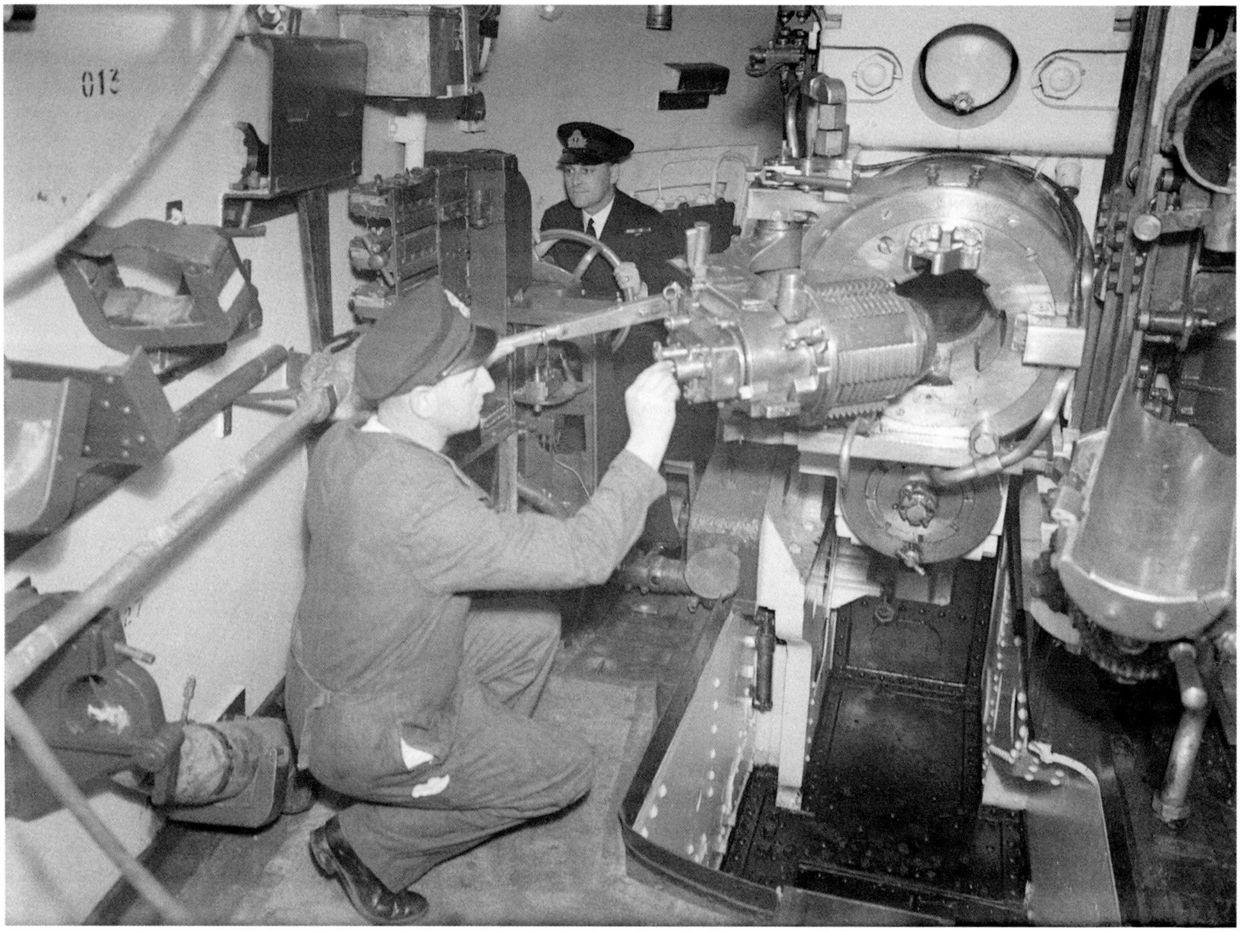

Gunner William West at the left-hand gun's gunlayer position and Petty Officer RH Johns adjusting the electric lock of the breech mechanism in one of HMS *Orion's* forward turrets, Rosyth, Scotland, 14 May 1944. A combined sponge and rammer is secured to its rack to the Petty Officer's left, likewise single racks for ready-use 6in projectiles. The deep gunwell allowing for 60-degree elevation is also visible. (Imperial War Museum, A 23469)

limited displacement, the protection scheme for these mountings was designed for splinter protection only.

Each gun was carried in a cradle, within which it moved during recoil and run-out, with an elevation arc fixed underneath. The cradle, with a 5in diameter trunnion pin on either side, was supported by two trunnion brackets, built up from the gunhouse floor, the pins being held in place by trunnion caps. A heavy balance ring fitted around the breech end of the gun, behind the cradle, balanced its weight about the trunnions to overcome the inertia of the long barrel and thus simplify achieving 60 degrees of elevation. The cradle also supported two semi-circular splinter plates mounted above and below the gun.

A recoil absorber, consisting of a liquid-filled recoil cylinder, was fitted to the underside of the balance ring with its piston rod secured to the fixed cradle, its function being to restrict the recoil distance and pull the gun up. Gun run-out was effected by a recuperator system featuring a compressed air run-out cylinder, fixed to the cradle and working in a recess in the moving balance ring, assisted by a liquid-filled intensifier.

Twin shell hoists entered the gunhouse between the guns close to the centreline abaft the trunnions while the two cordite hoists entered to the rear of the guns. Access to the gunhouse was through two doors at the rear using a folding ladder stowed under the tail of the turret, as well as an access hatch between the hoists from the working chamber and the ammunition lobby. Each door was fitted with ejection ports to manually expel used cordite cartridge containers.

A lookout platform for the Officer of the Turret (OOT) was located on the centreline to the rear of the gunhouse and fitted with a stool and periscope for external visibility. Directly in front was a telephone standard providing the telephone operator access to gunnery communications. Vent tube stowage lockers and ready-use racks were fitted close to the breechworkers' positions while a single 3pdr sub-calibre gun was stowed on the left side of the gunhouse with overhead lifting gear and guiderails above both guns. There were also 16 single ready-use rack stowages for 6in projectiles.

Unlike British heavy cruisers and many foreign light cruiser designs, the turrets were not fitted with

The guns of HMAS *Perth*'s 'A' and 'B' turrets at their maximum 60-degree elevation. Despite the original intention of long range controlled anti-aircraft fire being abandoned, the main armament of modern British cruisers, both 6in and 8in, was routinely employed in the barrage role throughout the Second World War. The contribution of main armament barrage to anti-aircraft defences steadily grew in importance during the conflict as efficacy improved through the development of specialised barrage directors working with automatic barrage units. (Australian War Memorial, 005237/10)

rangefinders or fire control instruments aside from the Local Director Sight (LDS). This was mounted between the guns in the front of the gunhouse in the local control cabinet. Capable of controlling the turret independently of the ship's fire control system, it could be segregated by a light partition door, the operators using two external sighting doors to observe and engage targets.

The turntable compartment was only 3ft 4in in depth with its bottom at weather deck level, while the working chamber beneath occupied the space between the weather deck and the ammunition lobby two decks below. The turntable compartment featured deep gun wells beneath the breech ends of the guns to allow them to elevate, as well as gun washout tanks holding fresh water for the combined hand sponge and rammers used both to cool the gun and to ram home the projectiles. The compartment, penetrated by the shell and cordite hoist casings, also held the turret's upper roller path, as well as the feed tank and several engines for the power system of the revolving structure. This system resided in the working chamber, which also supplied back-up batteries and the compressed air system, and acted as a conduit space for the hoists and other auxiliary services. The chamber also steadied the revolving structure against the fixed structure via guide rollers beneath its deck.

The power system used mineral oil for turret training and gunlaying, as well as shell and cordite ammunition supply. It was supplied by a pump located on the right-hand side of the working chamber, cooled by an oil-circulating pump and driven by a 65hp electric motor running off the ship's 220V direct current (DC) high power mains. Eight separate hydraulic motors powered two elevation and two cordite hoist engines in the turntable compartment, as well as two training and two shell hoist engines on the roof of the working chamber. The pump delivered pressurised oil to a pressure stack feeding control valves that allowed for the control of each engine from levers and pedals in the gunhouse.

Compressed air, fed from either the ship's system or three air bottles in the working chamber via a valve panel, provided the air blast system for the guns. Air was led to a valve at the rear of each balance ring and onwards into the chamber below the breech mechanism. The system engaged automatically as the breech was opened to blow air at 1,250psi through the bore to clear smoke and residue from the gun, the breechworker tripping a lever to shut it off manually as part of his drill. Both shell and cordite hoists projected below the working chamber into the ammunition lobby, the bottom of the hoist casings meeting the raised platform forming the light elevated false deck of the ammunition lobby with a cable gland penetrating it and the structural deck below.

Elevation drive for each gun was arranged around a two-way clutch enabling either power or hand control for each of the two gunlayers. The gunlayer's seat was behind the trunnions facing aft in the early turrets and inboard towards the guns from *Amphion* onwards. He controlled the elevation gearing via a power lever, connecting the elevation engine to one side of the clutch, or the elevation handwheel connected to the other side, via a foot pedal, the gearing acting upon the gun's elevation arc. Elevation speed in power control was nominally 10°/sec and hand control 2°/sec, while shafting from the arc provided elevation feedback to the gunlayer's elevation receiver.

Training arrangements centred upon two training pinions, mounted on opposite sides of the structure, which meshed and engaged with the training rack secured to the lower roller path of the fixed structure. The revolving structure's upper roller path rested and turned upon rollers fitted atop the lower roller path of the fixed structure. Power training was operated through the handwheel at the turret trainer's position in the local cabinet with a maximum speed of 7°/sec and a creep of 1°/sec. Manual training was facilitated via a clutch and hand training position in the working chamber; this required two men and a reasonable degree of effort to operate at a maximum of 1°/sec.

All ammunition hoists were of the endless chain type and were protected by flashtight casings. The revolving structure shell hoists carried 14 lifting heads. The lower hoists were loaded via flashtight scuttles in the ammunition lobby and operated by shell hoist workers in the gunhouse between the guns using power levers to cycle the hoists, one lifting head at a time, once the lower hoist doors were closed. Crank handles in the working chamber could be used to work the hoists manually, a lever being used to switch between power and hand operation. The height of the casing in the gunhouse complemented the gravity-assisted loading tray system delivering the projectiles to the rear of the guns.

Without mechanical assistance, the loading system was as simple as it was effective. Projectiles required virtually no manual handling in the gunhouse, thereby supporting and maintaining high rates of fire. The arrangement consisted of three shell trays. A small fixed tray received the projectile from a tilting bucket at the top of the hoist. A loading tray, pivoting on an arm bolted to the inboard side of the gun cradle, was swung manually behind the open breech by the tray worker, while an intermediate tray worked between the two. Projectiles arrived into the tilting bucket that tipped over to rest the base of the projectile on the fixed tray. When the intermediate tray was empty and in line with the fixed tray, the hoist worker pushed the projectile into it and the tray worker, releasing and bearing down on the intermediate tray, lowered the now-charged tray down level with the empty loading tray, where it was held by a stop unless the trays lined up. The projectile then slid into the loading tray to be swung behind the open breech by the tray worker, the gun's breech mechanism having been swung outboard. The projectile was then rammed by two ratings using the combined sponge and rammer before the tray worker withdrew the loading tray clear of the breech. Once empty, a balance weight forced the intermediate tray to rise for recharging from the fixed tray.

The lower cordite hoists were located opposite the revolving shell hoists suspended into the ammunition lobby, each featuring six open-ended cylindrical cannisters. A cordite cartridge container was loaded into an exposed canister and the lower door closed. With the upper hoist door closed, the hoist's engine automatically moved it through one cycle before opening the upper door on the gunhouse floor for the cordite hoist worker to remove the container by hand. He then stood on a pedal abaft the door to close it and start the next cycle,

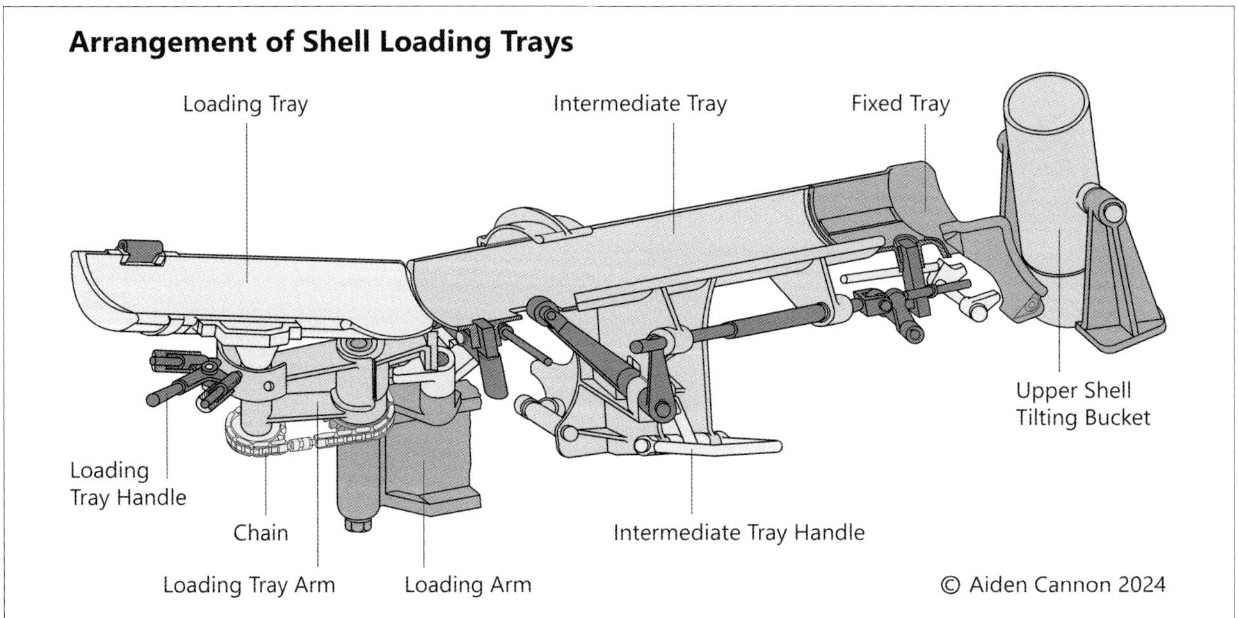

Mark XXIII gun shell hoist, fixed, intermediate, and loading tray arrangement. Drawn from BR 960B, Plate No 21.

so long as the lower door was also closed. Like the shell hoist, hand operation could also be engaged from the working chamber. The shell and cordite hoists were aligned to provide ammunition at a maximum of 12 rounds per minute, although 10 rounds per minute, easily exceeding requirements to sustain a normal firing cycle, was typical with all turret hydraulics running. Hand operation supplied only two rounds per minute.

The alternative supply method was via the secondary loading trunk. This was a plain trunk, sealed in the lobby by a sliding door and at gunhouse floor level by Miller's flaps for antiflash protection, used with a rope whip system for manual hoisting. The revolving structure was fitted with a seawater cordite drenching system, operated by handles in the gunhouse. When activated, this 'wet' (*ie* continuously charged), non-automated system redirected the oil cooler's water supply to fill the cordite hoists with spray.

Fixed Structure

The weight of the revolving structure was supported by stout bulkheads of high-tensile 'D'-quality (Ducol) steel, which was also required to resist the stresses acting on it during firing. A 40lb D1 cylindrical ring bulkhead extended from the deck above the ammunition lobby to the underside of the lower roller path seating, effectively covering the working chamber. The horizontal lower roller path sat atop the vertical ring bulkhead. The area between the weather deck and the gunhouse floor was further protected by a 40lb non-cemented (NC) armoured ring that enclosed the upper portion of the ring bulkhead and was sometimes referred to as the 'barbette'.

The shell rooms provided stowage (for the four-turret ships) of 1,600 6in service projectiles (1,440 CPBC and 160 HE: 180rpg and 20rpg respectively) as well as 248 practice rounds (31rpg) for a total of 1,848 projectiles. Both outfit and practice allowances were stowed in shell bins secured by wooden battens, 462 projectiles to a compartment. The centre of each shell room featured a single electric endless chain hoist within relatively easy reach of the bins. These hoists were operated from the ammunition lobby, where a 5hp electric motor delivered projectiles at 16 rounds per minute via the hoist casing. Projectiles were loaded by hand into a lower revolving flashtight scuttle and controlled by a rating at the upper hoist casing depressing a foot pedal. As projectiles reached the top of the casing in the lobby, they entered a tilting bucket that automatically tipped over before the projectile was mechanically rammed out onto a receiving ledge on the casing. It was then manually pushed into the lobby's circular loading chute that surrounded the hoists on the revolving structure. A crank handle was provided at the upper casing for manual hoisting, but this delivered only two rounds per minute.

The magazines, much larger than the shell rooms, stowed the cardboard cordite cartridge containers in sealed aluminium 'RN' cases, five containers per case. Cases were stowed on their sides in bays held fast by teak battens, their lids accessible and facing inboard for easy access. Each of the two magazines stowed 160 cases for a total of 800 cordite cartridges. Each magazine served a group of two turrets and housed two flashtight handling rooms, offset to port in the forward and rear sections of the compartment. These contained single electric cordite hoists, like those in the shell rooms, feeding the ammunition lobby. This arrangement provided extra separation to defend against explosive flash being transmitted from the gunhouse via the chain of supply to the main stowage.

The magazine crew, having removed the cordite containers from their cases, passed them one at a time into the handling room via revolving flashtight scuttles, two arranged on the bulkhead of each compartment. They were controlled by operating pedals in the handling room and fitted with mechanical indicators on each side. From there, containers were loaded into the cordite hoists to the ammunition lobby. The hoists were very similar in operation to the shell hoists, including the supply rate of 16 rounds per minute, except that they required a rating to catch the container at the upper casing. Containers were thereafter transferred by hand to the revolving hoists. Hand operation sustained eight rounds per minute.

'Y' turret used 'hand-ups' in place of hoists. The two hand-ups in 'Y' handling room were vertical flashtight steel tubes bolted to the deckhead. These were fitted with rotating flashtight doors, only one of which could be opened at a time, and fitted with indicators, the pair being capable of matching the hoists' nominal supply rate of 16 rounds per minute. Each of the other handling rooms was fitted with a single hand-up as a secondary source of supply in the case of hoist failure.[5]

The ammunition lobbies featured a raised false deck matching the depth of the revolving hoists that dominated the centre of the compartment as well as the upper casings of both fixed shell and cordite hoists, or hand-ups, as described above. The revolving hoists were in turn surrounded by the circular shell chute that forked from the top of the shell hoist casing, its gentle gradient down towards the opposite end of the compartment assisting ratings to push projectiles around the chute in either direction to suit the orientation of the revolving hoists. The lobbies also served as ready-use stowages for cordite and shell to complement the ready-use projectiles in the gunhouse, racks for 15 shell and 20 cartridge containers being provided. Secondary loading on the fixed structure entailed the use of rope whips hoisting projectiles in single rope shell bags through open ammunition hatches in the various decks before being manhandled through the lobby access doors while cordite used the same procedure in conjunction with the single hand-ups fitted in the handling rooms of 'A', 'B' and 'X' turrets for the purpose.

The scheme of protection for the fixed structure prioritised cordite over shell stowage while, as with the revolving structure, providing splinter protection only to the handling compartments and hoist casings – although this level of protection was superior to that of the earlier heavy cruisers. Magazines and handling rooms were protected by longitudinal sides of 140lb (3.5in) NC armour over 14lb (0.35in) D1 plating and 80lb (2in) NC over 14lb D1 crowns at the platform deck level above. Transverse bulkhead protection was 100lb (2½in) NC on 14lb plating. The scheme provided collective protection for the forward armoured 'box' of compartments including the magazine, 4in HA and small arms magazines as well as – coincidentally due to its location – 'B' shell room. 'A', 'X' and 'Y' shell rooms were not protected by armour except where they shared a bulkhead with a magazine. Bulkhead protection was 25lb (⅝in) D1 on 14lb plating with a 40lb (1in) single thickness on the crowns. Both shell and cordite hoist casings were constructed of 40lb D1.

Fire Control and Communications

Centralised fire control provided the elevation, training, range and deflection data turret crews needed to fight their mountings, the only provision for remote operation being electrical firing circuits. If central control was lost or impaired, the LDS allowed for alternative independent fire control. The principal control positions, manned by the fire control parties under the general direction of the Commanding Officer on the compass platform, comprised the DCT under the Control Officer, who directed the engagement, the TS, under the TS Officer in charge of the AFCT and associated equipment, as well as an after control position. The austere after control was substituted for a second DCT as a financial economy measure in 1932, a decision that reduced the efficiency of the fire control system and would haunt these ships in service. Over the course of six construction programmes, ships' fire control equipment underwent several iterative changes beyond the scope of this study, so a general description will suffice.[6]

Two director firing circuits, main and auxiliary, were fed from the ship's 24V DC low power supply and energised by the firing pedal in the DCT. Both circuits ran through the TS to the turret's Quarters Change-over Switch (COS) in the local control cabinet. With the COS set to 'Director', these circuits engaged the turret's main and auxiliary firing circuits through the interceptor, Breech Safety Change-over Switch (BCOS), 'expendable lead' cable and the electric lock to fire the vent tube. There was no provision for percussion firing. One side of the interceptor switch was secured to the gun's recoiling balance ring and the other to the cradle, the circuit being broken as the gun recoiled. The similarly arranged BCOS was secured to both breech and breech mechanism to break the circuit when the breech was open. Thus, the gun was prevented from firing unless fully run out and the breech mechanism screwed all the way home.

Having been jointly calculated in both the DCT and TS, gun training data was transmitted directly to the turret trainer's training receiver. Likewise, gun elevation was received on the elevation receivers at both gunlayers' positions, having passed through the Quarters COS. The turret trainer and both gunlayers worked their power drives and elevation handwheels to 'follow the pointer', matching the electrically-driven pointers with the mounting's own mechanically-driven pointers. Gun range

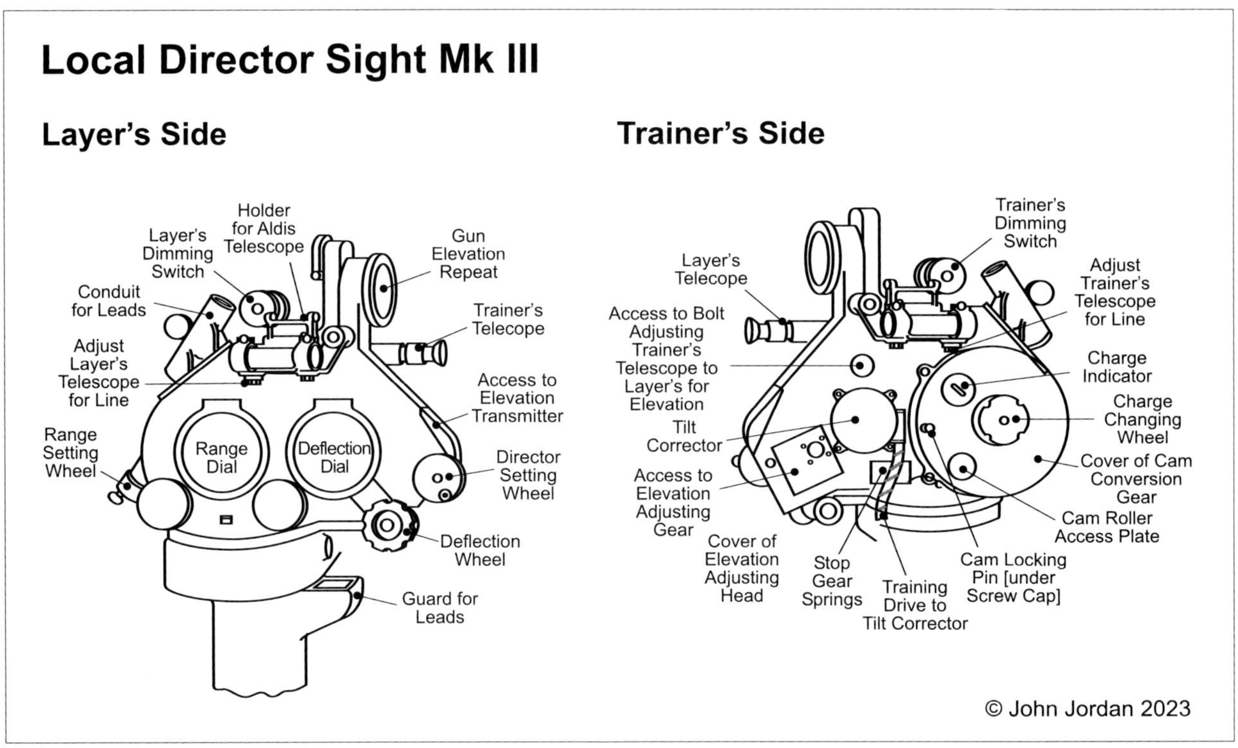

Local Director Sight Mark III as fitted to *Leander, Neptune, Achilles, Orion, Ajax* and *Amphion*. Drawn from BR 912(8), Plate No 3.

was electrically transmitted to the training receiver for convergence onto the DCT's line of sight as well as elevation receivers for MV corrections. Deflection was likewise sent to the LDS with verbal communications from the TS used as a backup. In action, the LDS crew kept the target under observation and their sights set with gun elevation, range and deflection, to enable them to take over rapidly if required.

The Local Director Sight was a pedestal mounting manned by the local sight layer on the left and turret trainer to the right, both using monocular laying and training telescopes aimed through the open sight doors in the gunshield. A sightsetter sat on the layer's left, facing the equipment and operating its range-to-elevation and deflection unit.

In director control, the turret trainer directed the turret in bearing with the layer practically an observer. To fight the turret independently, the turret trainer shifted the Quarters COS over to 'Quarters'. This disconnected all TS circuits. At the same time, the OOT moved his Local Control COS from 'TS' to 'Local Control', severing the fire control system's gun range transmissions to the elevation receivers as well as the fire gong. No longer following pointers, the trainer aimed at the target himself. The gunlayers' electrical elevation pointers were now driven by the LDS, as the local sight layer aimed his telescope via a director setting handwheel. Elevation and training values were now influenced by range and deflection settings applied by the sightsetter as directed by the OOT through spotting corrections. Furthermore, the LDS provided the firing pulses, which were fed from turret batteries from the layer's firing pistol to the guns. When in 'TS', the Local Control COS fed 'Gun Ready' lamp boxes in the DCT and TS as well as the local cabinet and OOT positions, but in local control the circuits were shifted to batteries and DCT and TS outputs disconnected.

Communications, both within the turret and to external fire control positions, were achieved by several means. 22V DC telephones connected the turrets with the fire control exchange in the TS. The telephone operator and OOT used two-way telephone headsets, sharing a jack box at the telephone standard, connected to the range and order phone operator in the TS via the gun order communications group, while the sightsetter used a receive-only headset on the same circuit. The telephone standard also carried two direct, non-exchange phones. One was wired to the adjacent turret in each group in the four-turret *Leander*s while the three-turret *Arethusa*s had 'A' and 'Y' both linked to 'B' turret. The other phone communicated either with After Control in the early ships, or the DCT in *Ajax* and later ships.[7]

The OOT used two voicepipes to communicate directly either with the local sight layer, or with the sightsetter, turret trainer and both gunlayers collectively. Two voicepipes were fitted between the RH gun's shell hoist and the LH gun's cordite hoist to the ammunition lobby. The lobby then used separate voicepipes to pass orders to both handling and shell rooms. Another voicepipe allowed the turret trainer to direct ratings operating the hand training gear in the working chamber. A multiphone electro-megaphone system mounted at the telephone standard was grouped with the adjacent turret as well as the TS and DCT and acted as a broadcast instrument. Other equipment included a speaker for the ship's general announcement system, the Warning Telephone, as well as the fire gongs operated from the TS and 'Cease Fire' (later 'Check Fire') bells from the compass platform. A turret training transmitter provided training bearing to a model turret training indicator at the OOT's position of both turrets in a group as well as a turret training telltale in the TS, allowing TS personnel to compare turret bearings with the transmitted output of the AFCT.

The turrets could be fought in director, group, local or gunlayer's control. If director control was unavailable, group control involved two turrets being fought by the OOT of the superimposed turret, either 'B' or 'X'. Local control saw each turret fighting alone under its OOT while gunlayer's control was completely in the hands of the local cabinet under the LDS layer, the OOT presumed to be incapacitated. However, these latter methods were the most inefficient means of directing the main armament and only used as a last resort. Divided control – fighting the forward and aft groups separately against different targets under director control – required two fire control computing instruments and was not provided for.

Turrets employed gyro, director, gunlayer's or quarters firing. Firing circuits were either closed by the director layer, using the DCT's stabilised gyro sight to coincide with the ship's roll, or instantaneously using the unstabilised sight in director firing. If communication with the DCT was lost, the after control position provided only training data through the AFCT. This required gunlayer's firing, the local sight layer using his telescope and firing pistol, while the turret trainer continued to follow electrical pointer bearings from the TS. *Penelope* and *Aurora* were slightly better off in this regard with combined HA/LA directors mounted aft, but these still supplied only training data when used in the LA role. The last method, quarters firing, saw both layer and trainer fighting the turret autonomously via the LDS.

Performance

These mountings, as examples of manually operated and directed gunnery systems, were reliant on highly choreographed and intensely practised drills to keep the guns in action and hitting their targets. The Royal Navy utilised its highly skilled and motivated gun crews to mitigate the automated mechanical refinements of many of their foreign contemporaries and increase reliability. However, anything less than perfect drill would result in missing salvos or reduced accuracy.

As previously noted, the Mark XXIII gun was a modern, well-balanced, powerful weapon that achieved everything expected of it. While nominally capable of eight rounds per minute, a relatively high rate of fire for hand-worked guns, this was achieved in local control with guns laid to loading elevations; the local sight layer pulling the trigger as soon as his two 'Gun Ready' lamps illuminated following the 'loading interval', the bare time it took guns crews to reload the gun. Reality in action was different, particularly under director control. This needed to account for the 'control interval', the time taken by the director layer to steady his aim (around three seconds on average) as well the need to depress and elevate the gun to facilitate loading angles while firing at longer ranges. Differing fire control procedures to ascertain line and range needed to be added to spotting corrections, also long times of flight at extreme ranges, change of point of aim, loss of sight of the target etc, resulting in multiple factors that slowed practical firing rates in action. With all of these in mind, six rounds per minute, relying on a good fire control solution while firing rapid broadsides, was considered the practical rate of fire of the mounting. However, during most actions such opportunities were rare.

Wartime concerns regarding excessive spreads for range affected several modern, high performance BL weapons including the Mark XXIII. These issues were not conclusively resolved during the war. However, precise gun drill served to minimise its effects while postwar quality control improvements in shell and propellant manufacture eventually redressed them.

Controlled main armament anti-aircraft fire was decided to be impractical while *Leander* was still under construction. This was due to the need for additional dedicated, costly and displacement-consuming high-angle control systems for guns that were, by necessity, primarily outfitted with ammunition for the anti-shipping role. This led to a reduced 45-degree maximum elevation in subsequent 6in cruisers. Thereafter, they were employed against aircraft in the barrage role using time fuzes, which were set by hand in the loading tray prior to ramming.[8]

Turret Manning and Gunhouse Drill

The complement of each gunhouse nominally comprised 20 men: two identical seven-man gun crews as well as six men to lay, train and control the turret.

Lieutenants or Sub-Lieutenants were allocated one to each group of turrets in 6in cruisers to fight their respective pair of mountings in group control if director control was lost. Hence 'B' and 'X' turrets were controlled by commissioned officers acting as Officers of the Turret, with Petty Officers in the role of Captain of the Turret supervising the gun crews. However, in 'A' and 'Y' turrets, these Petty Officers filled the OOT role, resulting in only 19 men in the gunhouse. Additionally, each turret system included a dedicated ammunition supply team of 24/25 men with an additional Petty Officer per magazine, where a total of eleven ratings supplied a pair of turrets with an additional three in the handling room for 'A', 'B' and 'X' turrets, and four in the handling room for 'Y' turret, which had hand-ups instead of a cordite hoist.

Table 2: **Gunhouse Duties**

Control Personnel

Officer of the Turret	OOT
Captain of the Turret	CT
Local Sight Layer	LSL
Turret Trainer	TT
Sightsetter	SS
Telephone Operator	TO

Guns Crews for Left and Right Guns

No 1	Gunlayer
No 2	Captain of the Gun, Breechworker, Tube and Interceptor Number
No 3	Rammer Number
No 4	Rammer Number (ejects cordite cartridge containers)
No 5	Tray Worker
No 6	Cordite Hoist Worker & Cordite Supply Number
No 7	Shell Hoist Worker

Each shell room was manned by six ratings, and the ammunition lobbies required ten for a total of 177 officers and ratings in a four-turret ship, or 134 in a three-turret ship.

Each gun was controlled by a Leading Seaman as Captain of the Gun. Qualified gunnery specialist Able Seamen undertook the more skilled roles within the turret and were supplemented by non-specialist Able Seamen and Ordinary Seamen, their roles being numbered one to seven. The gunlayer (No 1) elevated and depressed the gun from his seat with the power control lever and hand elevating wheel. The Captain of the Gun/breechworker (No 2) worked the breech mechanism with the BM lever to open and close the breech. He also inserted the vent tube into the electric lock and closed the interceptor to make the electrical firing circuits. The ramming numbers (Nos 3 and 4) manned the combined rammer and sponge to ram home projectiles into the breech from the loading tray. The tray worker (No 5) worked the shell loading tray and loaded the cordite cartridge into the gun. The cordite hoist worker (No 6) worked the cordite hoist bringing cartridge containers up from the ammunition lobby and supplied the cartridges to the tray worker. Lastly, the shell hoist worker (No 7), worked the shell hoist to lift projectiles, also from the ammunition lobby, up to the gunhouse and supply them to the tray worker (No 5).

Initially, the turret stepped through a procedure to close up and prepare for action including initiating

Table 3: **Ammunition Supply**

	'A' turret	'B' turret	'X' turret	'Y' turret
Magazine	—— 11 ——		—— 11 ——	
Handing Room	3	3	3	4
Shell Room	6	6	6	6
Ammunition Lobby	10	10	10	10
Total	99			

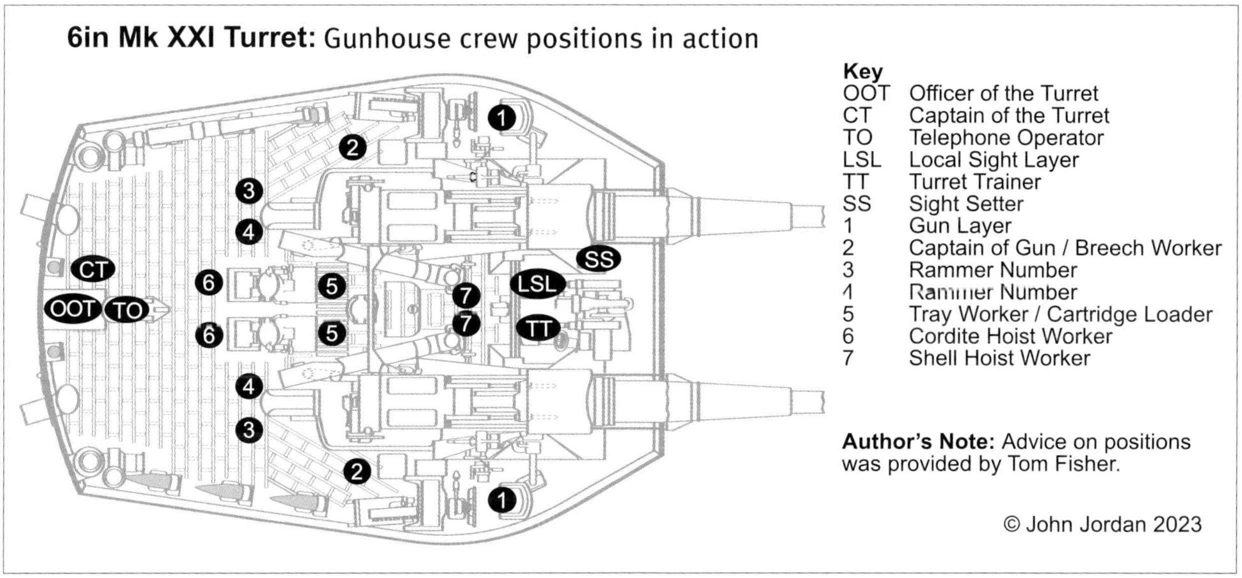

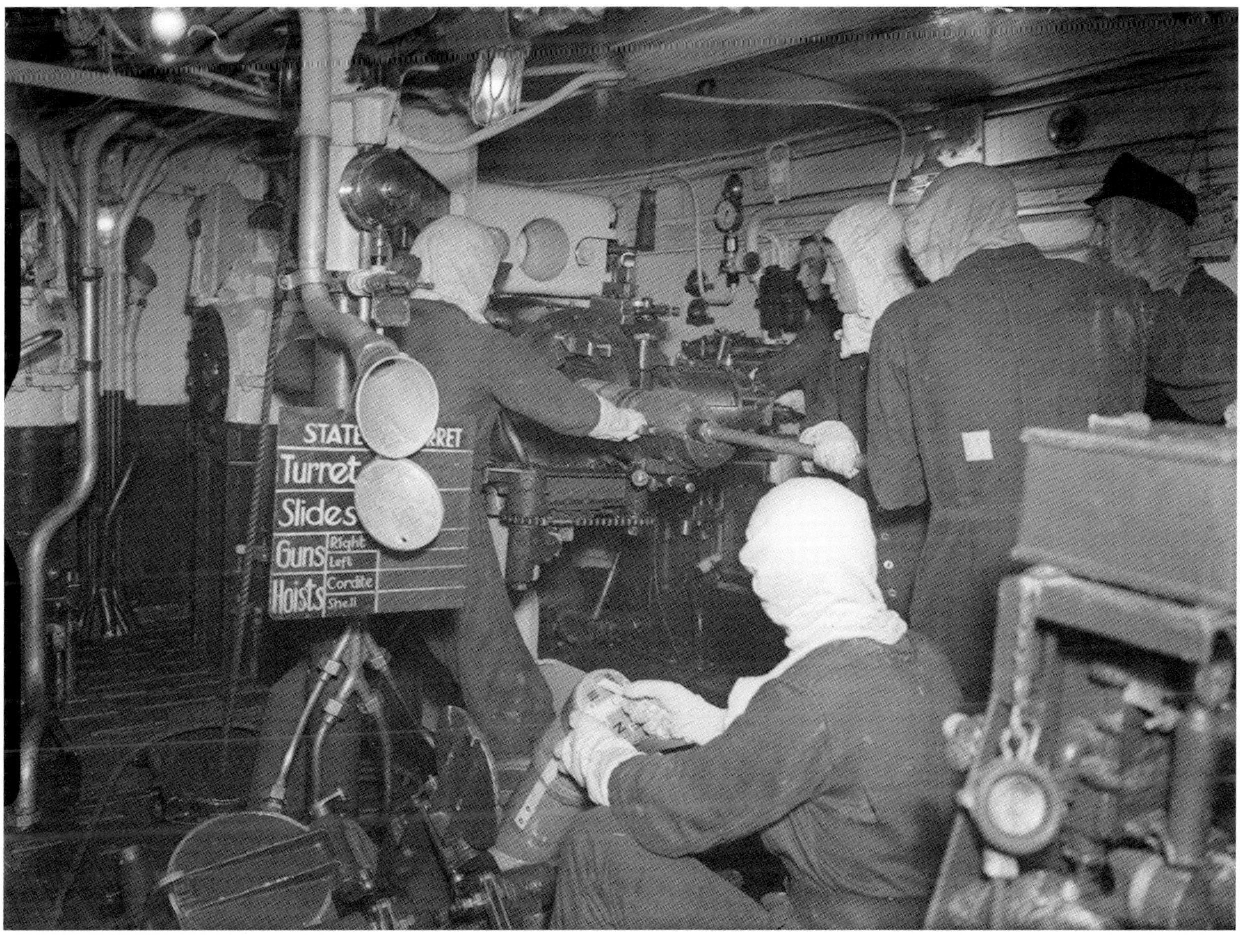

The crew of the right-hand gun in one of HMS *Orion*'s turrets, believed to be 'A', circa 1944; the photo was possibly staged for the camera. In the foreground, the cordite hoist worker has a cartridge container in hand with the right hoist door open. To the left, the tray worker has the charged loading tray in line with the open breech as the two rammer numbers to the right use the combined sponge and rammer to push the projectile into the chamber. In the background, the breechworker handles the breech mechanism while the Petty Officer Captain of the Turret looks on. The voicepipe in the foreground is for the revolving cordite hoists in the ammunition lobby, while the voicepipe in the background services the corresponding shell hoists. The rope on the left is for the auxiliary hoist, its Miller's flaps visible at deck level while the telephone standard is at bottom right. (Imperial War Museum, A 23468)

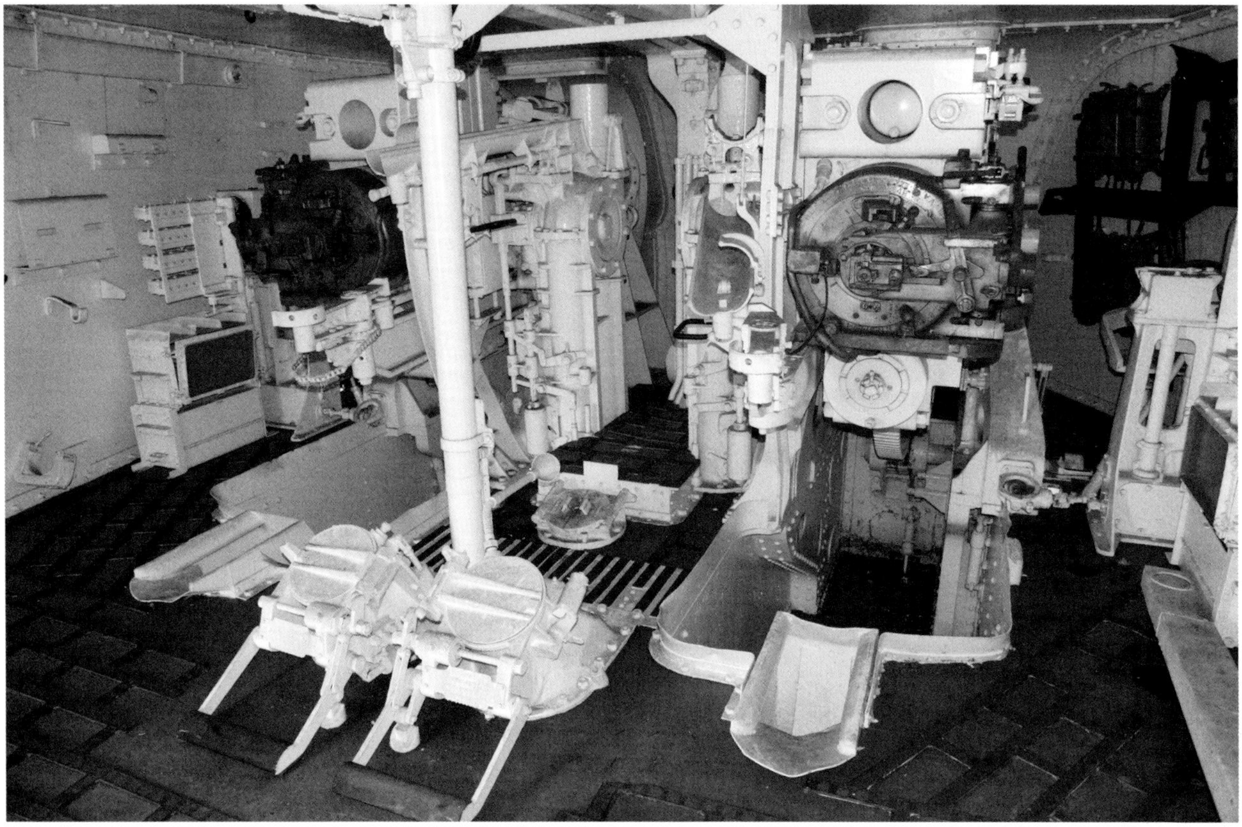

The interior of Ex-HMS/HMNZS *Achilles*/INS *Delhi*'s 'Y' turret. Comparatively more spacious than the later triple turrets, this gunhouse has been stripped of a range of equipment including voice pipes, telephone standard and Officer of the Turret position. The most notable omission in this view looking forward is the two loading trays. The upper cordite hoists and their operating pedals are prominent in the foreground with the Miller's flaps of the auxiliary hoist immediately behind. Shell hoists and their upper tilting buckets are visible in the middle of the image, with vent tube stowage locker and wooden ready-use racks visible outboard of the left-hand gun. Both guns have been changed out with 1939-manufactured weapons at some point in the ship's career, probably prior to her transfer to the Indian Navy. The overuse of white paint is attributable to mid-1990s preservation efforts. (Peter Cannon, Devonport, Auckland, 2 December 2016)

compressed air, starting the pump motor, removing the locking bolts keeping the turret and guns in place, and testing the operation of all circuits and machinery. The guns were loaded when ordered, the hoists being 'charged' by cycling them through until supply of both shell and cordite reached the gunhouse. Once a projectile appeared in the shell bucket at the top of the hoist, the shell hoist worker pushed it over into the fixed loading tray before tilting the bucket back to the vertical and hoisting the next projectile after the first entered the intermediate tray. The tray worker bore down on the intermediate tray, allowing the projectile to slide into the loading tray. As the intermediate tray rose back to its starting position, the shell hoist worker recharged it with the next projectile. The cordite hoist worker extracted the cardboard container and discarded the lid to the rear before swinging the open container onto his shoulder and pushing the cordite hoist pedal down to close the flashtight door.

The gunlayer laid the gun to the firing elevation, if not exceeding +12.5° or an otherwise convenient loading angle, before the breechworker opened the breech. The tray worker swung the charged loading tray in line with the bore and partially pushed the projectile into the chamber. The rammers rammed the shell into the bore until its copper driving band engaged with the rifled shot seating. The tray worker, having been handed a cordite cartridge by the cordite hoist worker, pushed the cartridge into the chamber behind the rammed shell, before swinging the loading tray inboard of the gun and away from the breech. The breechworker closed the breech and inserted a vent tube into the electric lock before closing it to secure the tube. With the breech closed, the gunlayer laid the gun to the elevation indicated by the pointers on his elevation receiver and reported 'On' before the breechworker closed the gun's interceptor, making the electric firing circuit and burning the gun's 'Ready' lamp in the control position. Meanwhile, the cordite hoist worker discarded the empty cardboard container towards the rear of the gunhouse for No 4 to expel via the ejection chute before withdrawing another container from the hoist. The gun was now at the 'ready' state before being fired remotely by director.

Ex-HMS/HMNZS *Achilles*/INS *Delhi*'s 'Y' turret on gate guard duty at HMNZS *Philomel*. The gunhouse and its turntable compartment, visible due to the absence of the armoured protection ring, have rested on this concrete plinth since 1994, having returned to New Zealand in 1979. The ceremonial tampions are in place. Minor fittings on both the sight doors and manual depression gear, the vertical rods with rollers beneath each gun port, are missing. The rollers originally ran on a cam rail secured to the front of the protection ring to prevent the guns being depressed past -5°. (Peter Cannon, Devonport, Auckland, 2 December 2016)

As the gun fired and recoiled, it automatically tripped the interceptor switch to open the firing circuits. As the gun ran out, it was re-laid to the loading angle, if required. The breechworker opened the breech before tripping the air blast lever on the balance ring after a couple of seconds and working the electric firing lock's hand lever, thereby ejecting the vent tube which fell to the deck. A fresh tube was inserted into the vent axial and the lock closed. As the breech screw swung outboard, the two rammers placed the soaking wet combined sponge and rammer, having been dunked in the gun washout tank, on the breech block's mushroom head to cool it. After the tray worker had partially inserted the next projectile into the chamber, the rammers dropped the combined sponge and rammer into the loading tray and rammed the shell, this action also serving to sponge the chamber free of burning residue. The cartridge was loaded, the breech closed, the gun re-laid to the firing angle indicated by the elevation receiver pointers, and the interceptor closed as before. The fire gong, initiated from the AFCT in the TS, would sound inside the gunhouse as the salvo (or broadside) was about to fire before the director layer in the DCT steadied his aim and depressed his firing pedal. The now-closed electrical circuit, running from the DCT firing pedal through the TS to the interceptor and BCOS to the electric lock, now fired the vent tube. As soon as the gun had run out, the loading and firing cycle repeated itself.

Battle Performance

The *Leander*s and *Arethusa*s fought many of the Royal, Royal Australian and Royal New Zealand Navies' most notable and celebrated actions of the Second World War. These included the battles of the River Plate, Calabria, Cape Spada, Matapan, Sunda Strait and many other actions, most notably in the Mediterranean, including Crete and the anti-convoy actions of the Malta-based Force 'K'. Numerous gunnery reports indicate that, in action, turret materiel was almost entirely reliable, the only exception being relatively frequent but minor hoist stoppages, cordite hoists often being jammed by the tapes

Table 4: **General Data**

Gun construction	steel: 'A' Tube, jacket, breech ring & breech bush
Length of gun	309in
Weight of gun	6 tons, 14cwt, 1qr, 5lb
Breech mechanism	Asbury type with Welin Screw breech & de Bange obturating pad
Weight of breech mechanism	3cwt, 3.5qr
Weight of gun (complete)	6 tons, 18cwt, 19lb (with mechanism in place)
Projectile weight	112lb (100lb in *Leander*)
Cordite cartridge weight	30lb
Design chamber pressure	20.5 tons/in^2
Working chamber pressure	20.5 tons/in^2
Pressure at MV 2,758ft/sec	19 tons/in^2 (new gun)
Service charge pressure	17 tons/in^2 (half-worn gun)
Muzzle pressure	6 tons/in^2 (approx)
Estimated life of gun	1,020 equivalent full charges
Gun elevation	hydraulic power via elevation control lever *or* manually worked by elevation hand wheel
Maximum elevation	+60°
Maximum depression	-5°
Extreme range	24,800 yards at 44° elevation (112lb shell)
Rate of fire (designed)	8 rounds per minute per gun (= 7.5 secs per round)
Rate of fire (director fire)	6 rounds per minute per gun (including control interval)
Loading angles	-5° depression to +12.5° elevation
Elevation rate (power)	10°/sec
Elevation rate (manual)	2°/sec
Training rate (power)	7°/sec max (1 engine), 5°/sec (2 engines) for superior control, 1° creep per minute
Training rate (manual)	1°/sec
Shell hoists (fixed structure)	two endless chain hoists driven by 5hp electric motors power operation: 16 rounds/min hand operation: 2 rounds/min
Shell hoists (revolving structure)	two endless chain hoists driven by hydraulic engines: 12 rounds/min (8 rounds/min normal) hand operation: 2 rounds/min
Cordite hoists (fixed structure)	two endless chain hoists driven by 5hp electric motors ('A', 'B' & 'X' turrets): 16 rounds/min (designed) hand operation: 8 rounds/min two vertical flashtight hand-ups ('Y' turret): 14 rounds/min
Cordite hoists (revolving structure)	two endless chain hoists, six cordite charge canisters: 12 rounds/min (designed) hand operation: 2–3 rounds/min
Mounting power	one 303rpm variable swashplate pump driven by 1,200rpm 65hp electric motor powering eight hydraulic engines (two shell hoist, two cordite hoist, two elevating engines, two training engines)
Ready-use shells (gunhouse)	8 rounds per gun
Ready-use shells (lobby)	10 rounds per gun
Ready-use cordite (lobby)	10 rounds per gun
Local fire control	Officer of the Turret periscope Local Director Sight (LDS) Mark III/IV/V
Emergency power	two 60 amp/hour 12-cell batteries for quarters firing, local control circuits & emergency lighting
Auxiliary shell supply	hand whip through ammunition hatches to ammunition lobby; hand whip up secondary loading trunk / auxiliary hoist along with cordite to the gunhouse
Auxiliary cordite supply	single hand-up in handing rooms for 'A', 'B' and 'X' Turrets to Platform Deck; hand whips through ammunition hatches to the ammunition lobby; hand whip up secondary loading trunk / auxiliary hoist along with shells to the gunhouse

on the cardboard containers. This rarely impacted practical rates of fire, shell supply being maintained by extremely well-drilled and efficient crews through secondary means. This reliability, backed by sound overall gunnery system design and performance, saw the turrets prove lethal at practical fighting ranges, allowing the *Leander*s and *Arethusa*s to establish their worth time and again in battle. In particular, the ships lived up to original staff requirements in their proficiency as destroyer killers, while several Axis Mediterranean convoys were torn apart at close range with ease. However, the main armament was let down on occasion by the fire control system's lack of redundancy, primarily the omission of an after DCT; both *Sydney* and *Perth* suffered great disadvantage in their respective final actions as a result.[9]

Conclusion

The ease of introduction and success of the Mark XXI system led to its further development into the triple Mark XXII and XXIII turrets fitted in the larger succeeding *Southampton* and *Fiji* classes. While ammunition supply arrangements differed, most notably the long trunk arrangement adopted in the Mark XXIII, other systems and equipment were virtually identical. Later British enclosed 5.25in and 4.5in mountings were far more cramped and unhandy, while the postwar 6in Mark 26 automatic mounting was a maintenance and performance nightmare compared to the simple, robust Mark XXI.

The Mark XXI and its triple-gun derivatives performed extremely well and arguably outclassed its foreign contemporaries for practicality and reliability with the exception of the US 6in/47 calibre Mark 16 guns on triple mountings, introduced in the *Brooklyn* class in 1937. The American system's high level of reliable automation, not to mention its impressive rate of fire, pointed the way towards the dual-purpose systems of the postwar era, when opportunities created by war-developed radar direction and proximity fuzing focused RN attention on the advantages of increased firing rates using automated QF guns. When finally withdrawn, the Mark XXIII was the last BL gun in British service.

During the Second World War, HM Ships *Neptune*, *Galatea* and *Penelope*, as well as HMA Ships *Sydney* and *Perth*, were lost to enemy action. The surviving ships had left service by 1948, with only *Achilles* and *Aurora* reprieved through transfer to the Indian and Chinese Nationalist navies respectively as *Delhi* and *Chungking* that same year. While *Chungking* was sunk by Communist air attack in 1949, *Delhi* served until 1978 with two of her turrets surviving today: one at the Indian Army Regiment of Artillery Museum in Nashik and the other serving as a 'gate guard' at the entrance to HMNZS *Philomel*, in Auckland, New Zealand.

Principle sources:

This article was compiled from primary sources in the National Archives of Australia, The National Archives (UK), RAN Sea Power Centre (Australia), HMS *Excellent* collection and interviews with ex-Leading Torpedoman Tom Fisher F2803, RANR (1921–2016) of HMA Ships *Sydney* and *Hobart*. The author would like to especially thank the Imperial War Museum's Andy Curran and National Museum of the Royal New Zealand Navy's Michael Wynd for special access to *Belfast*'s and *Achilles*' turrets respectively.

Endnotes:

1 While the preceding 8in turrets had completed grossly overweight, the design was expected to weigh approximately the same as the Mark XVIII in *Nelson*. However, despite the new Mark XXIII guns being lighter, the mountings were considerably heavier. The completed weight of *Leander*'s turrets, including guns, revolving and non-revolving weights, fixed shell and cordite hoists, DCT and all spares far exceeded Vickers-Armstrongs initial 1929 estimate of 391.1 tons, completing 61 tons heavier at 438.9 tons.

2 A 100lb shell weight had been adopted primarily to facilitate rapid manual loading. The provision of mechanical loading aids, which reduced the amount of manual handling, largely negated this concern.

3 Maximum range was achieved at an elevation of 43 degrees for a peak shell height of 20,000 feet and angle of descent of 58 degrees following a long time of flight of 75 seconds, time during which a high-speed target would invariably have moved a considerable distance.

4 Two 17lb 2oz half-charges of NF/Q/S fully flashless propellant could either be hoisted separately or be passed by hand to the ammunition lobby before being hoisted by hand into the gunhouse via the auxiliary cordite hoist. They were replaced by 32lb SC 150 NQ/F/P 128 full charge cartridges compatible with the powered hoists from March 1945.

5 Hand-ups were also used to supply cordite from a 'cordite gallery' in the revolving structure direct to the gunhouse in the subsequent triple Mark XXII turrets of the first two groups of *Southampton*-class cruisers.

6 Except for the 1934 programme *Aurora*, all ships featured the 'M' type synchronous transmission system to relay gun control data, while *Aurora* received the succeeding indicator Magslip, as opposed to power Magslip used in later generation follow-up systems. *Leander* was fitted with the AFCT Mark IV, *Neptune*, *Achilles* and *Orion* the Mark IV*, *Ajax*, *Amphion*, *Apollo*, *Sydney*, *Arethusa*, *Galatea* and *Penelope* the Mark V and *Aurora* the Mark VI. All *Leanders* and *Amphion* carried the Mark III LDS, and the remainder, except *Aurora* with the Magslip-compatible Mark V, carried the Mark IV.

7 Telephone Mark IV for two-way and Telaupad Mark I for receive-only fire control exchange communications and Mark V**, built for high noise environments, for direct telephone emergency communications.

8 Interest in high elevation in large-calibre guns was renewed from 1942 due to the development of the proximity fuze, leading to the pursuit of fully dual-purpose main batteries. Cruisers fired barrages from the DCT using much guesswork or more effectively when fitted with Type 283 radar-equipped barrage directors. These fed gun training and elevation to the turrets, while radar-fed automatic barrage units closed the firing circuits at precisely the right moment.

9 After losing her DCT seconds into the battle on 19 November 1941, during arguably the most brutal close-range cruiser action of the war against the German disguised raider HSK *Kormoran*, *Sydney*'s gunnery organisation was unable to switch to alternative methods of control before sustaining ultimately fatal damage. However, 'X' and 'Y' turrets, operating in either group or local control at point-blank range, likewise rapidly inflicted mortal damage aboard *Kormoran* before being knocked out. *Perth* was forced to split her main armament against an overwhelming number of Japanese targets in the Sunda Straight on the evening of 28 February/1 March 1942, putting 'X' and 'Y' into the highly-inefficient group control that achieved poor results against rapidly manoeuvring destroyers.

THE 'H'-CLASS COASTAL SUBMARINES OF THE ROYAL DANISH NAVY 1936–1950

The coastal submarines of the 'H' class were completed shortly before the Second World War. They saw no action due to Denmark's decision not to resist Nazi Germany's takeover in April 1940, and were scuttled in August 1943. **Tom Wismann** has provided a study of this interesting and well-developed design.

The Royal Danish Navy (RDN) was quick to to see the potential of the submarine. The first boat, HDMS *Dykkeren*, was ordered from the Fiat San Giorgio shipyard at La Spezia, Italy, in 1908. It was an all-electric boat that was meant to operate only in the Copenhagen area, defending the city and supporting the forts around Copenhagen. *Dykkeren* had to have its batteries recharged at Orlogsværftet (the Royal Naval Dockyard) between sorties, as it was incapable of charging the batteries independently.

To follow *Dykkeren*, three boats were ordered from Whitehead-Fiume in Austria in 1911, 1912 and 1913 respectively. These boats became numbers 1, 2 and 4 of the RDN's 'A' class. After assigning the construction of the first two boats to Austria, Orlogsværftet felt sufficiently confident to build the third, fifth and sixth boats without support. The following 'B' class, comprising five boats, were built at Orlogsværftet 1914–1917. Considering the newness of the technology, the rate of construction was impressive.

The 'B' class was followed by three boats of the 'C' class 1918–1922 and two boats of the 'D' class 1925–1927. However, after the First World War the financial means available to Danish politicians to allocate to the armed forces declined. This reached an all-time low with the catastrophic Defence Law of 1932, in which funding for the Navy was halved; nor did the Army fare much better.

The rise to power of Adolf Hitler in Germany in 1933 had little impact on Danish defence policy. However, rather than use the meagre funding from the law of 1932 for a modernisation of the elderly 'B' class, it was decided that it would be preferable to invest it in a new class of submarines.

Work on the specifications for these boats, initially designated the 'E' class, began in 1934. When it became clear that it was difficult to find suitable names starting with the letter 'E', the pragmatic step was taken to re-designate the class as the 'H' class, which was to comprise three boats. These were duly laid down at Orlogsvaerftet in 1936, and with the marked deterioration in the international situation a fourth boat of the class was laid down in June 1939; a fifth was authorised but had not been laid down when Germany invaded Denmark on 9 April 1940.

Table 1: **Characteristics**

Length:	
between perpendiculars	42.63m
overall	47.45m
Beam:	
waterline	4.45m
moulded	4.65m
Depth (moulded)	3.58m
Draught (surfaced)	2.84m
Displacement:	
surfaced	333.5 tonnes
in dived condition	$402m^3$
Metacentric height (GM):	
surfaced (light condition)	271mm
dived	259mm

Names

Orlogsværftet had built the vast majority of the ships – a total of 346 – for the Royal Danish Navy (RDN). The first, the 94-gun ship of the line *Dannebrog* (sometimes spelled *Dannebroge*) was laid down in 1692 with yard number 1, and Orlogsværftet built its last two vessels, the submarines *Narhvalen* and *Nordkaperen*, in 1965–70.

Havmanden: for 1,500 years the common term for a supernatural being that lives in the sea.

Havfruen: mermaid.

Havkalen: the Greenland or 'sleeper' shark is a little-known shark that lives in the seas around Greenland and the Faroe Islands. It swims very slowly and is almost blind because a small parasitic creature that lives in its eye.

Havhesten: a mythical creature with the body of a horse and the tail of a fish.

Probably the first sea trial on 17 May 1938 of the first boat of the class. The boat is seen here with the pennant number E1 and without any deck guns fitted. It is unclear when the pennant numbers changed from 'E' to 'H' and names starting with 'H' were adopted. E1 became H1 *Havmanden*. (Marinens Bibliotek)

The northern part of Dokøen (Dock Island) at Orlogsværftet. The submarines of the 'H' class were built on the three slipways to the right of the graving dock. Here torpedo boats of the *Dragen/Glenten* classes are seen under construction during the mid-1930s. (Marinens Bibliotek)

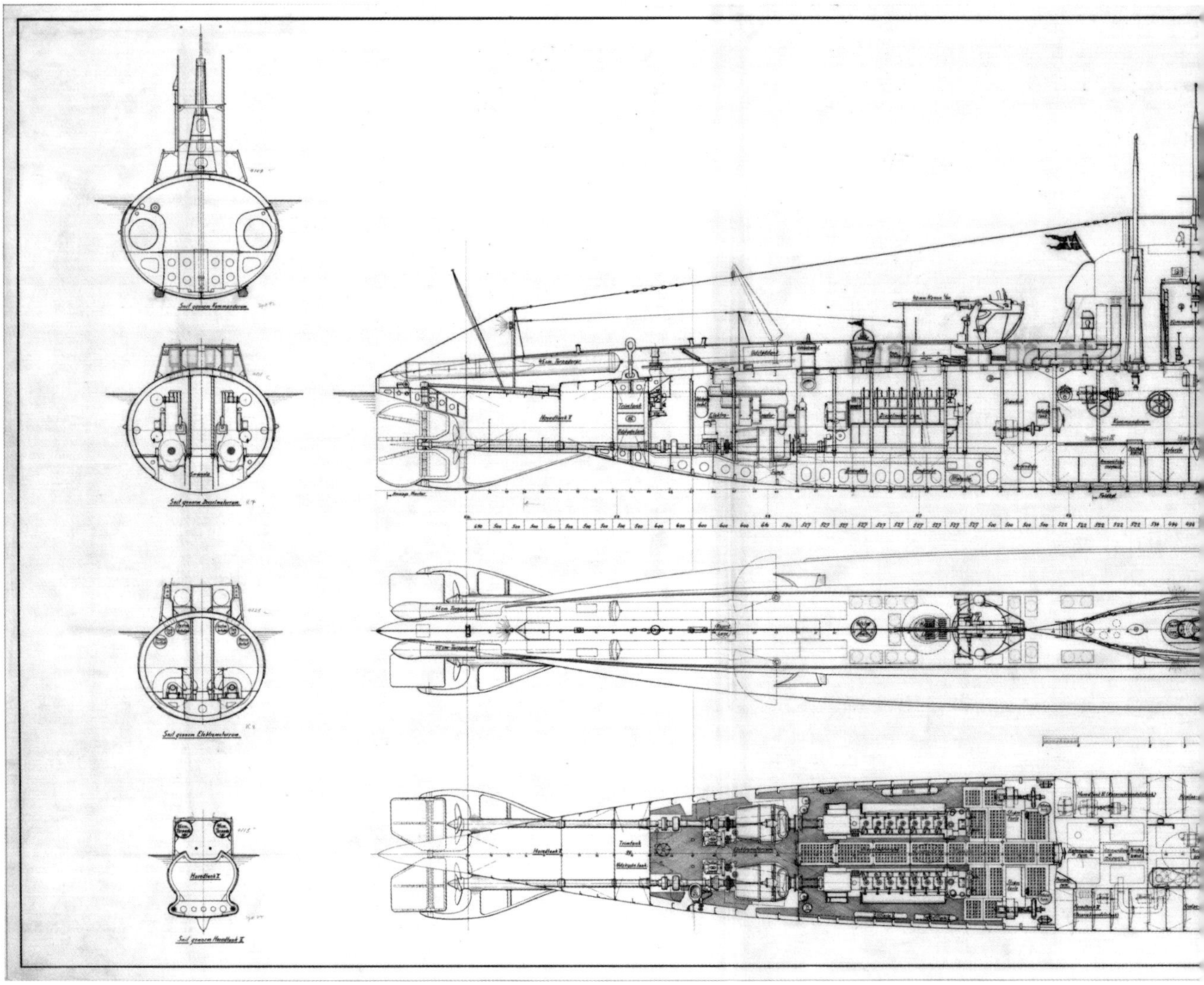

The 'H'-class boats had three 45 cm bow torpedo tubes with three spare torpedoes within the pressure hull and two stern torpedo tubes outside the pressure hull. The pressure hull was divided into five compartments: torpedo room, crew space, control room, diesel room and electric motor room. (Marinens Bibliotek)

Hull and Construction

The boats were built to the Danish regulations *Regulativ for Bygning …*, which translates in full as 'Regulations for the construction and examination of the Navy's ships and their supply with rescue equipment'. Forgings and steel castings have to satisfy Lloyds Register requirements.

The pressure hull had an elliptical cross-section and extended from frame 8 to 71; the main ballast tanks were fore and aft of these frames. Maximum beam amidships was 4.45 metres and maximum depth 3.58 metres.

The hull was constructed of special welding-capable Atlantes steel, which had a tensile strength of 52–60kg/mm^2. Electric welding was used extensively for the construction of the pressure hull, and the designed diving depth was 45 metres. The bow had a configuration that made it suitable for penetrating submarine nets

Table 2: **Building Data and Fates**

All built Orlogsværftet

PN/Name	Laid down	Launched	Commissioned	Scuttled	Raised	Broken up
H1 *Havmanden*	22 Aug 1936	19 Jun 1937	8 Oct 1938	29 Aug 1943	8 Nov 1943	1951
H2 *Havfruen*	12 Sep 1936	6 Nov 1937	6 May 1939	29 Aug 1943	8 Oct 1943	1951
H3 *Havkalen*	13 Nov 1936	3 Mar 1938	18 Jul 1939	29 Aug 1943	2 Nov 1943	1951
H4 *Havhesten*	15 Jun 1939	11 Jul 1940	22 Sep 1942	29 Aug 1943	6 Nov 1943	1951
H5	Cancelled – not laid down					

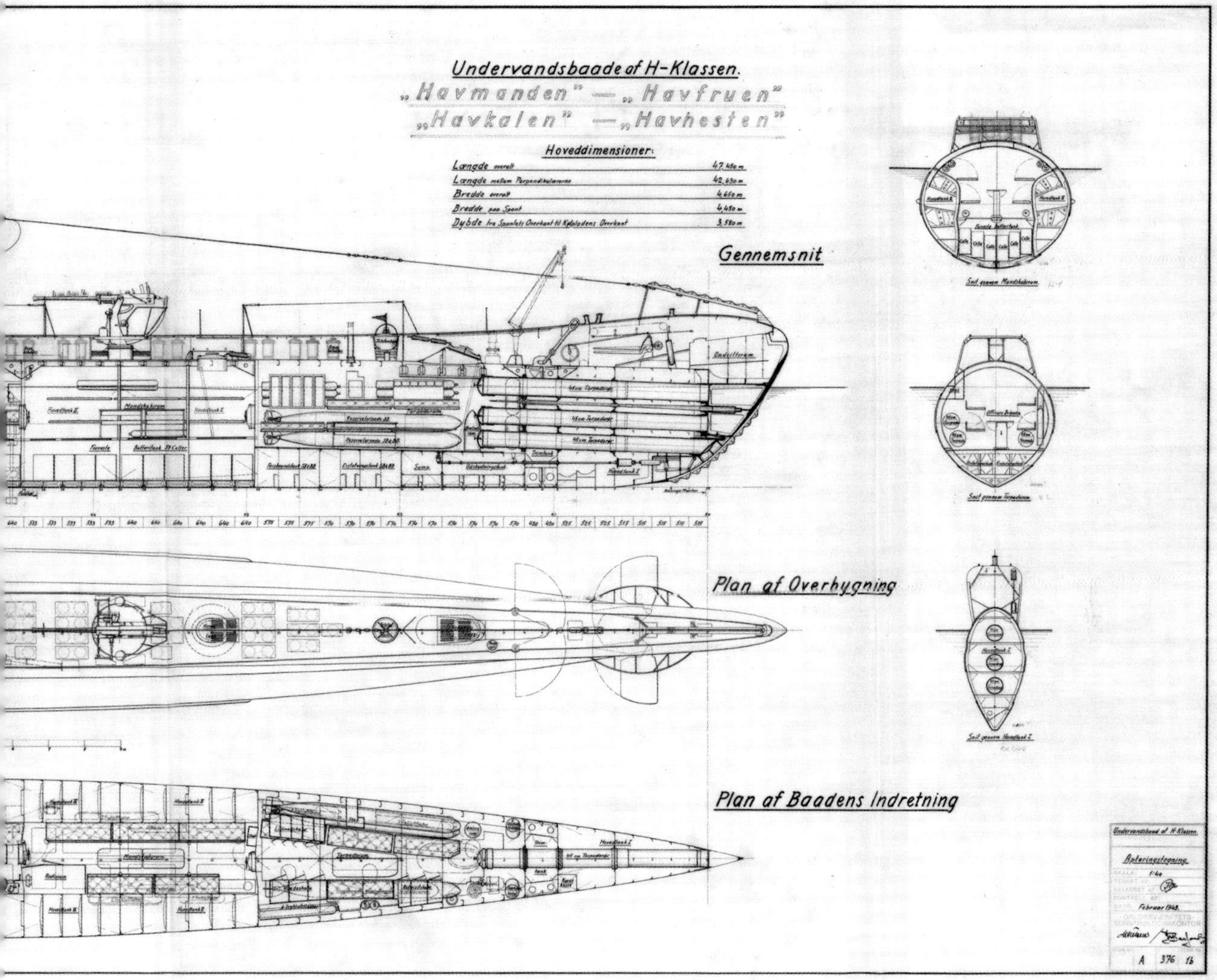

with cutters atop the outer casing and underwater fixed to the pressure hull. The conning tower was constructed of steel castings and plates welded together, and was riveted to the pressure hull.

The upper hull casing had a thickness of 10–8.5mm and extended from the bow to the stern. In order to ensure rapid diving and surfacing there was a significant stand-off between the pressure hull and the outer hull combined with multiple free-flood holes in the casing.

The bow and the stern were constructed as cast steel structures. The retractable forward diving planes were housed in the outer casing at the bow and protected by guard rails. The rudder and after diving planes at the stern were protected by both vertical and horizontal guard rails. The cast steel bow formed the framework for the three bow torpedo tubes, which were disposed one above the other on the boat's axis. The two upper tubes were opened or closed by a common shutter which was manually operated by a shaft located between the tubes. The lower tube was closed by a cast steel shutter turning on a shaft located below the tube and operated by compressed air.

Inside the outer casing there was space for a Hall's patent anchor weighing 335kg; the anchor cable had a length of 75 metres. There were also two salvage fittings each designed for a 100-tonne load, a torpedo davit and two telephone buoys as well as several pressure-proof containers for the ammunition for the two 40mm anti-aircraft guns.

Internal Layout

There were eight watertight bulkheads, of which six divided the interior of the boat into five sections. From the bow the foremost space was the torpedo room, which had a deck area of 14m^2. This compartment also served as the officers' and POs' mess and living space. There were six bunks, together with the requisite kit cupboards, as well as space for three spare torpedoes. The next section was the crew space, which had a deck area of 13m^2. It was fitted out with ten bunks, and on the port side aft was the galley – although 'galley' is possibly too grand a word; 'niche' is probably a more accurate description. Opposite the galley was the W/T Office. The

A submarine of the 'E'class (later 'H' class), probably H1 *Havmanden*, during construction at Orlogsværftet on 12 February 1937. This is the diesel room looking aft; the electric motors would be mounted on the raised platform abaft the diesels. (Marinens Bibliotek)

control room was in the centre of the boat, directly beneath the conning tower, the after two compartments comprising the diesel room and the electric motor room. All bulkheads, except that at frame 62 between the diesel room and the motor room, could withstand the water pressure at a 45-metre depth.

Access to the boat was via hatches above the torpedo room, the crew space (main hatch) and the diesel room plus a reserve hatch over the electrical motor room – this hatch was for emergencies only and was normally sealed. The hatches, except for the torpedo room hatch, were at the level of the pressure hull but had a heavy coaming to reduce the risk of their being damaged in the event of a collision. They could be opened from the inside or the outside. Access to the control room was via an upper and a lower conning tower hatch, both of which were designed for rapid closing.

There were rapid-closing hatches in the bulkheads between the torpedo room and crew space, between the crew space and the control room, and between the control room and the diesel room, but between the diesel and the electrical motor compartments there was only a standard watertight hatch.

There were five main ballast tanks, two trim tanks and two rapid-dive tanks. All except the forward and after main ballast tanks were inside the pressure hull. To improve sea-keeping on the surface, there was a pressure tank in the forward part of the outer casing which also served to strengthen the bow. The tank for the batteries was divided into two watertight compartments below the crew spaces and the forward half of the control room.

Machinery

The two Burmeister & Wain main diesel engines were built under licence by Orlogsværftet, and were of the two-stroke single-working reversible type. Each produced 600bhp at 550rpm, for a total horsepower of 1,200bhp, giving the boats a maximum speed on the surface of 15.75 knots. When cruising on the surface the main engines drove the propeller shafts directly.

The normal fuel capacity was 13.2 tonnes, but this figure could be increased to 26 tonnes by using one of the diving tanks for additional fuel. Range when running at 12 knots on the surface was 1,400nm with

Table 3: **Consumables (fully loaded)**

Fuel oil	13.2 tonnes
Fresh water	3.9 tonnes
Provisions	0.15 tonnes

The bow of an 'H'-class submarine in Orlogsværftets dock No 1 in 1942. The shutter for the two upper tubes and the one for the lower tube are seen in the open position. Also prominent in this photo are the net cutter mounted on the outer casing above the bow and the cutter attached to the lower shutter. The wire running aft from the forward net cutter also served as a W/T antenna. (Marinens Bibliotek)

The stern section of *Havmanden* at Orlogsværftet in 1942, showing the two external after torpedo tubes and the heavy guard rails protecting the diving planes and the propellers. (Marinens Bibliotek)

One of the two Burmeister & Wain diesels intended for a submarine of the 'H' class in the workshop at Orlogsværftet on 17 June 1937. (Marinens Bibliotek)

the normal fuel load, which allowed a transit from Copenhagen to Edinburgh and back.

The two main electrical motors for submerged operations were purchased from Brown-Boveri. Each produced 225hp, giving a maximum speed when dived of 8 knots – the contemporary German Type II submarine of the same size and build had a maximum dived speed of 7 knots. The batteries could supply 3,470Ah, giving the 'H' class a maximum range of 100nm at 3.25 knots when submerged.

The main bilge pumps had a capacity of 75 tonnes per hour at 10 metres pressure depth at 1,200rpm requiring 5hp, and 75 tonnes per hour at 45 metres pressure depth at 2,000rpm requiring 25hp.

Detection and Navigational Instruments

There were two periscopes. The forward periscope was supplied by Carl Zeiss, and had an optical length of 6,200mm and a stroke length of 3,750mm, which enabled it to be used from inside the conning tower as well as the control room. The after periscope was supplied by Barr & Stroud and had an optical length of 7,000mm and a stroke length of 2,640mm.

Along the sides of the submarine there were six hydrophones for the passive detection and monitoring

H3 *Havkalen* and H2 *Havfruen* during fitting out at Orlogsværftet in April 1938. The first boat, H1 *Havmanden*, had her first dock trial alongside and her first dive at Orlogsværftet on 12 May 1938, followed by sea trials on 17, 18 and 20 May. The first dive at sea to 10.5 metres was followed by a surfaced speed trial on 24 May during which the boat achieved 16.06 knots. The depth diving trials were carried out in Øresund (The Sound) south of the Swedish island of Ven, with the assistance of the Orlogsværftets floating crane. (Marinens Bibliotek)

of underwater noise emissions from ships or submarines.

A single Sperry gyro compass Mk VIII was installed as master in the control room with several repeaters. There was also a master compass as backup for the gyro compass located in the after part of the conning tower, with mirror transmission to the control room. The navigational outfit was completed by an electrical log and an echo sounder.

A slightly out-of-focus image of the torpedo room. The photo was probably taken on board H4 *Havhesten* and shows the three 45cm torpedo tubes with foldable bunks to the rear on either side. (Marinens Bibliotek)

The submarine's galley, which was on the port after side of the crew spaces. One of the submariners is happily smoking his pipe while stirring the pot. The watertight hatch leads to the control room. (Marinens Bibliotek)

Armament

The 'H' class had five torpedo tubes for 45cm torpedoes: three axial tubes disposed one above the other in the bow torpedo room (see section drawings), with three spare torpedoes stowed behind them, plus two stern torpedo tubes outside the pressure hull; the stern tubes had to be reloaded from outside the hull. The torpedoes were launched by compressed air.

A 45cm torpedo is loaded on board. (Marinens Bibliotek)

The crew space looking aft. On the tables can be seen bottles of 'small beer' from the King's Brewery. This beer had been served for hundreds of years in His Majesty's Ships, and was not allowed to be be stronger than 2.25% alcohol. In former times the beer came in casks and was served in place of water, which was of poor quality. (Marinens Bibliotek)

Part of the control room with the depth controls. The deck area of the control room was approximately 14m². In the forward part was the CO's accommodation with a bunk on the port side and space for a kit locker and a washstand. It was possible to install an extra bunk above the CO's bunk. There was a reserve bunk a little to the rear on the starboard side of the periscope well. The control room housed all the necessary dials, handles and valves necessary to dive or surface the boat. (Marinens Bibliotek)

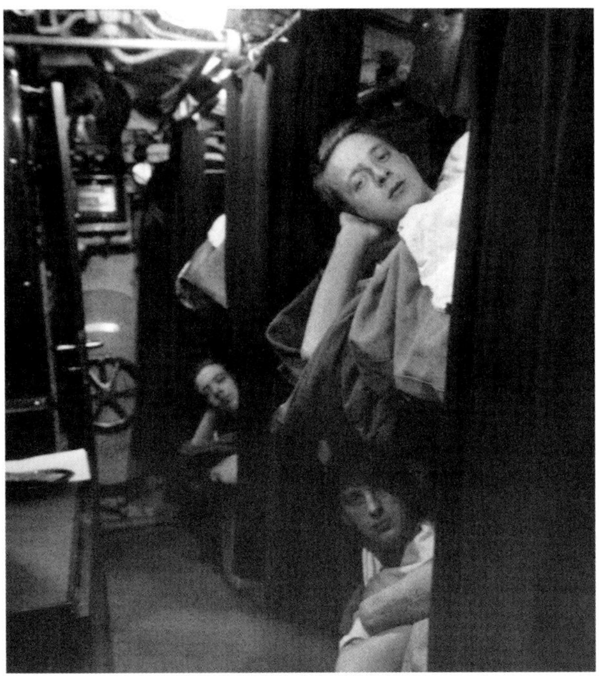

The photo shows four of the ten bunks in the crew space looking aft. The accommodation spaces in these small submarines were particularly cramped. All in all, there were 17/18 bunks in the submarine for a 25-man wartime crew; the peacetime crew was generally 18. The door on the left leads to the W/T Office. (Marinens Bibliotek)

The continued use by the RDN of 45cm torpedoes in submarines during the second half of the 1930s is perhaps surprising. All the major navies changed from 45cm torpedoes to 53cm (21in) torpedoes in their new-build boats after 1916. The rationale for the RDN's retention of the 45cm torpedo was largely economic. After the end of the First World War the political situation in Denmark was such that there was little enthusiasm for spending money on defence, so the RDN was compelled to stick with the 45cm torpedoes manufactured in Denmark, of which there were a large number in storage.

The newest Danish torpedo, the 45cm Type M2, had a length of 5.61m, a weight of 862kg, an explosive charge of 200kg, and a range of 3,000m at 40 knots, 6,000 metres at 34 knots, and 8,000m at 30 knots.

In the 'H'-class the former practice of fitting Danish submarines with a 57mm or 75mm gun for engagement of small surface targets was abandoned. The Danish Navy showed considerable foresight in fitting two single 40mm Bofors L/60 M/37 anti-aircraft guns instead. The magazine capacity was 1,184 fixed rounds, of which 208 were stored in a below-decks magazine in the after part of the control room and 976 in pressure-tight cylinders in the outer casing. This was arguably the most efficient and powerful AA armament seen in any submarine in 1938. The similarly-sized German Type II had only a single 20mm gun; the British 'S' class, which was twice as large, had only a single 0.303in machine gun. In the Danish 'H' class the 40mm guns were complemented by two single 8mm machine guns in the conning tower, which were supplied with 4.000 cartridges stowed below deck and in pressure-tight boxes in the conning tower.

The sanitary installations were almost non-existent. The only heads were at the rear end to starboard in the electric motor room. Here a submariner is enjoying the newspaper while relieving himself, probably enjoying the relative peace and quiet – at least if the diesels were not running. (Marinens Bibliotek)

One of the two 8mm MG mounted in the conning tower. (Marinens Bibliotek)

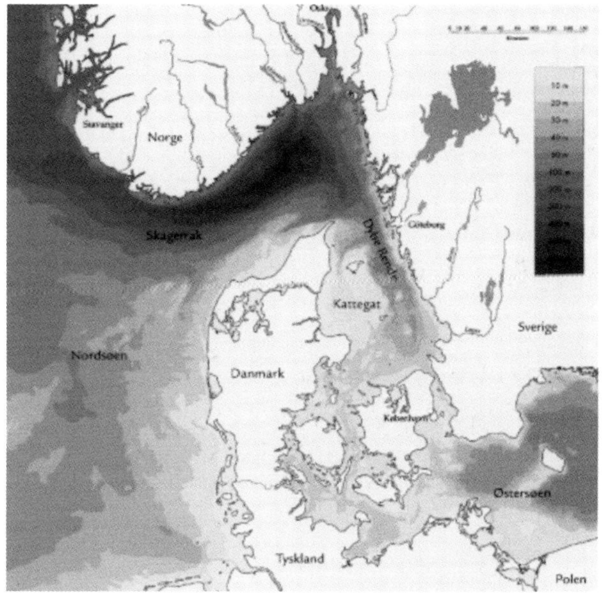

The Danish waters are very shallow, often no more than 20 metres deep, and the maximum diving depth of these boats was limited to 45 metres. (Courtesy of Søren Nørby)

In December 1939 both *Havmanden* and *Havfruen* were compelled to give up one of their two 40mm AA guns, and in January 1940 it was the turn of *Havkalen* and *Havhesten* (then building). The four 40mm AA guns were to augment the AA armament of the training cruiser *Niels Iuel* and the elderly Coast Defence Battleship *Peder Skram*.

Submarine Tactics

The Danish waters are very shallow, hence the boats' maximum diving depth of 45 metres. If the order 'Go deep 15 metres' were given, a boat could hit the bottom at a depth of 10–11 metres in the event of a slightly inaccurate position reading, or if the submarine's position was out due to its being attracted by objects on the seabed.

At a lecture in the *Søofficersforeningen* (Association of Naval Officers) given by Lieutenant-Commander C Weilbach on 9 November 1937 it was stated that a water depth of 15 metres was accepted for operating Danish submarines, as this depth was sufficient for the submarines to pass under all but the largest warships – a submarine of the 'H' class measured approximately 8 metres from the bottom of the keel to the top of the periscopes in the lowered position.

When a boat executed an emergency dive the rapid-dive tanks were constructed so that the boat would not

Table 4: **Cost Estimates 11 March 1940** (not including armament)

For the first three submarines of the 'H' class: *Havmanden – Havfruen - Havkalen*

Cost estimate Orlogsværftet 10 January 1936	6,000,000 Danish kroner
Surcharge: wage increase	59,000 Danish kroner
Surcharge: materials during building	150,000 Danish kroner
Surcharge: gyrocompass installation	18,000 Danish kroner
Surcharge: sandblasting of steel parts	20,000 Danish Kroner
Surcharge: *Selsyn* log vs S.A.L. log	60,000 Danish Kroner
Surcharge: additional compass repeater W/T Office	4,650 Danish Kroner
Surcharge: strengthening DF moving parts	3,000 Danish Kroner
Surcharge: four-inch rectifier	2,750 Danish Kroner
Surcharge: more advanced hydrophones	130,000 Danish Kroner
Total cost	6,447,400 Danish Kroner
Minus customs reimbursement	75,000 Danish Kroner
Final cost	6,372,400 Danish Kroner

Note:
1940 exchange rate: £1 sterling = 19.80d Danish Kroner, so total cost equivalent to £303,030.

H2 *Havfruen* as completed in 1937 with the two 40mm Bofors AA guns. (Marinens Bibliotek)

A superb overhead of three of the 'H'-class submarines at Århus after the outbreak of the Second World War in September 1939. From left to right: H3 *Havkalen*, H1 *Havmanden* and H2 *Havfruen*. The two external stern torpedo tubes are prominent in this view. Note that the two Bofors AA guns remain in place. (Ernst Chr Skabelund, Marinens Bibliotek)

Havkalen at Århus in the autumn of 1939. The forward 40 mm gun would be landed shortly after the photo was taken, in December. (Marinens Bibliotek)

A diving exercise at Mariager Fiord in 1942 during the German occupation. A daredevil is enjoying the ride. (Marinens Bibliotek)

The aftermath of a dive in Mariager Fiord: an 'H'-class submarine has dived at too steep an angle. The bow has struck the bottom, and the propellers and rudder are out of the water. As far as is known there were no serious consequences. (Marinens Bibliotek)

dive at a steeper angle than 8 degrees. During training there were experiments that involved the submarine operating on the surface at night, attacking with torpedoes on the surface and then diving, operating dived and again attacking with torpedoes. The optimal attack distance was set at 800 metres.

It was envisaged that the Danish submarines were to take part in the naval defence of Danish coastal waters and the Danish-laid minefields. Unfortunately, the Danish submarines were too small to be able to lay mines, which could have given an attacker an unpleasant surprise. Danish submarines could also be deployed to destroy an attacker's supply lines, as an attack on the Danish landmass would need shipborne resupply of materiel and/or troops.

The Second World War

Denmark had signed a non-aggression pact with Germany in the spring of 1939, in contrast to Norway and Sweden, which saw the German offer as a propaganda stunt on the part of Joseph Goebbels. Denmark was not at all prepared, neither militarily or psychologically, for a defence of Denmark.

Due to the tense political situation in Europe in the late 1930s the largest ship in the RDN, the cruiser *Niels Iuel*, which was due to represent Denmark at the World's Fair which opened in New York on 30 April 1939, had to remain in home waters. Most of the Navy's available ships were in commission.

The submarine flotilla and a number of minelayers joined the training squadron, which was one of the largest since the First World War. When Germany attacked Poland on 1 September 1939, *Havmanden*, *Havfruen*, *Havkalen*, together with three other Danish submarines, were in commission out of a total submarine force of eleven boats.

When the situation escalated on 3 September the three 'H'-class submarines, together with their depot ship *Henrik Gerner*, joined the First Squadron stationed at Aarhus, Jutland. From 3 October 1940 all Danish warships wore large Danish national markings

(*Dannebrog* – the Danish flag) painted on superstructures/conning towers and decks. Danish merchant ships wore the painted *Dannebrog* from shortly after the beginning of hostilities in September 1939.

After Germany invaded Poland, leading to a declaration of war by France and Britain, a Neutrality Guard Force was established comprising most of the ships of the RDN plus approximately 30,000 soldiers from the Danish Army. However, most of the soldiers from the Army were demobilised after only a couple of weeks in uniform, and daily life went on almost as before 1 September.

A second submarine group of two 'D'-class, one C'-class and one 'B'-class submarine, together with a depot ship, was formed and stationed at Elsinore at the northern entrance to the Øresund. The two submarine groups at Aarhus and Elsinore switched stations approximately every three weeks. The boats at Elsinore had only two hours sailing to Orlogsværftet, so it was quite easy to reach Orlogsværftet if yard assistance were needed.

Most of the Navy was kept on alert, and in December mine barrages for the defence of Copenhagen were established and minefields laid in the southern part of the Great Belt following a German request.

The German Occupation of Denmark

Although Denmark had declared its neutrality, and Germany had offered Denmark a non-aggression pact in February 1939 which Denmark had accepted, on 9 April 1940 Germany invaded Denmark. The Danish Army and the Navy had not been mobilised due to political indolence. A force of 30.000 German troops invaded Denmark over the common border in southern Jutland. Germany also landed troops in several Danish harbours, and a German battalion of 800 soldiers was landed directly at the Langelinie Pier in Copenhagen – the pier nowadays used by cruise ships. The reason the (admittedly small) Danish armed forces did not resist the German invasion was arguably the natural consequence of the previous 20 years of Danish disarmament. Ordering the few soldiers under arms to fight would have been tantamount to murder.

In 1940 Denmark was not a wealthy country, and there was little inclination among the population and within the political system to spend money on defence. Both the Navy and the Army were ready to fight with what means they had, but the Danish politicians declined to mobilise the armed forces. After some four hours of minor skirmishes close to the border with Germany in the southern part of Jutland, which tragically resulted in 14 deaths among the Danish soldiers, the Government and the King ordered the Danish troops to lay down their arms. The situation could be summed up by re-writing Winston Churchill's famous words from his speech of 18 June 1940 after the fall of France and the evacuation from Dunkirk: 'This was NOT their finest hour'.

The following three years were a strange period in Danish history. Denmark was allowed by Germany to keep its own government and part of its armed forces; this was termed the Cooperation Policy. Denmark received raw materials, including coal and oil, supplied from Germany, but this also meant that much of Danish industrial and agricultural production was delivered to Germany – a supremely unequal outcome.

The Years 1940–43

During the occupation years the submarines could to a certain extent undergo training at Isefjorden (the inland waters at Sjælland/Zeeland) and at Mariager Fiord, located in the northeast corner of Jutland where the waters were deeper (30 metres). It was an unusual situation in that Denmark was occupied by a hostile power but was still permitted to train its submarine crews. Deployment of the 'H' class during these years was as follows:

Havmanden

Assigned to the Neutrality Guard Force from 1 September 1939 to 9 April 1940. At the beginning of the occupation, from 17 September 1940 to 26 October 1941, *Havmanden* was laid up at Vejle, eastern Jutland. From 27 October 1941 to 8 December 1941 she was commissioned for a training cruise. From 9 December 1941 to 29 August 1943, she was again laid up at Naval Station Holmen at Copenhagen.

Havfruen

Assigned to the Neutrality Guard Force from 1 September 1939 to 9 April 1940. Decommissioned on 31 May 1940, she was recommissioned from 15 April 1942 at Holmen. She was in a floating dock at Orlogsværftet on 29 August 1943 and scuttled together with the dock.

Havkalen

Assigned to the Neutrality Guard Force from 1 September 1939 to 9 April 1940. Decommissioned 31 May 1940, she was laid up at Vejle. She was later sailed to Holmen, and was scuttled there on 29 August 1943.

Havhesten

Commissioned 29 April 1943 and assigned to *Niels Juel* as a target towing vessel. She was decommissioned on 25 July 1943 at Holmen and scuttled there on 29 August.

The Scuttling of the RDN

The Cooperation Policy ended on 28 August 1943. From 1941–43 relations between the Danish population and the German occupation forces had gradually become more and more strained. This culminated in massive strikes during which ordinary people took to the streets in the spring/summer of 1943, first at Odense on Funen and later also in Copenhagen and other cities. The strikes were spurred on by the Danish resistance movement, which from 1942 had become more and more active as weapons and explosives were dropped by parachute from British aircraft. On 28 August the occupation forces had given the Danish government an ultimatum that not even the politicians could accept: the conviction of Danes according to German law and the introduction of the death sentence for sabotage activity. The government rejected these demands and cooperation ended.

On the morning of 29 August 1943, 27 ships of the RDN were scuttled by their own crews at Holmen in order to prevent them from falling into German hands. When German forces entered Holmen, the crews either used demolition charges or opened the seacocks to make sure that the ships would not fall undamaged into the hands of the enemy. The scuttling was ordered by the senior officer of the RDN, Vice Admiral Vedel, not by the Danish government; the most the Minister of Defence was prepared to do was to instruct the Vice Admiral that the Navy was not to resist a German entry into the Holmen naval base by force of arms.

Vedel had, on his own initiative, pre-emptively given secret orders to the ships at Holmen and HDM ships outside Copenhagen. Should the German *Wehrmacht* attempt to seize the Danish warships, i) the ships should attempt to reach (neutral) Swedish waters, and if this were not possible, ii) the crews must scuttle their ships.

If Vedel had decided that the ships at Holmen should try to leave Copenhagen and fight their way out, the ships would have been led by *Havkalen* with the Chief of Staff and the naval staff on board. However, given the presence of several armed German warships in Copenhagen Harbour, scuttling was the only feasible course of action.

All four of the 'H' class were scuttled by their crews. *Havmanden*, *Havkalen* and *Havhesten* were all alongside at the submarine piers on the northwest side of Holmen. The CO of *Havkalen*, Commander Aage Linde, later reported to the Naval Command:

> On 29 August at approx 0407 I received the order to scuttle the boat, which was obeyed one minute later by opening the seacocks so that the sea entered the main ballast tanks and the bilgepipes could fill the interior of the boat. After this the crew was ordered to evacuate the boat and at approx 0412 the deck hatches were level with the sea's surface.

Havkalen's two sisters, *Havhesten* and *Havmanden*, were scuttled in the same calm and efficient way. No explosives were used; the seawater did the damage to the machinery, batteries, and electrical installations.

The fourth boat, *Havfruen*, was in Floating Dock No 2 at the Orlogsværftet some 100 metres from the submarine piers. Søren Nørby writes in his book *Sænk skibene*:

> On August 28, *Havfruen* was due for inspection in Orlogsværftet Floating Dock No 2, which was located on the north side of the dock island. The *Havfruen* was not in commission, and the submarine's crew was therefore not on board. Until the year 2004, the sinking of the submarine was attributed to some determined seamen who opened the hatches and valves in the submarine and, to complete the work, they simultaneously opened the bottom valves of the floating dock and punched a hole in the dock with a sledgehammer.

However, the real story turned out to be a little different. At a meeting of the submarine division in the afternoon of 28 August 1943, the responsibility for the sinking of the submarines that were not under command was divided between the various submarine commanders, and chief engineering officer first class Wilhelm C Tuxen was given the responsibility for destroying the *Havfruen*. When Tuxen turned in for the night on board the *Hekla*, the barracks ship for the submarine squadron, on the evening of the 28th, he gave the order that he should be awakened if the slightest thing happened. When he was woken up a little after 4 o'clock, however, it was with the message that the ships had already been scuttled! It had all gone so quickly that they had forgotten to wake Tuxen, and while the submarines at the submarine piers had been scuttled, the *Havfruen* was still intact in the floating dock.

Tuxen was told that all personnel were to head towards the Plan Building at the Mastekranen, but he decided to try to carry out the assigned task first. Tuxen crept from *Hekla* to the Dock Island and the two floating docks there. The Navy's general report on the events at Holmen on 29 August 1943 states that *Havfruen* was scuttled at 0415, but according to Tuxen's report it was closer to 0430 before he reached the floating dock. Upon arrival, Tuxen discovered that he had not brought the key to the floating dock's engine room. Access to this was necessary in order to lower the dock, and Tuxen therefore had to break the glass of the door with his elbow before he could gain access. Once inside, Tuxen tried to figure out which buttons and handles would cause the dock to sink. After some deliberation, he chose the simple solution of placing all the handles opposite from where they now stood, and when shortly afterwards he could hear escaping air, he knew that the dock was sinking. Tuxen now calmly left the floating dock and headed for the Plan Building.

Havfruen was scuttled together with the drydock in which she was undergoing maintenance. (Marinens Bibliotek)

Havfruen was the first of the four 'H'-class boats to be lifted by the German *Schwimmkran* floating crane; she is being lowered onto the quay on 8 October 1943, Note the *Dannebrog* painted on the fin. *Havkalen* was raised on 2 November, *Havhesten* on 6 November and *Havmanden* on 8 November. (Orlogsmuseet)

Raised (formerly scuttled) submarines in Sømingraven at Holmen in 1945. In the foreground is H3 *Havkalen.* (Marinens Bibliotek)

After the Scuttling

After the scuttling of the ships at Holmen, the Germans deployed an enormous floating crane, *Schwimmkran 350^t I/38,* with the nickname *Langer Heinrich* from Gotenhafen (now Gdynia, Poland) to Copenhagen; it arrived on 3 October 1943. The crane was able to lift 350 tonnes, which in 1943 was an enormous lift capacity. The *Schwimmkran* was one of four that would have been key to the building process had the German Navy's prewar 'Z' Plan materialised.

Now the *Schwimmkran* had to clear the basins of scuttled Danish ships, as the Germans needed the harbour and quays cleared for their own use; they were also keen to recommission those Danish ships that were not too badly damaged. In all, thirteen of the Danish ships were later in the war recommissioned into the *Kriegsmarine.* In 2024, two of the *Schwimmkräne* are still in operation: one in St Petersburg, Russia, and the other in the Panama Canal.

The End

After the war all four 'H'-class submarines, together with other wrecks of Danish submarines and surface ships, were found in the southern part of Søminegraven at Holmen, where there was a veritable ships' graveyard.

From the outset the Navy recognised that the submarines of the 'B' and 'D' classes were to be discarded, and they were sold for scrap in 1946. For some years it was debated whether to use the 'C' class for target practice or for experiments with depth charges. The reconstruction of the 'H' class for use as training boats was also considered. However, it was acknowledged that only the hull itself could be employed for this purpose, as all the technical installations in the boats had been destroyed.

The development of submarines and submarine technology, especially in the *Kriegsmarine* in the last years of the Second World War, had made such rapid progress that any proposal for reconstruction of the 'H' class had little traction. The mere fact that the 'H' class was built with 45cm torpedoes meant that the 45cm tubes would need to be enlarged to 53cm, which made a reconstruction almost impossible.

This situation was difficult to accept for Orlogsværftet and the Navy. In order to move forward, Denmark succeeded in 1946 in leasing three British submarines, one 'U' class and two 'V' class, so it was at least possible to ensure that the Navy had submarine crews and knowledge of how to operate submarines in the future.

In June 1950 the 'H'-class wrecks were stricken from the Navy list and on 18 November 1950 *Havmanden, Havfruen, Havkalen* and *Havhesten* were put up for sale and disposal in Denmark. The hulks were purchased by Petersen & Albeck and broken up at their Copenhagen South yard.

When the hulls were put up for sale, the following specifications were given: 120 tonnes of steel, 12.7 tonnes of other metals, 6 tonnes of copper, 4.4 tonnes of lead, plus 29 tonnes of assorted items which included cement, paint and linoleum, and 22.5 tonnes of water and dirt. All four boats were sold without their batteries.

Acknowledgements:

The author wishes to thank Lars Jordt and Søren Nørby for their assistance.

Sources:

Den Danske Ordhog (ordnet.dk).

Ubaade af 'H-Typen', Oplysningsbog for Baaden (Copenhagen).

Madsen, Kaj Toft, *Danske Torpedoer 1868–2008* (Copenhagen, 2008).

Marinens Bibliotek (MAB).

Nørby, Søren, *Danske Ubåde 1909–2004* (Copenhagen, 2005).

Steensen, Steen, *Vore Undervandsbåde gennem 50 år, 1909–1959* (Copenhagen, 1960).

Thiede, Sven Egil, *Dansk Søartilleri 1860–2004, Bind 2,* (Copenhagen, 2004)

Weilbach, Lt-Cdr, 'About the Use and Misuse of submarines, lecture Sølietenant-Selskabet (Copenhagen, 1937).

RUSSIA'S BATTLECRUISER DESIGN COMPETITION OF 1911

In the summer of 1911, the Imperial Russian Navy invited six Russian and seventeen of the most prominent foreign shipbuilders to participate in a competition for the design of a powerful battlecruiser. **Stephen McLaughlin** describes the process that eventually led to the four ships of the *Izmail* class, which were laid down but never completed.

The disastrous Russo-Japanese War had revealed deep structural and administrative problems in the Russian Naval Ministry, and one response was the formation of the Naval General Staff (*Morskoi generalnyi shtab*, or MGSh). It was tasked with the formulation of the Navy's war plans, the development of shipbuilding programmes to match those plans, and the drawing up of requirements for the ships to be built. As a starting point towards these goals, in 1906 it canvassed a large number of officers, seeking their views on the lessons of the recent war. The role of armoured cruisers was one of the more contentious topics considered. One respondent argued that this type of ship was 'absolutely useless. For squadron battle it is weak, for reconnaissance it is heavy and expensive', but most others believed that such ships would play an important role in future naval warfare. For example, Lieutenant Graf A P Kapnist believed that, in addition to traditional cruiser duties such as commerce raiding and scouting, they could also serve as 'a flying detachment, which attempts to reinforce the attack of the main forces, directed at a part of the enemy's squadron. They attempt to attack his flank, positioned ahead of his van, or astern of his tail…'.[1]

However, in order to carry out this role, these ships would need not only a speed superior to that of battleships but an armament to match them as well. The Imperial Japanese Navy was already building ships along these lines: the large armoured cruisers *Tsukuba* and *Ikoma*, laid down in 1905.[2] With a displacement of about 14,000 tons they were much larger than previous armoured cruisers and were armed with four 12in guns in twin turrets fore and aft. But by the time they entered service in 1907–1908 they had already been overtaken by the Royal Navy's *Invincible* class, then under construction. Like the *Tsukubas* these three ships were originally classified as armoured cruisers, but their powerful armament of eight 12in guns soon led to them being dubbed unofficially as 'battle cruisers'. The first unit of the class to enter service was *Indomitable* in June 1908, but information about these new turbine-powered ships was circulating well before this – a detailed report was published in the June 1907 issue of the Russian professional journal *Morskoi Sbornik*.[3] As a result, in August 1907 the MGSh developed a set of requirements for this new type of cruiser (Table 1, Column A). The desired characteristics were very similar to those of *Invincible*, although more extensive protection was specified. The intended roles of the ship were deep reconnaissance for the fleet, engaging enemy armoured cruisers, and forming a flying wing of the battle fleet in a general action.

However, in the wake of the war with Japan and the subsequent economic turmoil there was little money available for rebuilding the fleet, and no design work was

Table 1: The MGSh's Armoured Cruiser Requirements 1907 and 1910

	A August 1907	B May 1910
Draught:		
normal	27ft 4in (8.33m)	–
full load	29ft (8.84m)	29ft
Armament:		
main guns	8 – 12in/52 (4 x II)	10/12 – 12/14in guns (2 x III, 2 x II / 4 x III)
secondary	16 – 4in (16 x I)	20 – 4in (20 x I)
torpedo tubes	three	six
Protection:		
main belt	6in (152mm)	7.5in (190mm)
longitdinal bulkhead	–	2in (51mm)
upper belt	3in (76mm)	3–4in (76–102mm)
upper deck	1.75in (44.5mm)	1.75in (44.5mm)
middle deck	–	1.25in (32mm)
lower deck	1.25in (32mm) 2in (51mm) slopes	1.25in (32mm) 2in (51mm) slopes
conning tower	12in (305mm)	10in (254mm)
turrets	?	10in (254mm)
funnel uptakes	1in (25.4mm)	?
Machinery:		
speed	25 knots	28 knots, 30 knots with forced draught
endurance	?	48 hours at 28 knots (normal displacement)

Source: Kuznetsov, *Lineinye kreisera tipa 'Izmail'*, 7, 8-9.

initiated at this time. In May 1909 the armoured cruiser requirements were updated, perhaps in response to rumours that the next generation of British battlecruisers would be larger and more powerful than the first-generation ships of the *Invincible* and *Indefatigable* classes. Now the Naval General Staff wanted a 25,000-ton ship with a speed of 28 knots, armed with nine 12in guns in triple turrets and an anti-torpedo boat battery of sixteen 130mm guns. The more conservative Naval Technical Committee (*Morskoi tekhnicheskii komitet*, or MTK) thought a 23,000-ton ship with a speed of 23 knots was more realistic. However, the Navy's budget was still severely limited, and what funds there were had to be devoted to the four battleships of the *Sevastopol* class, laid down the following month. Once again the battlecruiser requirements remained a paper exercise.

New Specifications

A year later the MGSh drew up yet another set of desired characteristics (Table 1, Column B). Not surprisingly, these were more ambitious: the possibility of a heavier main battery was raised, the speed was greater, and the armour thicknesses were increased in expectation of engagement ranges in the order of 40–60 cables (8,000–12,000 yards). Another feature was the inclusion of a longitudinal splinter bulkhead inboard of the belt armour, as in the *Sevastopol* class. An icebreaking bow was also desired, as was a silhouette similar to that of the dreadnoughts, the idea being to confuse an enemy regarding what sort of ship it was up against.

The MTK's drawing office worked out some sketches to determine the feasibility of these requirements, and concluded that even with the battery limited to eight 12in guns in twin turrets the ship would be very large. This result was confirmed when the requirements were circulated to the German firm of Blohm und Voss and to Vickers in Great Britain, both of which were willing to draw up sketch designs in the hope of obtaining future contracts (see Table 2).

As had been the case with the MTK's effort, the two foreign shipyards were unable to fulfil all of the MGSh's requirements on a reasonable displacement, and the staff was forced to reduce its specifications: the range was halved and the desired 35-degree elevation of the main battery guns was abandoned. The new version of the requirements was now given to a colonel in the Constructor Corps, I A Gavrilov of the Admiralty Shipbuilding Works in St Petersburg. Gavrilov was fast becoming the Russian Navy's expert in cruiser design. In 1904 he had proposed a high-speed scout cruiser, in 1907–1908 he had drawn up a series of protected cruiser design studies for the Navy's leadership, and he would soon become heavily involved in the design of the light cruisers of the *Svetlana* class.[4]

Gavrilov submitted his armoured cruiser design sometime in the spring of 1911 (Table 3, Figs 1 and 2). As was common practice amongst Russian naval constructors, he gave it a name – '*Simmetriia*', after its almost symmetrical profile. There was never any intention of building a ship to this design; it was instead an investigation of the feasibility of the requirements. Gavrilov assumed a main battery of ten 12in guns, although he made allowance for replacing these with eight 14in, a change that would have required some reduction in the thickness of the armour belt. The most unusual feature of the design was the use of superimposed turrets at a time when the MGSh was fixated on the 'linear' arrangement, with all the turrets on the same level, as in the battleships of the *Sevastopol* class. The staff believed that separating the main battery turrets would prevent more than one being disabled by a single hit, and also reduce the chances of a catastrophic magazine explosion. Other features, however, mirrored those of the battleships, including the two identical conning towers, the extensive hull protection designed to resist high-explosive shells, a 2in longitudinal splinter

Table 2: **Proposed Armoured Cruiser Characteristics, Autumn 1910–Spring 1911**

	MTK Drawing Office	Blohm und Voss	Vickers*
Displacement	28,000 tons	26,420 tons	29,000 tons
Length	204m / 669ft 4in	210.92m / 692ft	?
Beam	27m / 88ft 7in	26.52m / 87ft	?
Draught	8.84m / 29ft	8.3m / 27ft 3in	?
Armament:			
main guns	8 – 12in (4 x II)	8 – 12in (4 x II)	8 – 13.5in or 14in (4 x II)
secondary	24 – 100 (24 x I)	16 – 120mm (16 x I)	14 – 6in (14 x I)
Protection	?	?	8in belt
Machinery:			
horsepower	80,000shp	95,000shp	?
speed	28 knots	30 knots	28 knots

* This design – if indeed it was ever developed into a design – does not appear in the Thurston Notebooks in the NMM.

Source: Kuznetsov, *Lineinye kreisera tipa 'Izmail'*, 9.

bulkhead inboard of the main belt, and two rudders on the centreline.

While this preliminary design work was underway, the Navy Minister, Admiral I K Grigorovich, had presented a draft naval programme covering the years 1911–1915 to the Emperor which included the construction of four armoured cruisers. Tsar Nikolai II, always a naval enthusiast, quickly approved the proposal, but under the semi-constitutional government created after the revolutionary disturbances of 1905–1907 the bill had to be approved by the new Russian parliament, the State Duma. Unfortunately, the bill was placed before the Duma too late to be considered before its summer recess, and therefore had to be carried over until the following session, delaying the start of the programme until 1912.

In the meantime the MGSh and the MTK were putting the finishing touches to the requirements. These were approved by Navy Minister Grigorovich on 9/22 August 1911 and would form the basis for an international design competition (See Table 4). [Dates are given here in both the Old Style used in Russia until 1918, and the New Style used in Western nations.] The Imperial Navy had been using such competitions since the late 1890s: the protected cruisers *Askold*, *Bogatyr*, and *Novik* had all been the products of contests for the best designs, as had the dreadnoughts of the *Sevastopol* class. In each case, however, the winning designs had been subjected to numerous changes before and during construction, resulting in a complex process that required repeated modifications to the characteristics of the ships. The same would be true in the case of the new armoured cruisers.

The technical specifications detailed a number of additional requirements. These covered not only fundamental matters such as hull strength and stability (the metacen-

Table 3: **Gavrilov's '*Simmetriia*' ('Symmetry'), *circa* Spring 1911**

Displacement	26,100 tons
Length	700ft (213.36m)
Beam	88ft (26.82m)
Draught	28ft (8.53m)
Armament:	
main guns	10 – 12in (2 x III, 2 x II)[1]
secondary	24 – 100mm (24 x I)
Protection:	
belt	8in (203mm)
upper belt	3in (76mm)
longitudinal bulkhead	2in (51mm)
upper deck	1.25in (32mm)
middle deck	1.25in (32mm)
slopes	2in (51mm)
turrets	10in (254mm)
barbettes	8in (203mm)
conning towers	10in (254mm)[2]
Machinery:	
boilers	18 watertube
engines	4 sets turbines
horsepower	72,500shp
speed	28 knots, 30 knots max
endurance	4,100nm at 13 knots

Notes:

1 Gavrilov made allowance for replacing the 12in guns with 14in, which would have involved a reduction in the main battery to eight barrels and some sacrifice of the armour protection.

2 Gavrilov's design had two identical conning towers, one forward and one aft.

Source: Kuznetsov, *Lineinye kreisera tipa 'Izmail'*, 9–12.

Table 4: **Armoured Cruiser Competition Specifications, August 1911**

Displacement	Not specified
Draught	8.84m (29ft)
Armament:	
main guns	9 – 14in (3 x III; 25°; 80rpg)
secondary	20 to 24 – 130mm (24 x I; 200rpg)
anti-aircraft	4 – 2.5in
saluting	4 – 47mm
torpedo tubes	six underwater
Protection:	
main belt	250mm amidships, reduced to 125mm fore and aft
longtdnl bulkhead	50mm inboard of main belt
upper belt	125mm over casemates, 75mm forward
upper deck	37.5mm
middle deck	25mm
lower deck	50mm slopes
conning towers	305mm sides
turrets	356mm faces, 305mm sides, 127mm roofs
barbettes	275mm
funnel uptakes	25mm between middle and upper decks
Machinery:	
boilers (recommended)	Yarrow three-drum watertube, mixed firing
engines	Parsons, Curtis-AEG, Vulcan, or Curtis-Vulcan turbines
speed	14 knots economical, 26.5 knots full, 28.5 knots forced
endurance	24 hours at 26.5 knots (636nm) normal 72 hours at 26.5 knots (1908nm) full load

Sources: Kuznetsov, *Lineinye kreisera tipa 'Izmail'*, 13–16; Vinogradov, *'Izmail': sverkhdrednout Rossiiskoi imperii*, 4; Usov, 'Lineinye kreisery tipa "Izmail"', 61.

Gavrilov 1911: Profile & Plan

Displacement	26,100 tons
Dimensions:	700ft oa x 88ft x 28ft
Horsepower:	72,500shp = 28 knots
Armament:	10 – 12in guns (2xIII, 2 xII)
	24 – 100mm guns (24 x I)

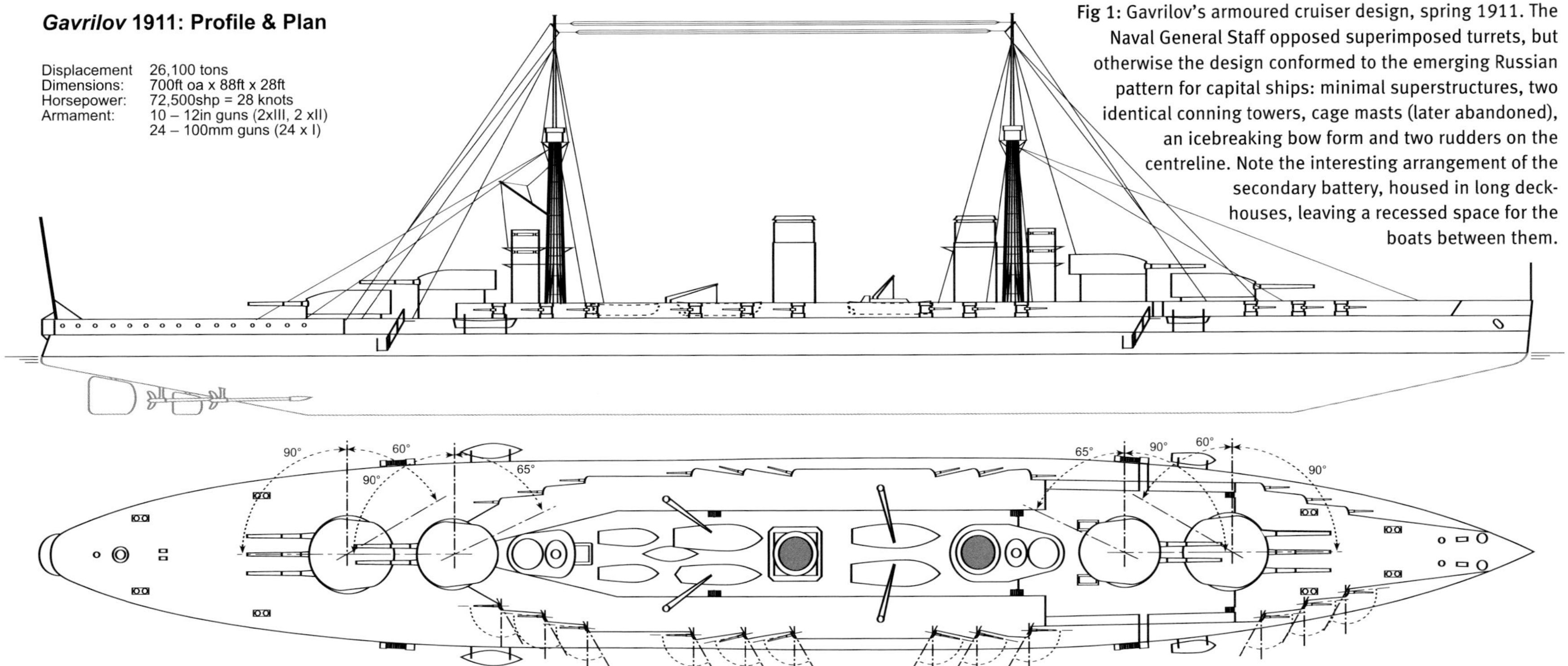

© John Jordan 2022

Fig 1: Gavrilov's armoured cruiser design, spring 1911. The Naval General Staff opposed superimposed turrets, but otherwise the design conformed to the emerging Russian pattern for capital ships: minimal superstructures, two identical conning towers, cage masts (later abandoned), an icebreaking bow form and two rudders on the centreline. Note the interesting arrangement of the secondary battery, housed in long deck-houses, leaving a recessed space for the boats between them.

Gavrilov 1911: GA & Protection Plans

Inboard Profile

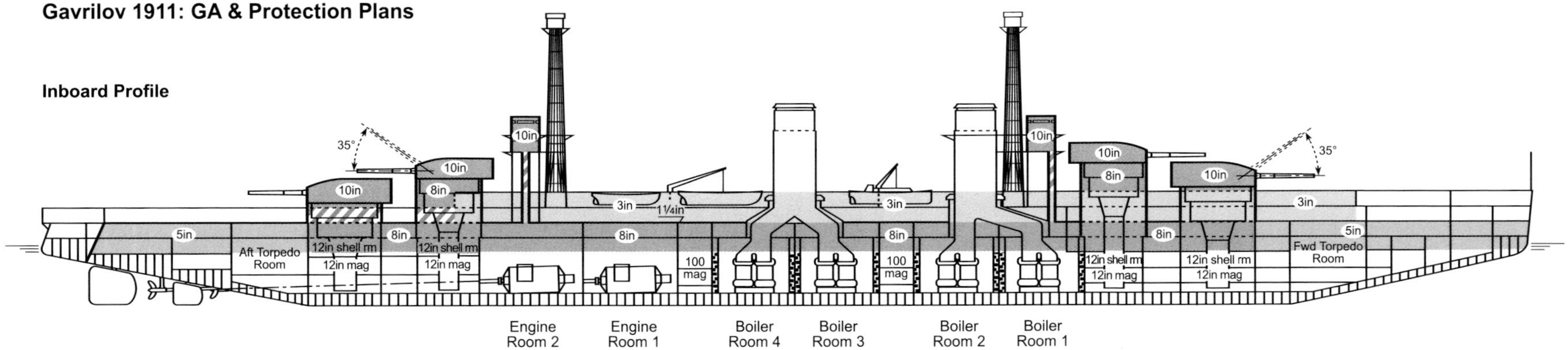

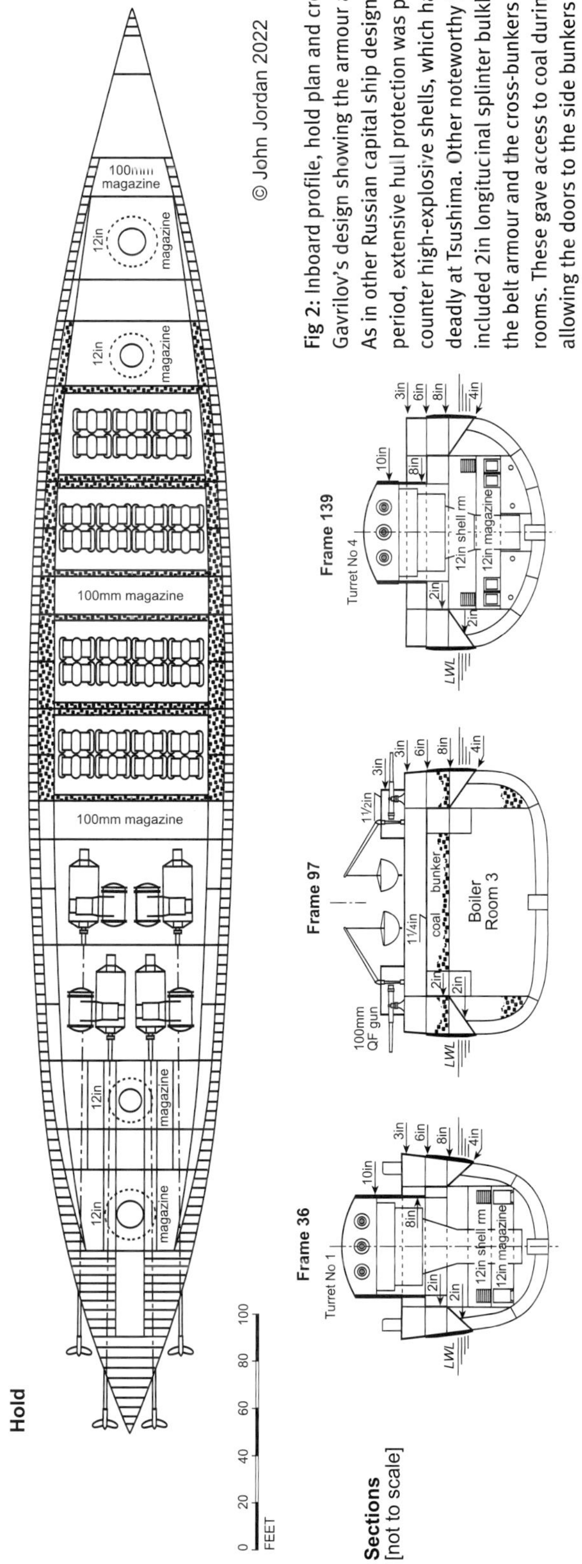

Fig 2: Inboard profile, hold plan and cross sections of Gavrilov's design showing the armour arrangements. As in other Russian capital ship designs of this period, extensive hull protection was provided to counter high-explosive shells, which had proven deadly at Tsushima. Other noteworthy features included 2in longitucinal splinter bulkheads backing the belt armour and the cross-bunkers between boiler rooms. These gave access to coal during battle, allowing the doors to the side bunkers to be kept shut, thereby helping to maintain watertight integrity.

© John Jordan 2022

tric height was to be between 1.7m and 2.1m), but also numerous demands related to individual aspects of the design. Thus the main battery turrets were to be 'equally spaced' along the centreline and the guns were to have a maximum elevation of 25 degrees. Moreover the Russian Navy's gunnery experts insisted that they were to achieve 'the highest permissible technical rate of fire, no matter what sacrifice that entails'.[5] In order to improve gunnery accuracy, any design submitted should include systems 'for automatic transfer of water loads from side to side', such as Frahm tanks, to reduce rolling. There should be two masts of 'light tripod or other construction', with spotting tops 30.5 metres above the waterline. The 26.5-knot speed was to be achieved using only 75 per cent of the boilers. Most importantly, no matter who won the competition, the ships were to be built in Russia.

The Competitors

On 26 August/8 September 1911 the MTK began issuing invitations to participate in the armoured cruiser design competition. Submissions were to be presented within six weeks – five copies of the plans were specified – and the companies invited were:

Britain: Fairfield Shipbuilding and Engineering (Govan), John Brown and Co (Clydebank), Swan Hunter and Wigham Richardson (Wallsend), Vickers Sons & Maxim (Barrow) and William Beardmore (Dalmuir);
France: Forges et Chantiers de la Méditerranée (La Seyne);
Germany: Blohm und Voss (Hamburg) and Vulcan Stettiner Maschinenbau AG (Stettin);
Russia: The Admiralty Works, Baltic Works, Nevskii Works and Putilovskii Works (all in St Petersburg); Naval and Russud (both in Nikolaev);
United States: Charles Cramp & Sons (Philadelphia) and New York Shipbuilding (Camden).

The Nevskii Works was forced to decline the invitation as it was located upriver, where the narrow bridges of the Neva River would make it impossible for such a large ship to reach the sea. The two shipyards in south Russia, Naval and Russud, also withdrew as they were fully occupied with the construction of the battleships of the *Imperatritsa Mariia* class. The Putilovskii Works and Blohm und Voss submitted designs jointly, an arrangement that benefited both companies: the Putilovskii yard was newly-established and lacked an experienced design staff, while the German firm gained an uncontroversial conduit to potentially lucrative Russia contracts. This was something it had lacked three years previously, when it had come within a hair's breadth of winning the competition for the design of the *Sevastopol*-class dreadnoughts – the news that a German firm had been selected had aroused anger in France, Russia's ally and chief creditor, and as a result the Blohm und Voss design was set aside and a domestic design chosen instead. So when the Putilovskii company decided to cash in on Russia's

Fig 3: Beardmore believed that the Russian competition's requirements were unrealistic, so modified them in order to limit the displacement. Its 12in main battery and 9in (229mm) armour belt failed to comply with the specifications, as did other elements of the design, and it was quickly rejected. One interesting feature is the two-tiered arrangement of the secondary battery near the forward and after turrets. Some elements of the protective scheme are conjectural due to a lack of definite information.

Beardmore 1911: GA & Protection Plans

Displacement: 29,500 tons
Dimensions: 670ft pp x 92ft 6inft x 29ft
Horsepower: 60,000shp = 26.5 knots
Armament: 9 – 12in guns (3xIII)
20 – 130mm guns (20 x I)

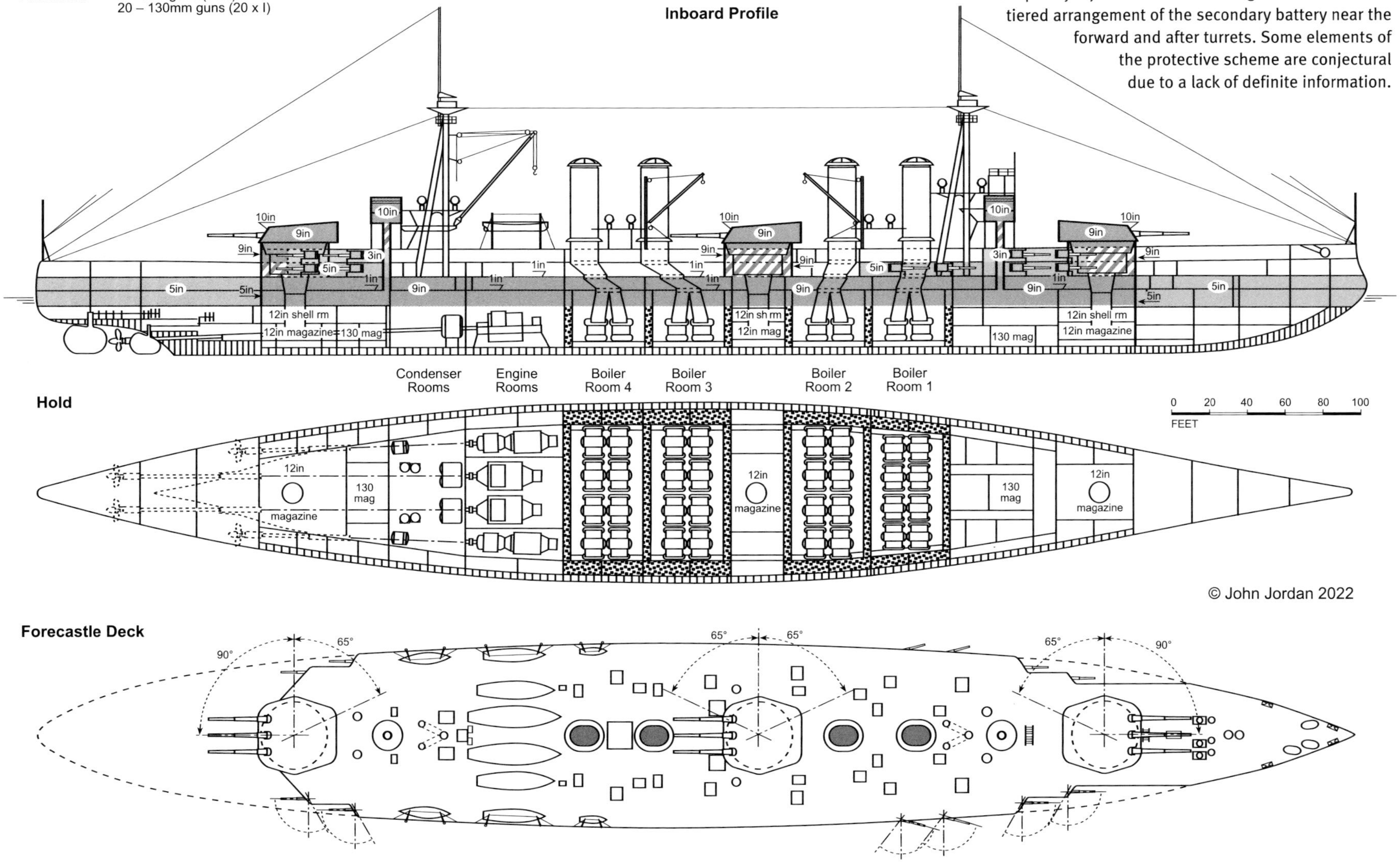

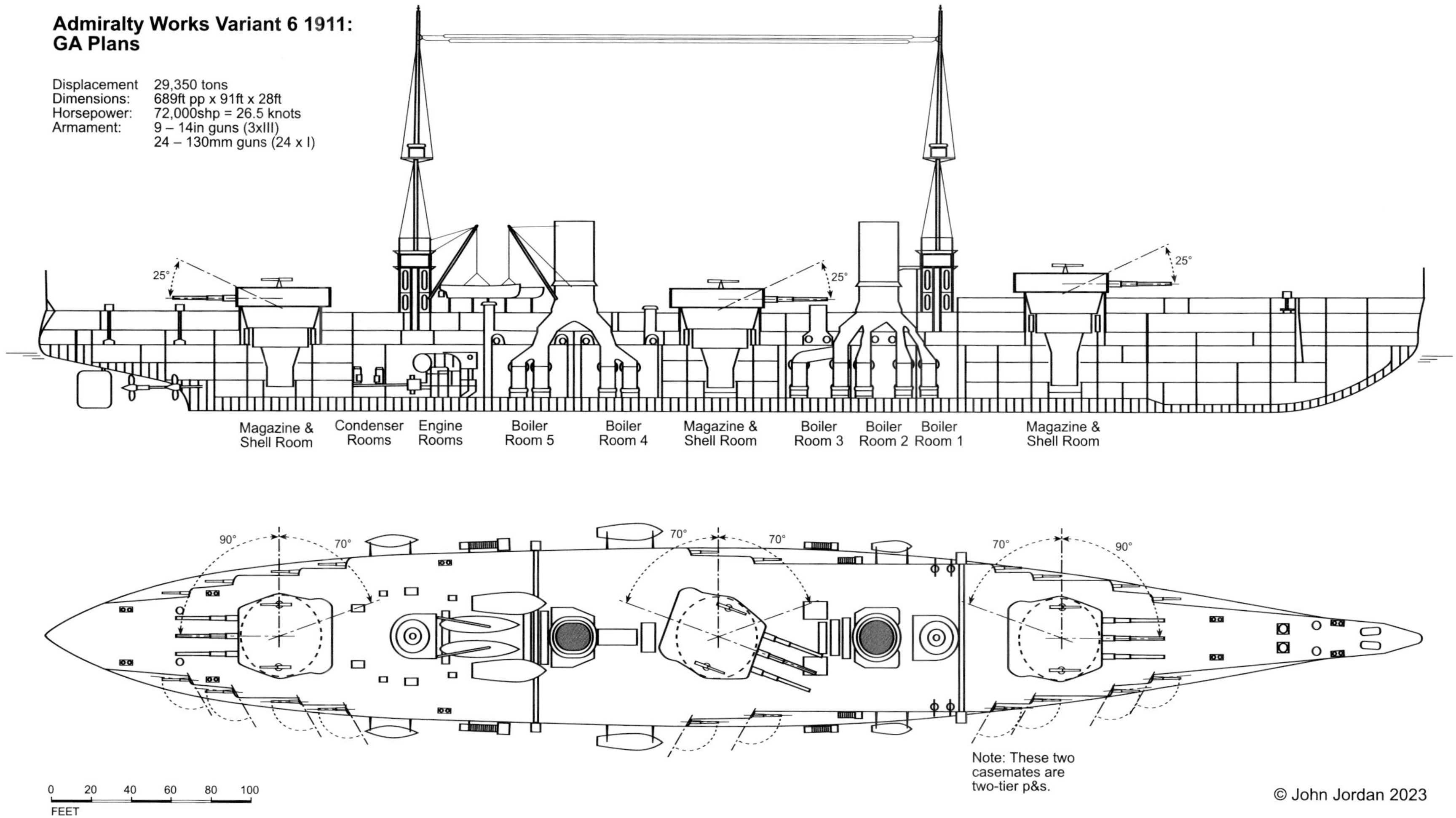

Fig 8: The Admiralty Works was one of the two largest state-owned shipbuilders in St Petersburg and had a large and experienced technical staff. Its design shows many of the features that would be incorporated in the final design, including a raised forecastle and two-storey casemates forward. It was an early favourite among the competitors

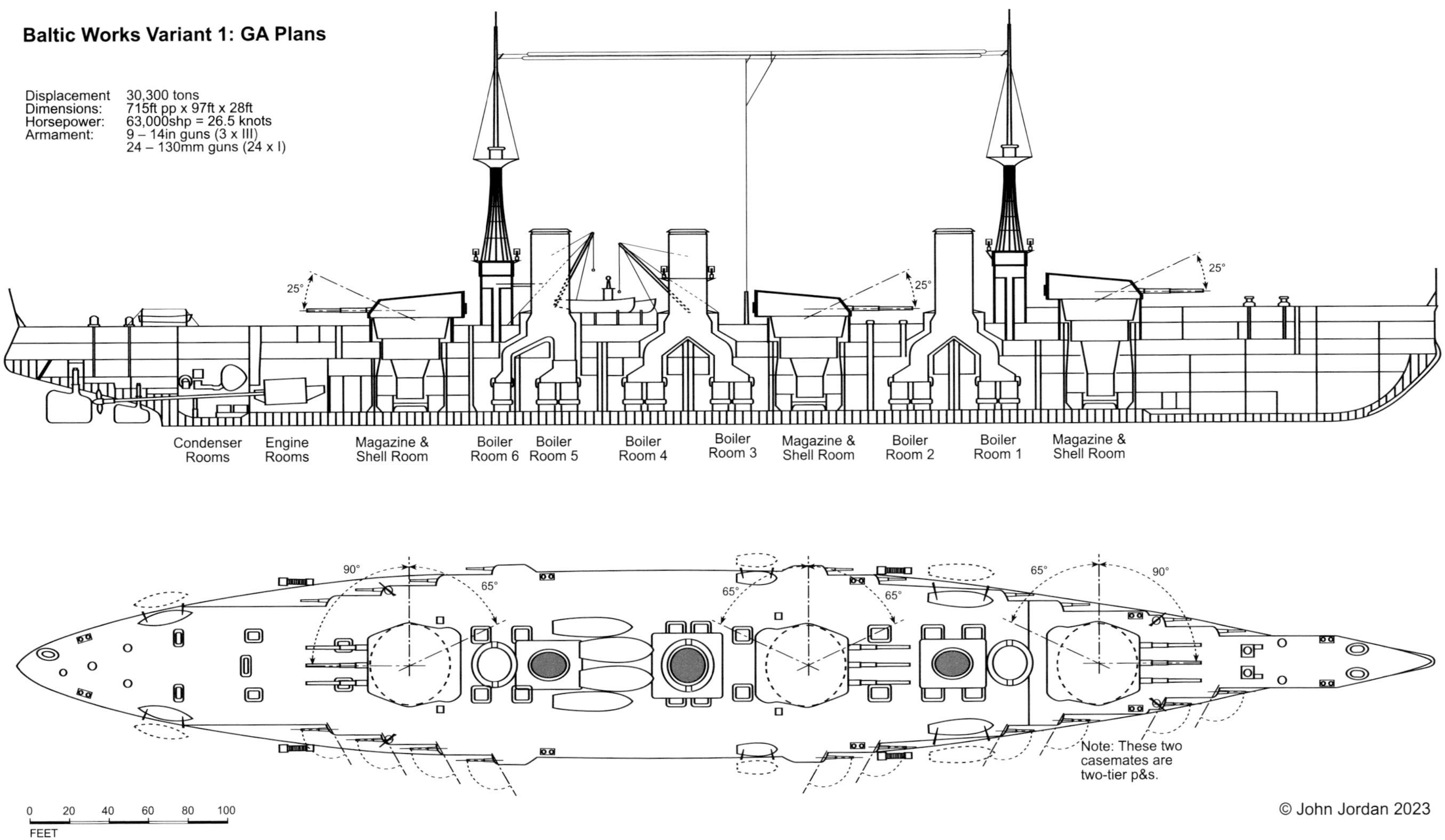

Fig 9: The Baltic Works was the other great state shipyard and its Variant 1 design shares some features with that of the Admiralty Works, but differed in having a longer and broader hull that required considerably less power to achieve the required speed. Note that the engine rooms are abaft the magazines for the third turret, leading to an unusually long quarterdeck which, combined with the spacing of the three funnels, gives the design a unique profile. Variant 3 of this design placed the engine rooms immediately abaft the boiler rooms, shifting the after turret toward the stern.

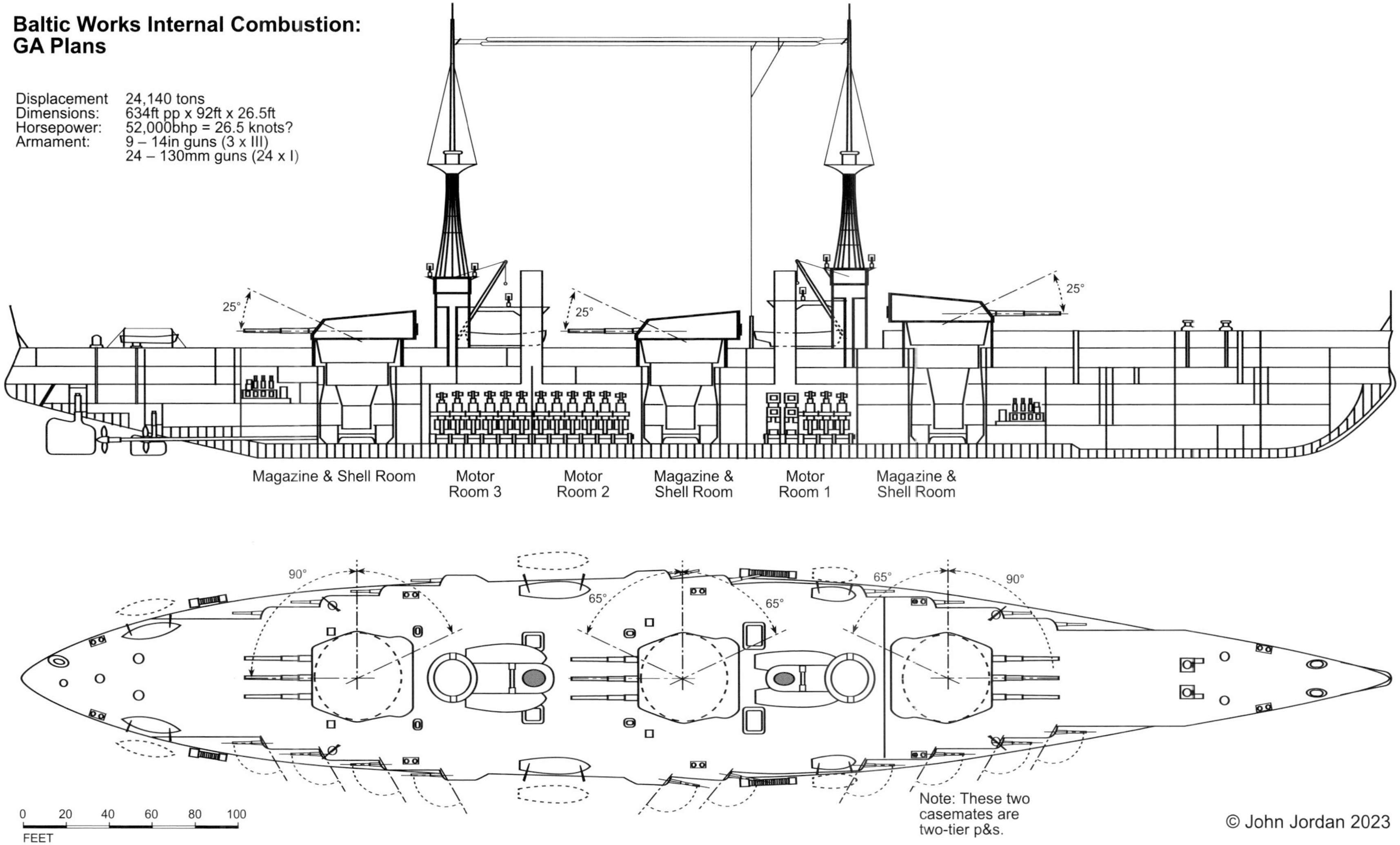

Fig 10: In parallel with the main competition, the Naval Ministry also explored diesel propulsion, hoping that this would make it possible for ships to travel from the Baltic to Vladivostok without the need to refuel. The Baltic Works drew up this design, which is basically a modified version of its competition proposal. The elimination of the boiler rooms allowed a shorter citadel, leading to a ship with a smaller displacement. Note the exhaust pipes for the motor rooms near the conning towers. It would be more than a decade, however, before the technology for such a ship was developed.

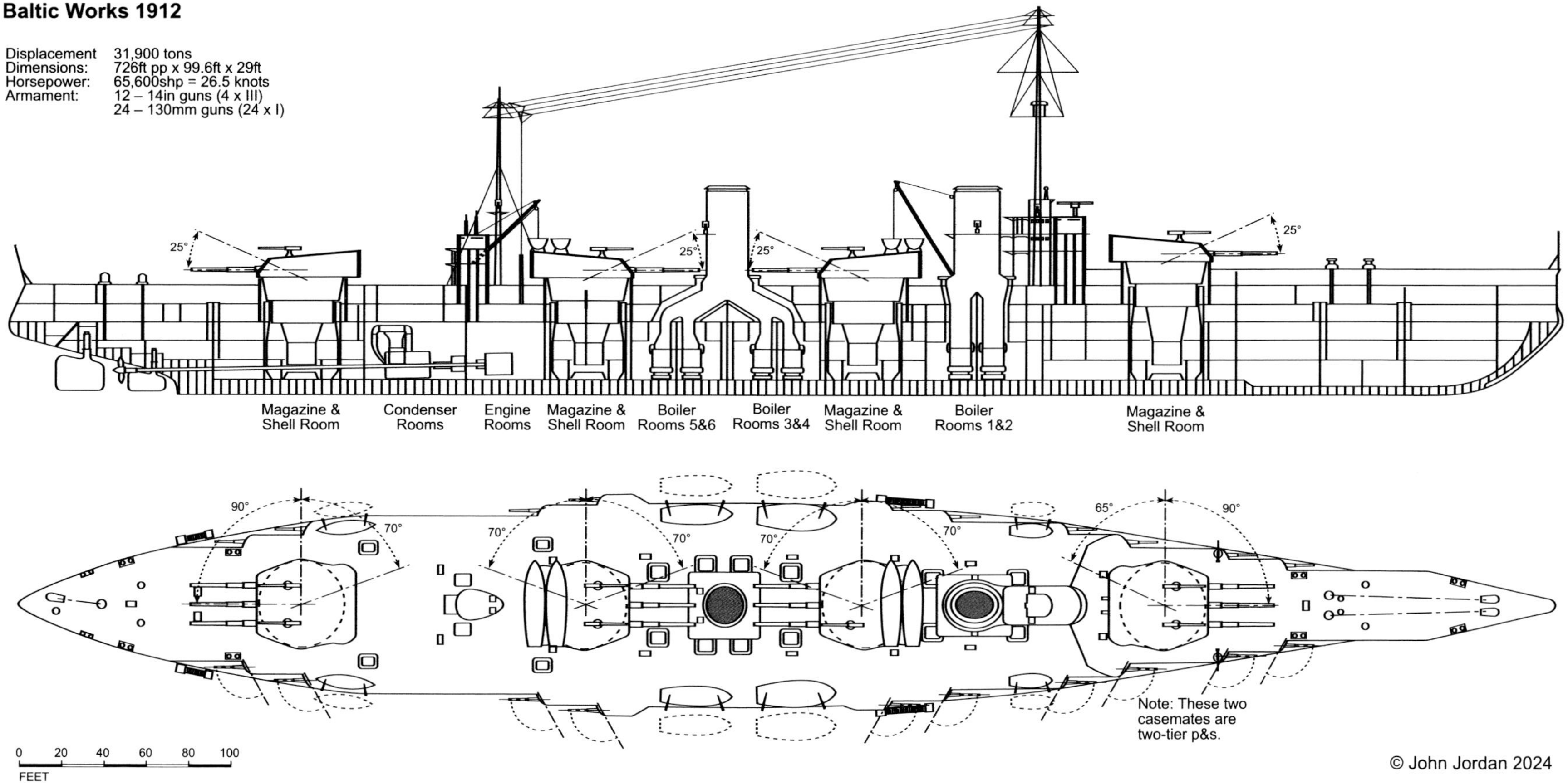

Fig 11: The Baltic Works had also submitted a design with twelve 14in guns in four triple turrets to the competition, which triggered the Naval General Staff's interest in larger and more powerful designs. In July 1912 Naval Minister Grigorovich opted for the bigger ship, and this design, dated 23 July/5 August 1912, was drawn up in response. Over the following months the combined efforts of the Admiralty and Baltic works would produce the final design. In the course of this work the after conning tower would be eliminated and the belt armour would be reduced, but the overall configuration otherwise remained similar.

expanding naval construction by establishing a shipyard, Blohm und Voss was only too willing to provide its technical assistance in order to acquire a 'Russian' conduit to potentially lucrative contracts. This partnership was thus already in place when the armoured cruiser competition was announced.[6]

Several of the foreign firms had requested that the deadline be extended due to the difficulties of translating the technical conditions, and this was granted, the new cut-off date being set at 7/20 November 1911. In the end only six groups submitted proposals: the Baltic Works (six variants), the Admiralty Works (seven variants), Putilovskii/Blohm und Voss (eleven variants), Vulcan (two variants), Beardmore, John Brown (three variants), and Vickers. But the British yards were quickly eliminated from the competition as the MGSh deemed they

Table 5: **Battlecruiser Competition Designs**

	Beardmore	John Brown	Vickers	Puilovskii/Blohm und Voss	
Designation		Variant A	Design 553	No 707-I	No 707-IX
Date	17 Nov 1911	3 Nov 1911	?	Nov 1911	Nov 1911
Displacement	29,500 tons	30,000 tons	30,000 tons	30,400 tons	32,350 tons
Length	204.2m/670ft pp	219.5m/720ft pp	195m/640ft pp	213.0m/699ft	219.0m/718.5ft
Beam	28.2m/92.5m	30m/95ft	29.6m/97ft	30.6m/100.4ft	31.2m/102.4ft
Draught	8.84m/29ft	8.5m/28ft	8.53m/28ft	8.65m/28.4ft	8.8m/28.9ft
Armament:					
main guns	9 – 12in (3 x III)	9 – 14in (3 x II, 1 x III)	9 – 14in (3 x III)	9 – 14in (3 x III)	10 – 14in (2 x III, 2 x II)
secondary	20 – 130mm (20 x I)	24 – 130mm (24 x I)	20 – 130mm (20 x I)	24 – 130mm (24 x I)	24 – 130mm (24 x I)
Belt	225mm/8.8in	225mm/8.8in	250mm/9.8in	250mm/9.8in	250mm/9.8in
Machinery:					
horsepower	60,000shp	70,000shp	63,000shp	64,000shp	82,000shp
speed*	26.5 knots	26.5 knots	26.5 knots	26.5 knots	28 knots, 29.5 knots max
endurance	3,350 tons coal 1,000 tons oil	5,750 tons coal max 9000nm at 14 knots	1,016 tons coal (trials)	7,100nm at 14 knots	6,830nm at 14 knots

Sources: Kuznetsov, *Lineinye kreisera tipa 'Izmail'*, 18–20, 22, 25; NMM MSS/72/017, Thurston Papers, Design Notebook 2.

	Admiralty Works	Baltic Works	Baltic Works	Baltic Works
Designation	Variant 6	Variant 1	Diesel Drive	Four-turret Design
Date	19 November 1911	10 October 1911	13 December 1911	23 July 1912
Displacement	29,350 tons	30,300 tons	24,140 tons	31,900 tons
Length	210m/689ft pp	218m/715.2ft pp	193.2m/633.9ft pp	221.27m/726ft wl
Beam	27.73m/91ft max	29.45m/96.6ft	28m/91.9ft wl	30.35m/99.6ft wl
Draught	8.61m/28.2ft	8.64m/28.3ft wl	8.1m/26.6ft	8.77m/28.8ft
Armament:				
main guns	9 – 14in (3 x III)	9 – 14in (3 x III)	9 – 14in (3 x III)	12 – 14in (4 x III)
secondary	24 – 130mm (24 x I)	24 – 130mm (24 x I)	24 – 130mm (24 x I)	24 – 130mm (24 x I)
Protection	250mm/9.8in	250mm/9.8in	250mm/9.8in	250mm/9.8in
Machinery				
horsepower	72,000shp	63,000shp	52,000bhp	65,600shp 85,400shp forced
speed	26.5 knots	26.5 knots	26.5/28 knots**	26.5 knots
endurance	?	3,000 tons coal 4,800 at 14 knots	?	?

* All designs, with the exception of Putilovskii/Blohm und Voss 707-IX, were to have a maximum speed of 28 knots with forced draught.

** The available sources do not quote a speed for this design; if it complied with the competition requirements, it would have had a speed of 26.5 knots or possibly 28 knots.

Source: Kuznetsov, *Lineinye kreisera tipa 'Izmail'*, 25, 28, 31, 33, 57.

had departed from the requirements. Beardmore, for example, argued that meeting all the specifications would lead to a ship of 36,500 tons, which would 'exceed not only all the vessels already built, but even those being designed at the present moment by any of the great powers'. It also noted that that the latest British battle-cruisers of the *Lion* class had a displacement of only 27,500 metric tons, yet carried an armament of eight 13.5in guns, arguing that '[t]he benefit of the ninth gun, in our opinion, absolutely does not compensate for the reduced speed, the considerably greater displacement and the reduced ability to manoeuvre in limited waters in view of its exceptional length, and also the great difficulties of placing in dock a vessel of such size. ... We are convinced that the above deficiencies would be so obvious to the technical members of the Ministry of the Navy that they would inevitably reject the vessel, even if we were to work out detailed drawings for it. Considering such a result to be completely inevitable, we prepared only a sketch plan for such a vessel...'.[7]

John Brown reached the same conclusion, but did submit three preliminary designs.[8] The cover letter began by arguing that a ship meeting all the specifications 'would not be much short of 35,000 Metric Tons', with dimensions of 780ft x 98ft x 28ft 6in, and that the cost of such a ship would be 'enormous'. The firm also believed that it would be a 'grave mistake' to rely solely on triple turrets for 14in guns, as these were 'as yet untried'. Moreover, the narrowing of the hull fore and aft made it difficult to arrange the magazines for triple turrets there. As a result, its preferred design, Variant A, showed three twin turrets – a superimposed pair forward and a third aft – with a triple turret amidships. The main belt was 229mm (9in) thick instead of 250mm, no provision was made for Frahm tanks, and the firm proposed running the full-speed trials with 83 per cent of the boilers on line instead of 75 per cent as specified. The firm did submit an outline design with three triple turrets, Variant B, and another that embodied all the requirements, Variant C. But given the more responsive offerings from other shipyards, the John Brown proposals were quickly eliminated from consideration.

The proposal from Vickers, which had recently built the armoured cruiser *Riurik* for the Russian Navy, was also rejected, but its perceived shortcomings are unfortunately not given in any available source. It may have

Izmail, lead ship of the class, takes to the water at the Baltic Works, 9/22 June 1915. The light-coloured stairway near the bow led to the viewing platform for distinguished guests, which included Tsar Nikolai II and other members of the imperial family. In the background can be seen the numerous cranes of the Baltic Works, and a submarine under construction can just be made out in the background to the left of *Izmail's* stern. (Courtesy of Sergei Vinogradov)

owed something to the recent political controversy stirred up by the Vickers-built armoured cruiser *Riurik* – upon delivery there were problems with the gun mountings, while in the Duma it was alleged that the Naval Ministry had revealed secret Russian information to the British company during her construction. The Ministry may simply have decided to avoid revisiting that squabble.

The Putilovskii/Blohm und Voss combination submitted no fewer than eleven variants, all designated by the number 707 followed by Roman numerals (707-I, 707-II, etc). Variants I–VIII and XI all featured three triple turrets, Variant X had twin turrets, and Variant IX had twin turrets superimposed above triples for a ten-gun broadside. In the latter design the magazines of each pair of turrets were well separated by intervening compartments, perhaps in an attempt to meet the MGSh's concerns about multiple turrets being disabled by a single hit.

Vulcan's two designs followed the Russian requirements closely and varied from one another only in detail – the primary variant, A-1366, featured a long forecastle deck, while A-1366A was flush-decked and its machinery was arranged differently. Neither seems to have received much attention from the Russian evaluators beyond the fact that the arrangements of the officers' accommodations in A-1366 were regarded favourably.

Although not part of the main competition, the Naval Ministry had also expressed interest in exploring the potential of diesel propulsion, the goal being to create a ship able to cruise from the Baltic to Vladivostok without refuelling. Diesel engines of sufficient power did not yet exist, but it was hoped that the required technology would be available by the time construction of the new ships began. In December 1911 the Baltic Works therefore drew up a 24,140-ton design (193.2m pp, 194.5m oa x 28.0m x 8.1m, 52,000bhp, range 14,000nm at 14 knots) powered by diesels. It otherwise seems to have complied with the competition's requirements, but the expected speed is not given, and the proposal was probably not feasible with existing technology. An engineer named A F Bushev also offered a pair of designs, one of 20,000 tons with an armament of nine 14in guns, the other an unlikely 30,000-ton fantasy with five triple 14in turrets. Both were clearly impossible, and nothing more came of this interest in diesel propulsion.

A Restructuring of the Navy

In October 1911, while the various shipyards were drawing up their proposals, the Naval Ministry's shipbuilding and design organisations were restructured. The MTK, which had previously supervised ship designs, was merged with the Main Administration of Shipbuilding and Supply, which had been responsible for financial matters and construction work, forming the Main Administration of Shipbuilding (*Glavnoe upravlenie korablestroeniia*, or GUK). The head of the new entity presided over the Council on Shipbuilding, which included the various department heads: shipbuilding,

Table 6: *Izmail* Class Final Characteristics

Displacement	32,500 tons normal
Length	222.4m wl, 223.85m oa
Beam	30.05m wl
Draught	8.81m design
Armament:	
main guns	12 – 356mm/52 (4 x III; 80rpg)
secondary	24 – 130mm/55 (24 x I; 200rpg)
anti-aircraft	4 – 63mm/35 AA (4 x I; 220rpg)
saluting	4 – 47mm (200rpg)
	4 – MG (7,200rpg)
torpedo tubes	6 – 450mm (broadside, u/w)
Protection:	
main belt	237.5mm, reduced to 125mm fore and aft
upper belt	100mm amidships, 75mm forward
longtdnl bulkhead	50mm
upper deck:	37.5mm
middle deck:	20mm + 40mm
lower deck:	50mm slopes
turrets:	300mm sides, 200mm roofs
barbettes:	247.5mm
conning tower:	400mm sides, 250mm roof
Machinery:	
boilers	25 Yarrow
engines	4 sets turbines
horsepower	66,000shp, 70,000shp forced*
speed	26.5 knots, 28 knots forced
fuel	1,904 tons oil, 1,974 tons coal
endurance	2,280nm at 26.5 knots
Complement	42 officers, 1,132 other ranks

* This is the forced-draught horsepower given in all available sources, but it seems too low; in comparison, the Baltic Works four-turret design in Table 5, of similar displacement and dimensions, was designed for 65,600shp for 26.5 knots, and 85,400shp for 28 knots.

Sources: Vinogradov, *'Izmail': sverkhdrednout Rossiiskoi imperii*, 17; Usov, 'Lineinye kreisery tipa "Izmail"' Part 2, 56–57.

artillery, torpedoes and mines, machinery, etc. At the same time, the MGSh was elevated to equal status with the GUK; as before, it was responsible for strategic and tactical matters.

In November 1911 the Council on Shipbuilding began receiving the responses from the participating shipyards and undertook the lengthy process of evaluating the designs and analysing their accompanying calculations. Meanwhile, the MGSh was reconsidering the ship's armament. Among the Baltic Works' submissions was a variant of its primary design with twelve 14in guns in four triple turrets, which the staff found interesting. As a result, in April 1912 the Admiralty and Baltic works were asked to calculate the cost per ton of an armoured cruiser displacing 32,500–34,000 tons. The impetus towards the bigger ship received an additional boost when the Putilovskii/Blohm und Voss partnership submitted a new proposal, 707-XVII – essentially an enlarged version of 707-IX with the superimposed twin turrets replaced by triples, making for a twelve-gun broadside.[9] Despite its previous opposition to designs with superimposed turrets, the MGSh was quite taken with this proposal and tried to have it included in the Technical Council's deliberations, but at its session on 12/25 May the Council rejected it for failing to meet the competition's requirements.

The MGSh continued to press for a four-turret design, receiving support from the Council's own Artillery Department, and as a result, in July 1912 when the Council met to make its final selection of the winning design, it made not one, but two recommendations: the Baltic Works' three-turret design, which had overtaken the earlier favourite from the Admiralty Works, and the Baltic Works' four-turret design. The Council noted, however, that the latter design would require numerous changes. This decision was laid before Navy Minister Grigorovich, who was much influenced by the views of the MGSh, holding its 'young and energetic personnel' in 'great esteem'.[10] He therefore ordered that preference be given to the four-turret design.

However, this created a problem. The 'Law on the Definition of the Cost of the Construction of Vessels of the Navy', approved by the Duma on 23 June/6 July 1912, laid out in precise terms the allocations for the various warships to be built under the 1912–1916 programme, and this budget was based on armoured cruisers with three triple turrets, not four. Grigorovich, when testifying in support of the bill, had promised that the Naval Ministry would present 'no additional demands' for money during the course of the programme and, having gained the trust of the Duma, he was loath to go back on his word.[11] Redefining some of the armoured cruiser's characteristics would allow some savings: the forced-draught speed was reduced from 28.5 to 27.5 knots, and the main belt was thinned from 250mm to 242mm (in the final design it would be further reduced to 237.5mm), but this still left a shortfall of 28 million roubles. The only option was to make cuts somewhere else in the construction programme, and as a result, Grigorovich compelled the shipyards to reduce the cost of the *Svetlana*-class light cruisers – as he put it, he was 'snatching as much as possible from the light cruisers for the armoured ones'.[12]

A further major factor in rescuing the programme's finances came from an unexpected source: the German

Izmail afloat in the Neva River immediately after launch, the cradles still attached to the hull. Note in particular the two-tiered casemates forward and the ice-breaking bow, features that had appeared in many of the competition designs. As the hull came afloat the imperial flag was broken out amidships, with the Russian naval ensign at the stern. (Courtesy of Sergei Vinogradov)

Another view of *Izmail* shortly after launch. The steam pinnace in the foreground is returning to shore, having just taken the imperial suite, including the tsar, Naval Minister I K Grigorovich and the head of the Baltic Works, A I Moiseevich, on a cruise around the new battlecruiser. (Courtesy of Sergei Vinogradov)

firm of Schichau, which was angling for the contract to build a pair of light cruisers. It offered a package deal: if awarded the cruiser contract, it was willing to reduce the cost of the nine destroyers to be built at its subsidiary, the new Russko-Baltiiskii Shipyard in Revel, for a saving of more than five million roubles.[13] This would in turn force the other shipyards building destroyers to reduce their prices. So desperate was Grigorovich to accept this offer that he overrode the objections of the MGSh and the GUK, neither of which liked Schichau's cruiser design.[14]

In the meantime the Baltic and Admiralty shipyards were directed to resubmit their designs, incorporating the latest demands for changes. In late July/mid-August 1912 the two shipyards submitted their revised four-turret designs. The Admiralty Works' submission was judged better overall, although in a few areas the Baltic Works' design was considered superior. On 27 July/9 August the GUK's Technical Council proposed that the two yards jointly develop the general design, which was given the force of a command when Admiral Grigorovich approved the decision on 4/17 August.

Work continued on the design even after an order was issued on 5/18 September for the construction of four ships, two at each shipyard, with the first two ships to be completed by July 1916, the second pair by September 1916. The final characteristics of the design are given in Table 6.

Conclusion

When war broke out in August 1914 the completion status of the ships was as follows: *Izmail* 43%, Kinburn 38%, *Borodino* 30%, and *Navarin* 20%. However, the pace of work soon slowed as urgent war needs took priority, while components ordered abroad (including in some cases from Germany and Austria-Hungary) had to be reordered from new suppliers. In July 1915 the ships were redesignated 'cruisers of the line' (*lineinye kreisera*). Some progress was made during the war, and by 15 April 1917 their completion status was *Izmail* 65%, *Borodino* 57%, *Kinburn* 52% and *Navarin* 50%. *Izmail*, it was now hoped, would join the fleet in late 1919. However, the revolutions of 1917 and the ensuing civil war put an end to these hopes and the incomplete hulls were laid up. In 1922 the cash-strapped Bolshevik government sold *Borodino*, *Kinburn* and *Navarin* to a German firm for their scrap value. *Izmail* lingered for several years while the Red Navy considered plans to arm her with 16in guns or convert her into an aircraft carrier, but in the early thirties it was finally decided to scrap her as well.[15]

The *Izmail* design was in many ways an impressive one, combining a very powerful armament with reasonable armour protection and speed, but these qualities were acquired at the cost of an almost complete absence of underwater protection. The gestation period of the design – twenty months from the finalisation of the specifications in August 1911 to the laying down of the ships in March–April 1913 – was so long that in its aftermath the head of the GUK, Vice Admiral P P Muravev, submitted a memorandum in which he wrote that:

> ... I consider it necessary to turn over the development of preliminary designs to the GUK because, as experience has shown, entrusting this work to shipyards through a competition leads to significant delays in the development of the final preliminary design, and also entails the possibility of errors in calculations that hinder and delay the preparation of complete ship designs. ... It is high time we did this work ourselves.[16]

As a result the core of what was to be a central design bureau was formed, based on the existing technical staff at the Admiralty Works. Unfortunately, it was too little, too late, as war and the subsequent collapse of the tsarist regime prevented this new organisation from demonstrating its potential.

Acknowledgements:

As always, the author is grateful to his friend of many years Sergei Vinogradov for supplying most of the material used in the preparation of this article. Ian Johnston generously provided the John Brown letter from the archives of the University of Glasgow. John Jordan once again created the superb drawings. And as so often before, my wife Jan Torbet read through the article and thereby saved me from innumerable errors and infelicities.

Sources:

Kuznetsov, Leonid, *Lineinye kreisera tipa 'Izmail': Izmail', 'Borodino', Kinburn', 'Navarin'*, Iauza (Moscow, 2013).

Usov, V Iu, 'Lineinye kreisery tipa "Izmail"', *Sudostroenie*, Part 1 No 7 (1986), 61–64; Part 2 No 8 (1986), 56–59.

Vinogradov, S E, *'Izmail': sverkhdrednout Rossiiskoi imperii*, Modelist-konstruktor (Moscow, 2004).

Glasgow University Archive Services, John Brown & Co (Clydebank) Ltd, engineers and shipbuilders, 'Letter to Their Excellencies the Ministry of Marine', 3rd November 1911.

National Maritime Museum, Greenwich. MSS/72/017, Sir George Thurston papers, Design Notebook 2.

Endnotes:

1 Kuznetsov, *op cit*, 7.

2 For details of these ships, see Jiro Itani, Hens (Hans) Lengerer, and Tomoko Rehm-Takahara, 'Japan's Proto-Battlecruisers: The Tsukuba and Kurama Classes', *Warship 1992*, 47–79.

3 'Spusk angliiskago bronirovannago kreiser Inflexible', *Morskoi Sbornik* Vol CCCXL No 6 (June 1907), unofficial section, *Morskaia Khronika*, 1–6.

4 Adam Smigielski, 'A Scout for the Imperial Navy', *Warship* Vol VII No. 27, 204–207 (1983). For more on Gavrilov's involvement in cruiser design work, see Stephen McLaughlin, *Russian Cruisers of the Tsarist Era*, Seaforth Publishing (forthcoming), Chapters 30 and 31.

5 Vinogradov, *op cit*, 4.

6 S E Vinogradov, *Poslednie ispoliny Rossiiskogo imperatorskogo flota: Lineinye korabli s 16" artillereriei v programmakh flota 1914–1917 gg*, Galeia Print (St Petersburg, 1999), 208–209.

7 Kuznetsov, *op cit*, 17. Note that these quotes are translations from the published Russian versions, and may not correspond exactly to the original English of the cover letter.

8 John Brown, 'Letter to Their Excellencies the Ministry of Marine' dated 3 November 1911, Glasgow University Archive Services, courtesy of Ian Johnston.

9 In February 1914 the Putilovskii Works drew up a series of large battleship designs with 16in guns, one of which was an enlarged version of its 1912 Variant 707-IX; see Vinogradov, *Poslednie ispoliny Rossiiskogo imperatorskogo flota*, 211–219.

10 I K Grigorovich, *Vospominaniia byvshego morskogo ministra*, Deva (St Petersburg, 1993), 85, 97.

11 Kuznetsov, *op cit*, 56.

12 *Ibid.*

13 N N Afonin, 'Malye kreisery "Murav'ev-Amurskii" i "Admiral Nevel'skoi"', *Gangut* No 26 (2001), 31–42, at 34.

14 The two Schichau cruisers were named *Muravev-Amurskii* and *Admiral Nevelskoi*; still under construction in Danzig when the First World War began, they were taken into the German Navy as *Pillau* and *Elbing* respectively.

15 For details on the construction of the ships of the *Izmail* class and subsequent conversion schemes, see Stephen McLaughlin, *Russian and Soviet Battleships*, Naval Institute Press (Annapolis, 2003; reissued 2021), Chapters 29 and 38.

16 S E Vinogradov, *Poslednie ispoliny Rossiiskogo imperatorskogo flota*, 141.

THE IJN PROTECTED CRUISERS OF THE *CHIKUMA* CLASS

The three protected cruisers of the *Chikuma* class were the among the first ships built for Japan to be powered by steam turbines instead of the traditional reciprocating engines. They were also among the first cruisers in the world to have a uniform main armament. **Kathrin Milanovich** provides a detailed study of this class, which served until the end of the 1920s.

The three protected cruisers (*Bōgojunyōkan*) of the *Chikuma* class were the last of their type built by the Imperial Japanese Navy (IJN) and the first second-class cruisers to be driven by turbines. Superheaters were also used for the first time in the name-ship, *Chikuma*. Two different types of turbines were fitted (Curtis impulse in *Chikuma* and *Hirato*; Parsons reaction turbines in *Yahagi*), and two- and four-shaft configurations were trialled in order to compare characteristics. The introduction of turbines was instrumental in attaining the designed speed of 26 knots – an increase of three knots over their immediate predecessor, the piston-engined protected cruiser *Tone*, on which the design was based.

In a paper titled 'On the Ships Built for the Imperial Japanese Navy during the Past Ten Years' delivered to a general meeting of the Japanese Institution of Naval Architects held at Kōbe in March 1914, Rear Admiral Kondō Motoki explained that the increase in speed in the *Chikuma* class was necessary to match the speed of the latest battleships. He noted that parallel developments were taking place in the Royal Navy ('Town' class) and the Imperial German Navy (*Mainz* class) and pointed out the similarity of these ships to the *Chikuma* class, both in size and speed, irrespective of the fact that they were developed independently. Turbine-driven protected cruisers were also being built in the USA and Italy, and a

The turbine-powered cruiser *Chikuma* running a full power trial in February/March 1912. All three ships of the class underwent trials without their armament. Fukui Shizuo states that the turbines developed 24,940shp during *Chikuma*'s full power trial; in *The History of Naval Machinery* (*Kaigun Kikan Shi*) the figure given is 24,700shp. *Chikuma*'s 10/10 trial speed was 26.83 knots, which exceeded the designed speed by 0.83 knots; shaft revolutions were 357rpm vs 340rpm as designed. (Lars Ahlberg collection)

common feature of all these ships was a move to four funnels; the high performance of the turbine required the supply of more steam which, in turn, resulted in an increase in the number of boilers.

While the turbine propulsion system was the outstanding characteristic of these ships, distinctive features included a clipper bow, a cruiser stern and two vertical masts. The principal feature they had in common with *Tone* was the pronounced clipper bow; the latter ship had only three funnels, and the funnels and masts were raked. The stern of the *Chikuma*s had a very different configuration, principally to avoid the problems experienced in fitting the armour of the stern torpedo room but possibly also to reduce resistance.

General

On 28 September 1906 Navy Minister Saitō Makoto presented the Supplementary Warship Expansion Plan at a Cabinet meeting chaired by Prime Minister Prince Saionji Kinmochi. He explained the rationale for the new ships as follows:

- They were to incorporate lessons learned from the sea battles of the Russo-Japanese War (RJW) and the latest developments in naval technology.
- The ships that had fought in the RJW had suffered losses and were fast becoming obsolescent.
- The warships captured from the Russians were of doubtful fighting value, even after reconstruction.
- The new programme would assist in the development of Japanese shipbuilding and industry.
- They were needed as a counterweight to the naval forces of foreign powers with interests in the Far East.
- They would sustain the Japanese strategic position achieved by victory in the RJW.

In order to attain this goal, Vice Admiral Saitō proposed the construction of three battleships, four armoured cruisers, three 2nd class cruisers of about 4,500 tons, six large and 24 small destroyers and six submarines. However, in view of the parlous condition of the state finances only a single battleship, three armoured cruisers and twelve small destroyers were to be built in the short term, with priority given to the 2nd class cruisers, the large destroyers and the submarines. The plan was presented to the 23rd Diet in December 1906 and approved in early March 1907; it was signed by the Emperor on the 16th and published in the official gazette (*Kanpō*) on the 18th.[1]

The budget procedure was complex, but in essence was as follows:

- The budget for two 5,000-ton ships in the Third Period Expansion Programme scheduled for FY 1911 was postponed until after 1914.
- Two 4,000–5,000-ton ships scheduled to be laid down in 1908 and 1909 under the *Kantei Hosoku Hi* were postponed to 1910.
- The two 4,500-ton cruisers were to be laid down in 1909. However, the funding for two large destroyers (*Ro Gō* & *Ha Gō*) and two standard destroyers (*I Gō* & *Ro Gō*) in the Supplementary Warship Expenses (*Hojū Kantei Hi*) was to be used to add a further 2nd class cruiser, making it possible to lay down three 2nd class cruisers at the same time.
- In order to replace the destroyers, one 2nd class cruiser in the Third Period Programme (No 1) was replaced by one large destroyer (*Kō Gō*) and two standard destroyers, and the two 2nd class cruisers of the *Kantei Hosoku Hi* (*Otsu Gō* & *Hei Gō*) were replaced by a second large destroyer.
- The implications of the above decisions were that the three 2nd class cruisers of the *Chikuma* class would be laid down in the same year, but that the overall number of 2nd class cruisers was reduced in favour of destroyers.
- Estimated building costs for *Chikuma* were ¥3,720,000, to be funded from FY 1909 to FY 1912, while those for *Hirato* and *Yahagi*, which were to be built in private shipyards, were ¥4,870,000 for each ship, to be funded from FY 1909 to FY 1913. No explanation for the marked difference in cost has been found. However, the budget allowances in the *Hojū Kantei Hi* were revised in November 1908; total and annual expenditure is shown in Table 1.

On 6 July 1909 the Navy Ministry duly placed the orders for the construction of three 2nd class cruisers with the budget stated above. Seven days later Sasebo Navy Yard received the order to build *I Gō* with funding from 1909 to 1911, and on 14 August protected cruisers *Ro Gō* and *Ha Gō* were contracted to Japan's two largest private shipbuilders, Kawasaki Kōbe and Mitsubishi Nagasaki. On 23 December 1909 the temporary budget names were replaced by *Chikuma, Yahagi,*[2] and *Hirato* respectively.

The Sasebo Navy Yard had built *Tone*, and following the order for *Chikuma* specialised in this type of ship. For the two private shipyards this would be their first order for a major vessel; previously they had built only small destroyers and torpedo boats. The front-loading of the budget in favour of the Sasebo NY reflected its position as the lead yard; it would be responsible for the detailed design and most of the working drawings – the design divisions of the two private yards produced only the

Table 1: Costings

Name	Total budget	1909	1910	1911	1912	1913
Chikuma (*I Gō*)	¥3,790,000	¥80,000	¥1,190,000	¥1,680,000	¥840,000	–
Yahagi (*Ro Gō*)	¥3,800,000	¥10,000	¥100,000	¥100,000	¥1,270,000	¥2,320,000
Hirato (*Ha Gō*)	¥3,800,000	¥10,000	¥100,000	¥100,000	¥1,270,000	¥2,320,000

Yahagi on trials off Nagasaki on 10 March 1911. *Yahagi* was the only one of the three ships with Parsons turbines and, unlike her sisters, had four shafts. In her 10/10 full power trial on 25 May 1912, on a displacement of 4,998 tons. she achieved 27.14 knots with 29,246shp, which was a marked increase on the designed figure of 22,500shp. (NHHC, NH 111693)

drawings for the machinery, which differed in all three ships.

The Design

The configuration of the *Chikuma* class resulted from the deliberations of the Investigation Committee into Ship Configuration (*Kankei Shosei Iinkai*) chaired by Admiral Yamamoto Gonnohyōe (Gombei). The committee was established in 1906 and finished its work the following year. It was tasked with determining and recommending the configuration of future warships. Sub-committees investigated the configuration and characteristics of corresponding types in foreign countries, while other key factors in the discussions were:

- the lessons of the RJW
- the strategic principles of defence in the Pacific against any advance of the US Navy towards Japan
- technical progress in shipbuilding, engine and weapon production, including turbine propulsion (for speed), a uniform armament of large-calibre guns, new models of torpedo and mine and the development of the appropriate tactics for use of the latter.

The conclusions of the committee exerted a considerable influence on shipbuilding policy, and in particular on the characteristics of the battleships of the *Kawachi* class (see the author's article in *Warship 2017*), the protected cruisers of the *Chikuma* class, and the destroyers of the *Umikaze* class. Not all of the Navy's ambitions could be realised due to a lack of funds, but with the *Chikuma* class the key characteristic – experimental turbine propulsion – was translated into action.

The *Chikuma* class was designed in 1908–09 on the basis of the recommendations of the committee. The preliminary design was drawn up in the Third Division of the Naval Technical Department (NTD) by a team headed by naval architect Kondō Motoki, whose rank (*Zōsen Sōkan*) corresponded to that of Rear Admiral, while the Fourth and First Divisions were responsible for the machinery and the armament respectively. The final draft, bearing the basic design number C18, was submitted on 25 May 1909. However, the Naval General Staff (NGS) required a revision of the armament from a mixed to a uniform main gun calibre, as proposed by the committee, and also queried the fuel stowage (hence endurance), citing the contemporary cruisers of the Royal Navy as an example.

The design was developed from *Chikuma*'s immediate predecessor, the protected cruiser *Tone*. Similarities with the 'type ship' were unmistakable. They included a forecastle and a poop deck with a bulkhead 3 feet high between them, tumblehome of the hull sides and a clipper bow; the latter, however, was slightly less pronounced

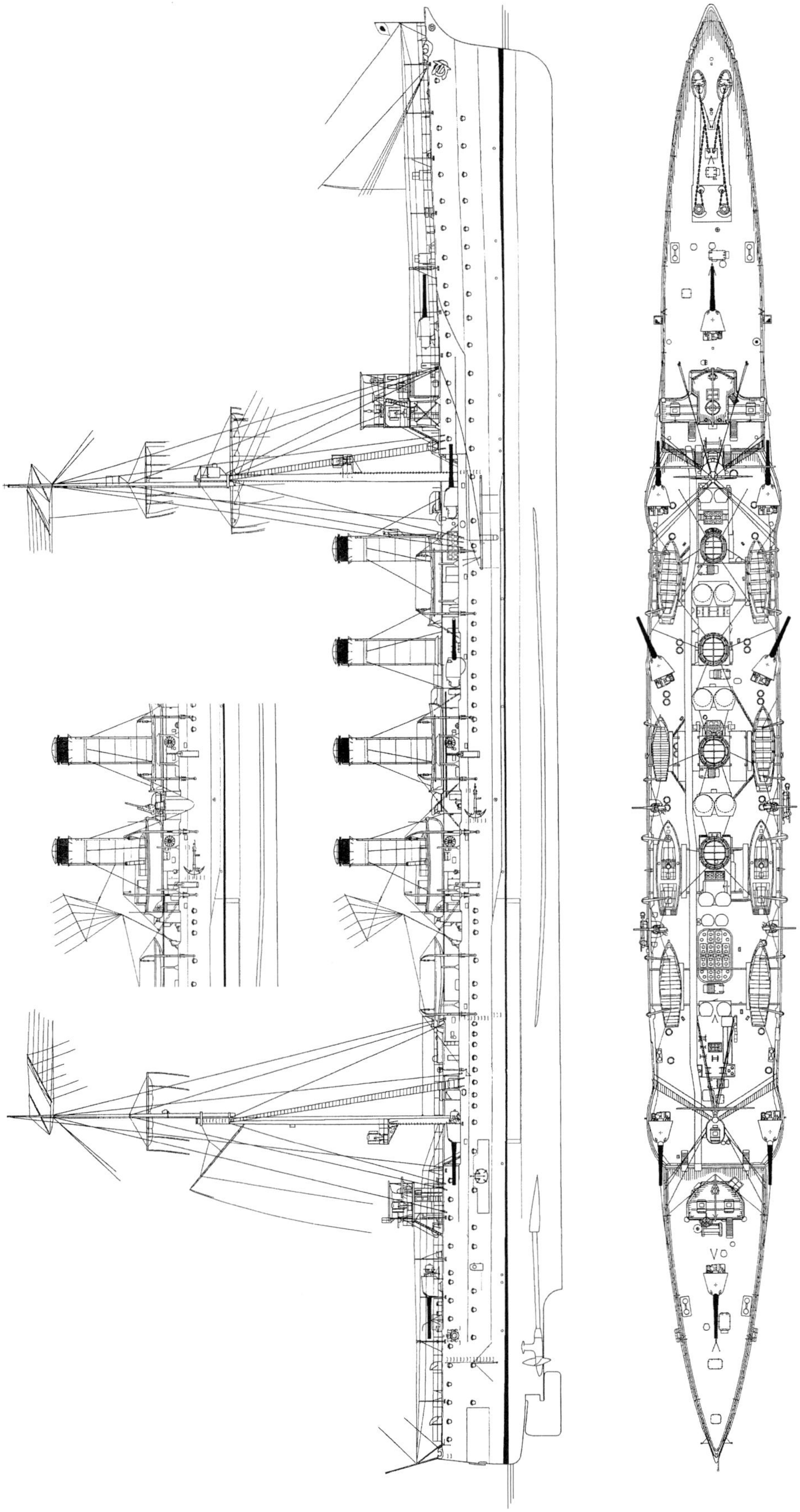

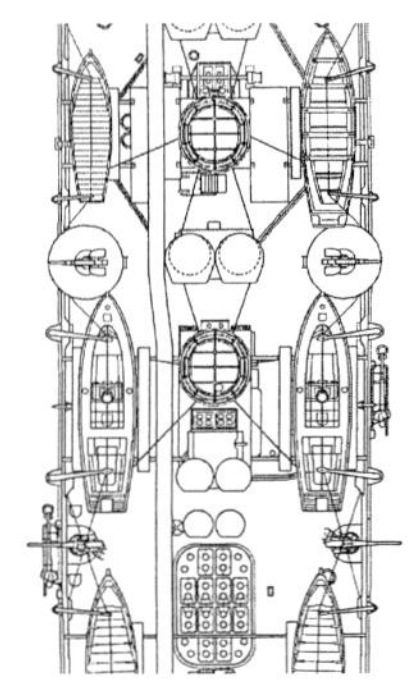

Chikuma as completed in 1911. The insets show the ship post-1919 following the replacement of the forward pair of 3in Type 41 guns by two Year 3 40-cal high-angle guns. (Drawn by Waldemar Trojca)

Table 2: **Building Data & Characteristics**

	Tone	*Chikuma*	*Hirato*	*Yahagi*
Builder	Sasebo NY	Sasebo NY	Kawasaki Kobe	Mitsubishi Nagasaki
Laid down	27 Nov 1905	23 May 1910	10 Aug 1910	20 Jun 1910
Launched	24 Oct 1907	1 Apr 1911	29 Jun 1911	3 Oct 1911
Completed	15 May 1910	17 May 1912[1]	17 Jun 1912	27 Jul 1912
General:				
Length pp	360ft (109.7m)	440ft (134.1m)	[as *Chikuma*]	[as *Chikuma*]
Beam, max	47ft 3in (14.4m)	46ft 7in (14.2m)	[as *Chikuma*]	[as *Chikuma*]
Depth (moulded)	28ft (8.6m)	28ft 6in (8.59)	[as *Chikuma*]	[as *Chikuma*]
Displacement:				
designed	4,150 tons	4,950 tons	[as *Chikuma*]	[as *Chikuma*]
normal	4,113 tons	5,040 tons	4,973 tons	5,027 tons
Machinery:				
Engines	2-shaft 4-cyl VTE engines	2-shaft Curtis turbines[2]	[as *Chikuma*]	4-shaft Parsons turbines[2]
Boilers	16 Miyabara,	16 Kanpon / *Gÿ*,	[as *Chikuma*]	[as *Chikuma*]
	225psi	275psi, 212°F	[as *Chikuma*]	[as *Chikuma*]
	(8 large, 8 small)	(12 large[3], 4 small)	[as *Chikuma*]	[as *Chikuma*]
Shaft revs	160rpm (des)	340rpm (des)	340rpm (des)	470rpm (des)
	168rpm (actual)	357rpm (actual)	302rpm (actual)	??
Propellers:				
diameter	n/a	10ft 6in (3.2m)	10ft 2in (3.1m)	7ft 3in (2.22m)
pitch		9ft 3in (2.82m)	9ft 4in (2.85m)	6ft 10in (2.08m)
Rudder area	n/a	174ft² (16.18m²)	[as *Chikuma*]	[as *Chikuma*]
Horsepower	15,000ihp (des)	22,500shp (des)	[as *Chikuma*]	[as *Chikuma*]
	15,215ihp (trials)	24,940shp (trials)	26,126shp (trials)	29,246shp (trials)
Speed	23 knots (des)	26 knots (des)	[as *Chikuma*]	[as *Chikuma*]
	23.37kts (trials)	26.82kts (trials)	26.79kts (trials)	27.14kts (trials)
Coal/oil fuel	900/124 tons	974/388 tons	991/300 tons	1006/391 tons
Endurance		5,000nm at 16kts; 10,000nm at 10kts [all ships]		

Armament:		
	Tone	*Chikuma* class[4]
main	2 x 6in (15cm) 45-cal Type 41	8 x 6in (15cm) 45-cal Type 41
	10 x 4.7in (12cm) 40-cal Type 41	4 x 3in (8cm) 40-cal Type 41
MG	2 x Maxim 6.5mm	2 x Maxim 6.5mm
TT	3 x 18in (45cm) Armstrong	3 x 18in (45cm) Armstrong
	6 x Type 42?	9 x Type 44 No 2[5]
S/L	6 x 90cm Siemens-Schuckert Type	6 x 90cm Siemens-Schuckert Type
Protection:		
deck	1½in MS flat	⅞in NS (flat)
	+ 1½in NS (slopes)	+ 1⅜in NS (slopes)
CT	4in Krupp NC	4in Krupp
TT	1in + ½in Krupp NC (sides)	3½in + ½in Krupp (sides)
Complement	370 (designed)	414 (designed)

Sources: NTD (then *Kaigun Gijutsu Honbu*) document dated Dec 1919 and all sources stated under 'Protection'; also *Kaigun Suirai Shi* and Navy Ministry's annual report for 1919.

Notes:

1 Sasebo NY was ordered on 25 September 1909 to complete cruiser *I Gō* (*Chikuma*) in 1911, but this was changed (see revised distribution of the budget).

2 For details of the Curtis and Parsons turbine installations see text and Table 6.

3 The twelve large boilers of *Chikuma* were equipped with superheaters – see text.

4 All ships ran trials without armament: *Chikuma* Feb–Mar 1912 off Mie Point Nagasaki, *Hirato* Apr–May 1912 off Awajijima, and *Yahagi* on 25 May 1912 off Mie Point.

5 Endō Akiro gives 'Type 43 18" torpedoes x 15' but all other sources state Type 44 (confirmed by the late Rear Admiral Takasu Kōichi).

Table 3: **Breakdown of Weights**

	Tone	*Chikuma*	*Hirato*	*Yahagi*
Hull	1,778t (43.3%)	1,849t (36.7%)	1,843t (37.1%)	1,840t (36.6%)
Fittings	[incl in hull]	424t (8.4%)	408t (8.2%)	374t (7.5%)
Armour	[incl in prot plat^{g}]	115t (2.3%)	104t (2.1%)	113t 2.3%)
Protective plating	458t (11.2%)	323t (6.4%)	305t (6.1%)	305t (6.1%)
Armament	261t (6.4%)	326t (7.2%)	375t (7.4%)	375t (7.5%)
Machinery	1,059t[2] (25.8%)	1,172t (23.3%)	1,170t (23.5%)	1,179t (23.5%)
Fuel (Coal)[1]	300t (7.3%)	500t (9.9%)	500t (10.1%)	500t (10.0%)
Equipment	249t (6.1%)	285t (5.7%)	307t (6.2%)	307t (6.1%)
Unknown	–	11t (0.2%)	39t (0.8%)	33t (0.7%)
Total	4,105t (100%)	5,040t (100%)	4,973t (100%)	5,027t (100%)

Sources: Hand-written tables of Technical Vice Admiral Niwata Shōzō; Fukuda: *Gunkan Kihon Keikaku Shiryō*, 55 & 151 (on page 151 the weight of *Hirato's* machinery is given as 1,147t, horsepower as 26,149shp and the shp/weight ratio as 22.8)

Notes:

1 Only 500 tons of fuel was included in the normal displacement. This was roughly 35% of the total bunkerage.

2 Note the comparatively heavy weight of *Tone*'s piston engines compared with the turbines of the *Chikuma* class; power was 15,000ihp for the former vs 22,500shp for the latter.

than in *Tone*. Other common features included teak planking on the upper deck, a double bottom and bilge keels. The watertight subdivision of *Tone*'s hull had been designed on the basis of the lessons of the RJW and was further improved in the *Chikuma* class.

In order to obtain experience and provide the possibility of comparing various types of turbines, Curtis impulse-type built by the Japanese licensee, Kawasaki, were fitted in two ships, while the third (*Hirato*) had Parsons reaction turbines built by Mitsubishi Nagasaki; the latter installation had four shafts rather than two, while the name-ship of the class had superheaters for her twelve large Kampon boilers. As Kawasaki was the licensee for Curtis, *Chikuma*'s turbines were built by the Kawasaki Engine Division; her boilers were manufactured by Sasebo's Engine Division. The adoption of the turbine for the latest capital ships built for the IJN, which enabled them to steam at 20–21 knots, made it inevitable that the 'eyes of the fleet', as these reconnaissance vessels were often christened, would likewise require higher speeds, implying a similar move to turbine propulsion. The length to beam ratio of the new cruisers was also increased over *Tone* to improve straight-line speed.

High-tensile steel (HTS) instead of mild steel (MS) for the hull and fittings was adopted to save weight, thereby compensating for the increased weight of the propulsion system.

Protection

According to the authoritative *Japanese Cruisers of the Pacific War* by Eric Lacroix and Linton Wells (see pages 7–8 and drawing page 10) the protection system followed standard practice for protected cruisers, and was in principle similar to that of *Tone* except for the employment of nickel steel instead of mild steel for the protective deck. The latter was located beneath the middle deck with the flat section a foot (0.6m) above the waterline and the sloped sides connected to the shell plating some 3ft (1m) below the waterline. The deck was of 35lbs (⅞in) nickel steel (NS) with reinforcing plates of 55lbs NS (1⅜in) on the slopes. Japanese-language secondary sources (notably Kondō Motoki and Fukui Shizuo) give a figure of 57mm, but this appears to be the total thickness on the slopes (35+55lbs = 90lbs = 2¼in = 57mm). The midship section drawing of *Hirato* published here

Table 4: **Block Coefficient and Stability Data**

Block coefficient (Cb)	0.513 (*Hirato*)
L/B ratio	9.46 (*Hirato* & *Yahagi*)
d/B ratio	0 .36 (*Hirato* & *Yahagi*)
D/d ratio	1.69 (*Hirato*), 1.72 (*Yahagi*)
Transverse metacentric height (GM)	2ft 4in/0.72m (*Chikuma*)
Height of metacentre above centre of buoyancy (BM)	10ft 9in/3.27m (*Chikuma*)
Distance of centre of buoyancy (CB) abaft midship	18ft 2in/5.55m (*Chikuma*)
Distance of CB below waterline	6ft 8in/2.03m (*Chikuma*)
Immersed midship area	624ft^2/57.94m^2 (*Chikuma*)
	641ft^2/59.55 m^2 (*Hirato*)
Necessary weight to immerse 1in (25.4mm)	34.7 tons (*Chikuma*), 34.8 tons (*Yahagi*)
Moment to alter trim 1in (25.4mm)	834.3 metric tons (*Chikuma*)

confirms the arrangement depicted in Lacroix and Wells, whose section drawing of *Chikuma* is likewise based on the official plans.

The conning tower was protected by Krupp armour 4in (102mm) thick and there were 3½in (89mm) plates for the outer faces of the two torpedo rooms on the middle deck and for the sides of the torpedo room in the stern. Although some Japanese-language sources claim the latter thickness to be 'side armour', the reduced weight of protection (408–430 tons vs 458 tons for *Tone*) and the increased length of the hull (440ft vs 360ft) clearly precluded a waterline belt.

Machinery

A key feature of this class was the adoption of two different types of turbine, a different arrangement of the propeller shafts, and two steam temperatures; this would provide the opportunity for competitive evaluation. *Teikoku Kaigun Kikan-shi* Vol 2 claims that this had a considerable impacr on engine design for the warships that followed (see page 461). The *Chikuma* class would not only be the IJN's first cruisers powered by turbines, they would be the last to have direct drive, subsequent ships having their shafts driven through reduction gearing to secure greater efficiency and range.

Chikuma and *Hirato* were fitted with two sets of Curtis impulse turbines. Built under licence by the Kawasaki Kōbe Shipyard, the turbines for *Chikuma* and *Hirato* were the first of their type to be manufactured in Japan. They were essentially the same turbines fitted in the batteships *Kawachi* and *Settsu* but with the pitch circle reduced from 12ft (3.66m) to 9ft (2.74m).[3] The shape of the three nozzles of the first ahead stage of the LP turbine – there were fifteen stages in total – was adopted with a view to maximising steam expansion to give economy at cruising speed; consequently these were called 'cruising nozzles'. Besides the fifteen ahead stages the casing also housed two astern wheels. The IJN christened this the 'simple body model' (*tandō gata*). Each set of ahead turbines was designed to develop 11,250shp,

***Hirato:* Midship Half-section**

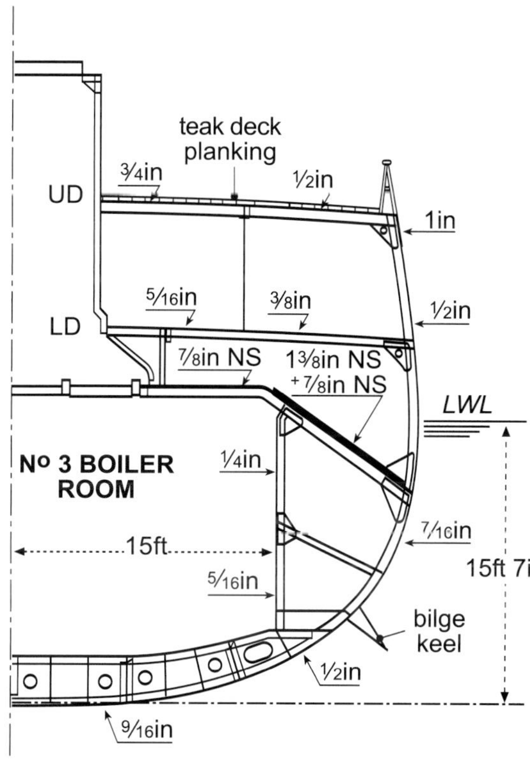

Drawn by John Jordan using material supplied by the author.

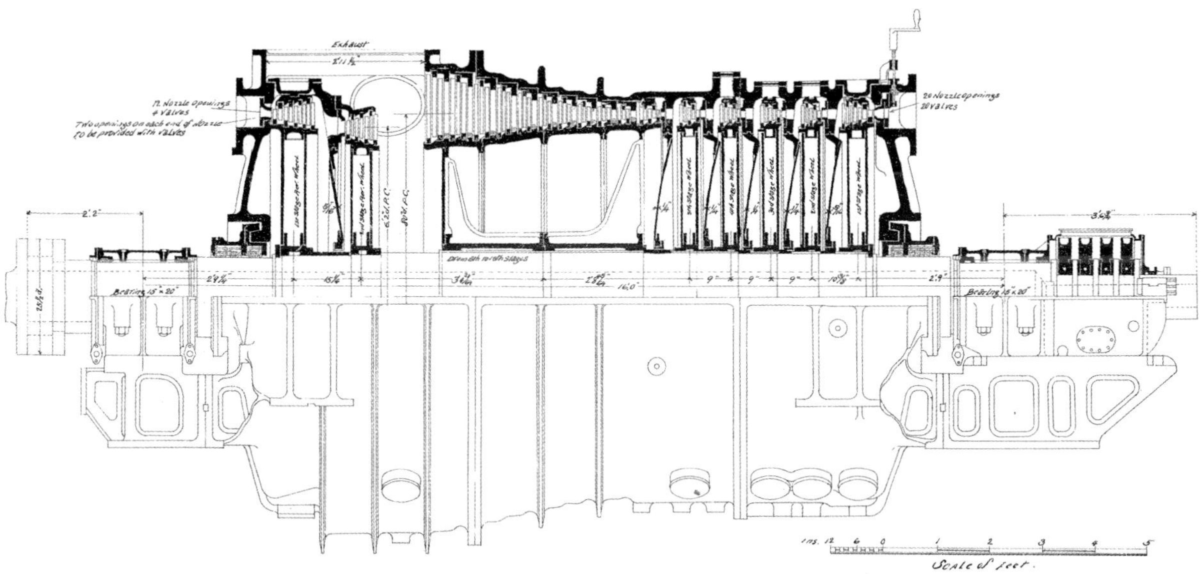

The Curtis LP turbine installed in *Chikuma* and *Hirato*. (Courtesy of Takagi Hiroshi)

One of the main Curtis turbines is lifted on board *Hirato* during fitting out at Kawasaki, Kōbe. (Courtesy of Takagi Hiroshi)

A rare (if poor-quality) image of the protected cruiser *Hirato* fitting out at Kawasaki, Kōbe, over the winter of 1911–12. Note the cranes – the symbol of happiness – on this New Year greetings card. (Courtesy of Takagi Hiroshi)

making for a total of 22,500shp; the two shafts were designed to turn at 340rpm, and rotated inwards.

The accompanying diagram, which is based on *Teikoku Kaigun Kikan-shi* Vol 2 (page 462), shows the steam flow at cruising speed. The downside of the initial arrangement was that steam from the cruise turbines had to pass on both sides of the longitudinal centreline bulkhead separating the engine rooms, which meant that the flooding of one of the engine rooms would have impacted on the steam supply for the other. Fortunately this potential problem was spotted early; the system was changed, and the steam was delivered direct to the HP cruise turbine (port side) and the exhaust from the IP cruise turbine could be led to the ahead HP turbines in both engine rooms. This arrangement required a complex system of valves which blocked the steam supply from the cruise turbines when running at high speeds, and there was concern that this could result in handling errors and a reduction in efficiency. However, these fears were not realised during trials, when the performance of the main turbines was found to be outstanding.

In contrast to her two sisters *Yahagi* was powered by a four-shaft Parsons turbine installation. The turbines were built by Mitsubishi under licence and were of the classic reaction type. The layout featured a HP turbine on each of the wing shafts and a LP turbine on each of the inner shafts. As with the Curtis turbines, the astern stage was housed within the main LP casing.The cruise turbines were connected to the inner shafts forward of the LP turbines, with the HP turbine to port and the IP turbine

Curtis Turbines: Steam Flow [Schematic]

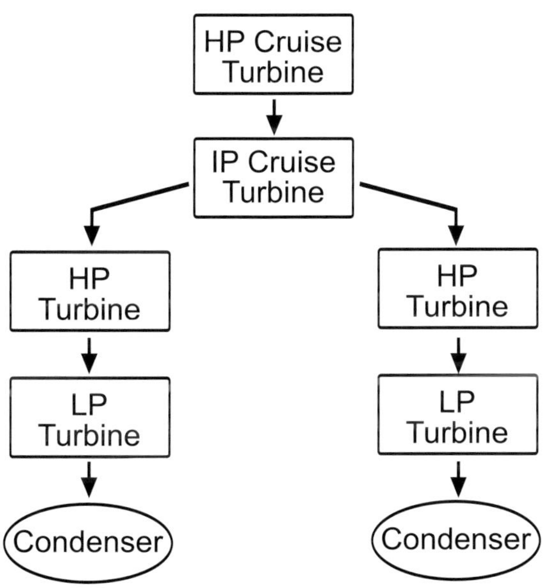

Note: Drawn by John Jordan using material supplied by the author.

***Yahagi*: Parsons Turbine Arrangement [Schematic]**

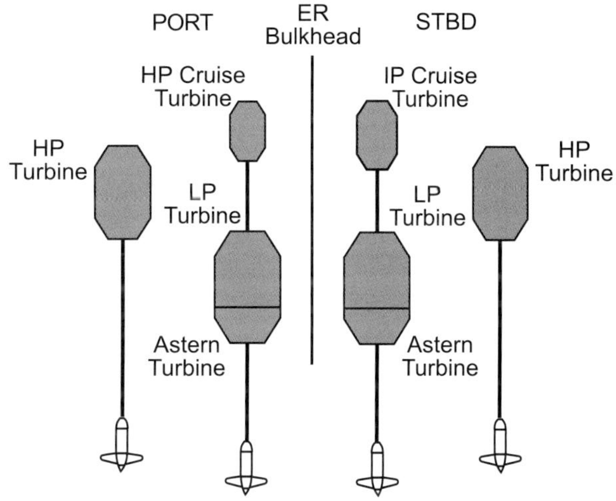

Note: Drawn by John Jordan using material supplied by the author.

to starboard (see accompanying diagram, adapted from *Teikoku Kaigun Kikan-shi* Vol 2, pages 401 and 161). Total horsepower was 22,500shp, as in the other two ships, but the four shafts revolved at higher speed: 470rpm vs 340rpm. The two inner shafts rotated inwards, the two outer shafts in the opposite direction.

According to Fukui Shizuo (*Japanese Naval Vessels Illustrated, 1869–1945*, Vol 2; *Cruisers, Corvettes & Sloops*, 147) it was initially planned to fit Miyabara large watertube boilers, as in *Tone* but, in a further technological leap, this was changed to a new small-tube boiler developed by the IJN, the Kampon *I Gō* type, which had a steam pressure rating of 275psi. Each ship had sixteen Kampon boilers in four groups of four, each group exhausting through a single funnel. *Chikuma* trialled superheated steam: twelve of her sixteen boilers were fitted with superheaters by reducing the rows of evaporating tubes – the ratio of the area of superheating tubes to that of evaporating tubes was 1:5.8, so roughly one-sixth of the heating area was used for superheating. The four smaller boilers in the forward BR and all sixteen of the boilers in the other two ships employed saturated steam. The cast steel tubes of the superheater headers used in *Chikuma*'s twelve larger boilers had to be ordered from the American Babcock & Wilcox Co because they could not be produced in Japan.[4] The valve rods had to be made of nickel steel and the valves of Ni brass (Takagi Hiroshi states that 'only first class materials were used').[5] However, the designed steam temperature was not attained during trials, so the experiment had to be considered a diappointment.

Despite the adoption of two different types of main engine with two or four shafts, the total weight of the machinery was almost identical across the three ships (see Table 3). However, a closer inspection reveals that the number of valves in *Yahagi* was almost twice as many as

The full power trial of *Hirato* took place in April/May 1912. She had the same American Curtis turbines as *Chikuma*. The maximum speed of 26.78 knots made her the slowest of the three ships despite a relatively high power rating of 26,126shp with shaft revolutions of 362rpm. Note the moderate bow wave of all three ships on trials, suggesting that the pronounced clipper bow was particularly successful. (Hans Lengerer collection)

This photo shows *Chikuma* at Yokosuka in July 1914. In contrast to *Tone*, *Chikuma* and her sisters had vertical masts and funnels. (Hans Lengerer collection)

in her sisters. The width of the engine rooms was identical, but the adoption of Parsons turbines on four shafts for *Yahagi* meant that they had to be lengthened by 4ft (1.22m, see Table 5).

The trial results are listed in the table of characteristics. Each ship exceeded the designed 26 knots; *Yahagi* was the fastest with 27.14 knots. Note that *Chikuma* recorded a speed of 26.83 knots with only 24,940shp (other sources give 24,700shp), whereas her sister *Hirato* attained only 26.87 knots with 26,126shp (other sources range from 26,000–26,500shp). The difference in performance is not significant, and may relate to the different locations chosen for the trials: Mie Point, Nagasaki for *Chikuma*, the Awajijima mile posts

Table 5: **Machinery Spaces**

	Engine Room	Boiler Rooms
Length	60ft (18.3m)	136ft (41.5m)
Width	40ft 3in (12.3m)	30ft (9.1m)
Height	19ft 1in (5.8m)	15ft 4in (4.7m)
Area	2,460sq ft (228.5m^2)	4,016sq ft (373.1m^2)

Source: Fukuda, *op cit*, 147; figures for *Chikuma*.

An NTD document dated 2 December 1919 gives additional data about the machinery spaces of *Hirato* and *Yahagi* as follows:

	Hirato	*Yahagi*
Engine room area	2,463sq ft (228.8m^2)	2,699sq ft (250.7m^2)
Boiler room area	4,020sq ft (373.5m^2)	4,011sq ft (372.7m^2)
Total area machinery spaces	6,483sq ft (602.3m^2)	6,710sq ft (623.4m^2)

These figures show the impact of the longer engine rooms in *Yahagi*, which had Parsons turbines in a four-shaft arrangement.

Hirato at Auckland, New Zealand some time between 1914 and 1915. The three cruisers of the *Chikuma* class operated in the South Pacific and in the Indian Ocean during the first year of the Great War. *Chikuma* and *Yahagi* searched the South Seas as far as the Fijian Islands for the German East Asia Squadron of von Spee. In 1917 *Hirato* and *Chikuma* patrolled the seas off Australasia to intercept German mercantile sea traffic, while *Yahagi* operated in the Indian Ocean. Due to the distant operational areas new W/T equipment was fitted and modifications made to the antennae and yards at this time. (Hans Lengerer collection)

for *Hirato*. Sea conditions may also have played a part.

The graph showing steam consumption is taken from *Teikoku Kaigun Kikan-shi*, Vol 2, page 460. The consumption of *Hirato* and *Yahagi* was almost the same for a given speed. *Chikuma* consumed a little more when running below 20 knots but showed greater economy above 21.5 knots. At full speed (26.75 knots) she consumed approximately 8% less than her sisters, thereby illustrating the benefit of using superheating over saturated steam.[6]

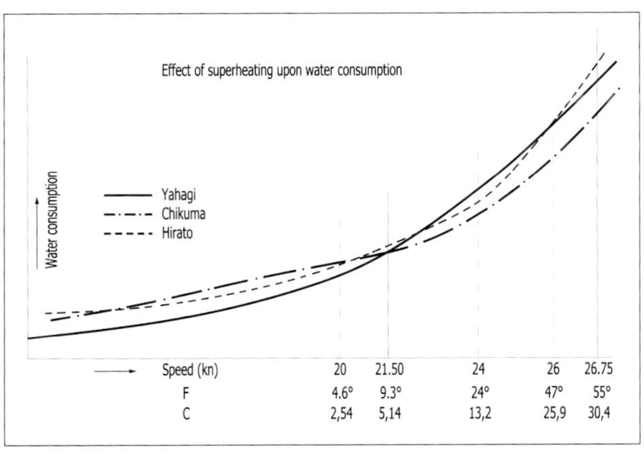

Armament

The calibre of the main guns was a uniform 6in (15cm) compared to the mixed battery of 6in and 4.7in guns in *Tone*. *Chikuma* and her sisters had eight 6in 45-cal Type 41 QF guns in shielded centre pivot mountings: two were located on the centreline fore and aft with the others in the waist. The latter guns were mounted on sponsons on the upper deck abreast the foremast, the second funnel and the mainmast, and fired over the bulwark connecting the forecastle to the poop. This arrangement permitted three guns to fire forward and

Table 6: Machinery

Performance on trials for *Yahagi* (Parsons turbines) was as follows:

	10/10 speed	8/10 speed	10 knots
Displacement	4,998 tons	4,984 tons	5,010 tons
Speed	27.14 knots	25.68 knots	10.14 knots
Horsepower	29,536shp	21,750shp	930shp
Fuel consumption:			
coal	15.73t/hr	13.64t/hr	2.24t/hr
oil	7.90t/hr	2.86t/hr	n/a
Range	1,526nm	1,527nm	6,468nm

Source: Endō Akira, *Outline of Shipbuilding History of the IJN* (*Nihon Kaigun Kenkan Ryūkushi*) No 19.

Note: The trial was run with 1,062 tons of coal and 354 tons of oil.

Machinery Plant of *Hirato*

The data which follow were supplied by Takagi Hiroshi, who also assisted the author in the translation of the technical terms, supplied the drawing of the Curtis LP turbine, and kindly allowed the reproduction of two previously unpublished photographs taken at the time the turbines were fitted. The arrangement of the data follows Mr Takagi's original presentation.

Main engine

Completion of the main engine (land)	1 December 1911
Builder	Kawasaki Shipyard
Type of turbine	Curtis horizontal
Number of turbines	two main turbine sets (HP + LP), two cruise turbines
Number and type of rotor shaft	one rotor shaft for ahead, two for astern
LP combined turbine	ahead and astern turbines in same casing
Number of turbine rotor shafts	two
LP turbine structure	6 blade wheels and one blade drum for ahead + two blade wheels for astern on same shaft and in same casing.
Maximum steam pressure:	
steam room	255psi (18 kg/cm^2)
boiler	270psi (19 kg/cm^2}
Full power (official trial):	
ahead	26,000shp = 26.8 knots (360rpm)
astern	10,500shp = 19.0 knots (280rpm)

Boiler

Completion of the boilers (land)	
11 April 1911	
Builder	Kawasaki Shipyard
Type	Kampon Type *I Gō* mixed-fired boiler
Number	twelve large, four small
Type of burner	Kampon Type 04 burner Model 4
Cone type	Kampon type 04 cone Model 1
Grate surface area:	
large type	63.5sq ft (5.90kg/cm^2)
small type	50sq ft (4.63kg/cm^2)
Heating surface:	
large type	3,295sq ft (306.20kg/cm^2)
small type	2,900sq ft (269.24kg/cm^2)
Number of tubes per boiler	1,386 (large type)
Tube cross-section (between plates):	
large type	6ft 5in (194.30cm)
small type	6ft 1in (186.34cm)
Length of the tubes:	
large type	6ft 7in (201.20cm)
small type	6ft 3in (191.20cm)

three aft, while five guns were available on the broadside (see Profile and Plan).

The secondary guns were four 12pdr (8cm) Type 41 QF guns; these were mounted on either side of the upper deck between the third and fourth funnels and aft of No 4 funnel. According to the authoritative *Japanese Cruisers of the Pacific War* (*op cit*, 788), the forward pair of mountings were replaced by 8cm Type 3 40-cal HA guns in 1924. *Sekai no Kansen*'s *Nihon Junyōkan-shi*, 61, claims that this substitution took place in 1919, and backs its claim with a photograph.

The gun armament was completed by two single 6.5mm Maxim MGs; these were replaced during the 1920s by two domestically-produced 6.5mm Type 3 MGs.

The arrangement of the three above-water 18in (45cm) torpedo tubes reflected that of *Tone*. They were mounted on the middle deck, one on either side of the after bridge structure and the third on the centreline at the stern. The configuration of the stern was modified to avoid the problems experienced in *Tone*. According to the authoritative *Kaigun Suirai-shi* (page 259) the torpedo was the

Auxiliary machinery

Item	Description	Power Rating	Number
Generator	4-pole compound winding aux shunt winding, driven by vertical reciprocating engine	upper deck: 44kW lower deck: 88hp	2 2
Steering engine	inverted 2-cylinder single expansion reciprocating steam engine	192hp	1
Main condenser	uniflux surface condenser	total cooling area: 9,600sq ft	2
Aux condenser	tubular type surface condenser	total cooling area: 1,500sq ft	2
Main air suction pump	Weir single acting	–	4
Main (cooling) water circulating pump	centrifugal type	3,525 tons	4
Aux air suction & water circulating pump[1]	combined type driven by 1-cylinder condensing reciprocating steam engine	water: 636 tons	2
Forced lubrication oil pump	Weir type	82 tons	4
Main (boiler) feed water pump [BR3–4]	Weir type	large type: 55 tons	12
		small type: 50 tons	4 (BR1)
Aux (boiler) feed water pump	Weir type	[as main]	[as main]
Forced draught fan	Siemens type vertical 1-cylinder reciprocating steam engine	–	16[2]
Heavy oil pump for burner	Weir type	100lbs(?)	4
Evaporator	Care drainer type(?)	40 tons	2
Cooling machine	Hall type	34,000Btu	2
Refrigerating machine	Hall type	53,000Btu	2
Capstan	vertical 2-cylinder simple expansion reciprocating steam engine	100hp	1
Air compressor	Brotherhood V type 4-stage compression	–	2
Boat hoist	vertical 2-cylinder simple expansion reciprocating steam engine	50hp	1
Fresh water pump	Worthington type	10 tons	1
Sea water pump	Worthington type	30–35 tons	5

Notes:

1 Pumps located side by side: water pump driven by steam piston rod directly, while air pump driven by see-saw lever.

2 Four per boiler room.

domestically-produced Type 44 No 2. The same source has additional data about the development of the torpedo (pages 16–17). The Type 44 was Japan's first 'heater' torpedo; length of the No 2 model was 17ft 8in (5.39m) and total weight was 1,650lbs (750kg) including a 240lb (110kg) Shimose explosive warhead; air pressure was 2,135psi (150kg/cm^2), and the torpedo was driven by a four-cylinder radial engine and was capable of running for 4,400 yards (4,000m) at 35 knots. According to the NTD document of 2 Dec 1919 the torpedo tubes were of the Armstrong type.

The tactical value of the stern torpedo tube was considered very low. When the ship was steaming at high speed launch was successful only in smooth water; water turbulence tended to affect the function of the gyroscope, so straight running was unlikely. Following the grand manoeuvres of 1912, Vice Admiral Yoshimatsu Shigetarō, C-in-C 2nd Fleet, recommended that the tube be removed and the rudder post moved aft, the weight thereby saved being used for the protection of the steering compartment. However, this does not appear to have been implemented.

In order to be able to illuminate targets at night, paired 90cm searchlights manufactured by Siemens-Schuckert were mounted in the forward and after bridge wings on either side, with a further two on the foremast and mainmast platforms; the latter were remotely operated using eletrical cabling.

First Impressions of Tactical Value

A few months after commission all three ships took part in the grand manoeuvres of 1912. These were both offensive and defensive in nature, and were focused on the defence of Tōkyō Bay. The primary tactical purpose of these exercises was to test the characteristics of the ships built under the naval expansion programme of 1906.

The operational aspect was divided into three phases (14–20 October, 21–27 October, and 28 October–1 November) and was remarkable for the large number of warships which took part. The three *Chikumas* were assigned to the 5th Squadron of the 2nd Fleet (other ships in the squadron were the protected cruisers *Otowa* and *Tsushima*) assigned to the 'Blue' (friendly) Fleet, and participated in all three phases against the 'Red' (enemy)

Fleet. The exercises took place south of Honshū, Shikoku and Kyūshū with the exception of Kii and Bungo Straits, and it was in this phase that the 'turbine cruisers' excelled with their high speed. The report of the C-in-C of the Blue Fleet, Vice Admiral Baron Dewa Shigetō, dated 8 Nov 1912, records:

> The performance of the three *Chikuma* class cruisers during the third phase ... was very useful for the Blue Fleet, and their achievements were far superior to that of the 1st class cruisers of the *Izumo* class. However, three ships can only satisfy a small part of the role [as the eyes of the fleet], and we neeed at least seven ships of this type to perform it effectively. Studies are also required with a view to reducing their high coal consumption and pronounced roll.

A Brief Chronology

1910
Laid down in January (*Yahagi*), May (*Chikuma*) and August (*Hirato*) respectively.
1911
Chikuma's launch ceremony at Sasebo NY on 1 April was attended by the Crown Prince Yoshihito, the later Taishō Tenno. *Hirato* (Kawasaki Kōbe) and *Yahagi* (Mitsubishi Nagasaki) were launched on 29 June and 3 October respectively.
1912
Chikuma was completed 17 May and assigned to Maizuru Naval Station. (When Maizuru was declassified to Important Naval Port as a result of the Washington Arms Limitation Treaty she remained assigned). Sisters *Hirato* and *Yahagi* followed on 17 June and 27 July respectively.
1913
Chikuma participated in the grand manoeuvres of 1913 and embarked the Emperor for the naval review off Yokosuka.
1914–1918
All three ships participated in the various operations of the Great War. *Chikuma*, together with the battlecruiser *Ibuki*, participated in the hunt for the German light cruiser *Emden*, ranging as far south as the Java Sea via Hong Kong and Singapore. *Hirato* and *Yahagi* were assigned to the Second Southern Branch Fleet which, together with the First Southern Branch Fleet, was tasked with guarding the Australian Route; this duty would subsequently be superseded by the occupation of the German island colonies in the Marshalls, Carolinas and Marianas, after which both fleets returned to Japan, being dissolved in February 1915. From 27 February 1917 to 26 January 1918 *Yahagi* operated with the First Special Duty Fleet tasked with escort duty in the South China Sea, Indian Ocean and Sulu Sea, while *Hirato* and *Chikuma* were assigned to the Third Special Duty Fleet to escort troop and supply transports around Australia from 13 April to 12 December 1917.
1920
Chikuma operated off the Russian coast, then Tsingtao and North China, conducting patrols in Chinese waters before returning to Yokosuka in 1924.
1922–1924
Chikuma was flagship of the 1st Flotilla of the 1st Submarine Squadron from 1 December 1922 until 15 May 1924. *Yahagi* was flagship of the 2nd Flotilla of the 2nd Submarine Squadron from 1 December 1922 until 1 December 1923; she was replaced as flagship by *Hirato* from 10 April until 1 December 1924.
1924–1930
Chikuma was 4th reserve ship at Yokosuka from 15 May 1924, and used as a moored training ship and naval barracks until stricken. *Yahagi* was reserve ship at Kure from early 1924 until early 1927. She operated in Chinese waters from 7 April 1927 to 1 May 1930. *Hirato* was reserve ship at Kure from 1 December 1924 for exactly one year. Subsequently she operated in Chinese

Yahagi at Kure on 30 January 1916. In the background, beyond her bow, can be seen her sister *Chikuma*. (Lars Ahlberg collection)

Hirato at Sasebo on 24 May 1918 after her return from Tsingtao. (Lars Ahlberg collection)

waters from 1 December 1925 until 10 December 1928 and again from 30 Nov 1929 until 1 May 1930, being assigned to either the 1st or the 2nd Expeditionary Force. She was dispatched again to China on 1 December 1930 and remained there until April 1933.

1931

Chikuma was stricken from the register of Imperial warships on 1 April and listed as Disabled Ship (*Haikan*) No 3. She was employed as a training ship for the Naval College until expended as a target ship in about 1935. She was decommissioned prior to her two sisters due to frequent engine breakdowns.

1933–1940

Yahagi stayed at Kure as 4th reserve ship, and *Hirato* was reclassified 4th reserve ship on 20 April 1933, also at Kure, where both ships were employed as stationary training ships until stricken.

1940–1945

Yahagi and *Hirato* were stricken from the register of Imperial warships on 1 April 1940; the latter became

Yahagi during the late 1920s. Note the 8cm (3in) HA gun on its elevated platform between the 3rd and 4th funnels. (NHHC, NH 111699)

Hirato dressed overall at Shanghai in 1930, towards the end of her active service career. (NHHC, NH 111733)

Disabled Ship No 11, her sister No 12. The former *Hirato* was used as a barracks ship at Kure–Etauchi until 5 August 1942, then at Kure–Kawaraichi. In December 1943 she was brought to Iwakuni and used as a barracks ship in place of Abolished Ship No 2. Now designated Abolished Ship No 12, she was used as a moored training ship for the Naval College at Etajima, and from the beginning of 1944 she served the Submarine School at Ōtake as a barracks ship. According to Fukui Shizuo, some of her funnels & masts had been removed.

1945 & after

No 12 remained afloat and was towed to Kure–Etauchi, while No 11 heeled over due to gradual flooding and sat on the bottom after the termination of the Pacific War. Both ships were broken up in 1947: No 12 from 31 January to 8 July by Kasado Dock, No 11 from 5 January to 29 April by Tōkyō Salvage. Her hull was used as a breakwater for Iwakuni harbour.

Notes

1 Details of the so-called 1906 Expansion Program can be found in *Outline of the Naval Armaments* (*Kaigun Gunbi Enkaku)* Vol 1, 132–146.

2 Named after the Yahagi River, which ran close to where Toyotomi Hideyoshi, the feudal lord (*Daimyō*) often referred to as the second 'Great Unifier' of Japan, was born.

3 A drawing of the Curtis turbine fitted in the battleships can be found in *Warship 2017*, 16. The Curtis turbines of *Chikuma* and *Hirato* had the mixed disk and drum structure typical of the Brown-Curtis type; however, because the basic design was from the US Fore River Company the word 'Brown' was omitted. The Brown-Curtis turbine differed from the original by (i) replacement of the last step of the rotor disk by a drum, (ii) an increase in the blade rows on the rotor to improve heat efficiency and reduce fuel consumption. In 1911 a contract was concluded between John Brown, the Kawasaki Kōbe Shipyard and the IJN according to which warships built by Kawasaki were to be propelled by Brown-Curtis turbines. In May of the same year an analogous contract was concluded for the manufacturing rights to the Parsons turbine between the Parsons Company, the Mitsubishi Nagasaki Shipyard and the IJN according to which warships built by Mitsubishi were to be propelled by Parsons turbines.

4 One source gives the reason as the non-availability of the special steel in Japan.

5 *Sekai no Kansen* Special Issue 1/2012, 'A Technical History of Japanese Cruisers: (3) Machinery', 183.

6 In his article in *Sekai no Kansen* (see Note 5) Takagi comments on the minimal effects of the superheater with the low degree of superheating ('to prevent condensation'), and *Kaigun Kikan-shi* refers to the reduction by one sixth in the heating surface in the large boilers – see text.

THE *300-TONNES*: THE FIRST FRENCH TORPEDO BOAT DESTROYER

The development of the torpedo boat destroyer (TBD) in the British Royal Navy and the French *Marine Nationale* followed different paths but resulted in ships with similar capabilities. **John Jordan** looks at the origins of the 300-tonne destroyer, 55 of which were completed for the French Navy between 1899 and 1910 with a further 20 built for export.

The first British destroyers were a response to the French 'sea-going torpedo boats' (*torpilleurs de haute mer*, or THM) that entered service from 1888 and, in particular, to the three high-speed torpedo boats informally designated *très rapides* ordered in 1891. Prior to the completion of these boats, the security of the battle fleet had been threatened only by small, short-range torpedo craft operating out of harbours for coastal defence; the new French torpedo boats were designed to operate in the open waters of the Channel, threatening attempts at even a distant blockade. Admiral 'Jackie' Fisher, Controller and 3rd Sea Lord from February 1892, proposed as a counter an enlarged torpedo boat with the weight of coal minimised, and requested from the Director of Naval Construction (DNC) a 27-knot boat with a powerful gun armament. The result was a series of prototype 'torpedo boat destroyers' (TBD): the first, the Yarrow-built *Havock*, was completed in January 1894; a sister, *Hornet*, was followed by pairs of similar (but competitive) boats from Thornycroft and Laird. These prototypes were followed by orders for a large number of 'production' boats, informally known as the '27-knotters'.

The French counterpart to the TBD, initially designated *contre-torpilleur d'escadre* (CTE), was developed for different reasons, but was likewise an enlarged torpedo boat. The French had long been conscious of the need to defend their own battle fleet against British and Italian sea-going torpedo boats and torpedo cruisers. On 17 March 1886 Minister of Marine Admiral Théophile Aube had proposed a conversion of a torpedo boat of the No 65 type[1] to perform this role. By the time the *Conseil Supérieur de la Marine* (CSM) published its 1891 Programme in July of the same year, it was planned to use torpedo cruisers of the *Condor* and *Wattignies* types as fleet torpedo boats and to supplemented them by five *avisos-torpilleurs* (AT – *D'Iberville* class and later), which would land their torpedo tubes and be reclassified as *contre-torpilleurs*.

The programme of December 1894 proposed the construction of two new types: a 220-tonne *torpilleur de haute mer* (THM) for the Squadrons plus a 300-tonne TB leader (*chef de groupe de torpilleurs de défenses mobiles*) for the coast defence fleet: 16 of the former and 30 of latter were to be built. On 20 April 1895 the Minister of Marine, Admiral Besnard, asked the *Conseil Supérieur* to draw up specifications, and on 28 May 1895 the *Conseil* submitted a proposal for a THM of 225–245 tonnes with a designed speed of 24 knots for the former role, and a TB leader (AT) of 320–375 tonnes with a speed of 26 knots.

On 7 June Jacques-Augustin Normand, the preeminent designer of torpedo boats for the French Navy, submitted a design for a *grand torpilleur* of 285 tonnes able to operate with the fleet, armed with light guns and two torpedo tubes and capable of 25 knots; he was asked to rework the design to meet the CSM specifications. Normand reported on 7 September that although the sea-going torpedo boat was feasible on 244 tonnes, in order to meet CSM specifications for the TB leader a displacement of 403 tonnes was required; moreover he expressed a strong preference for a single type to perform both roles. On 14 October 1895 Normand duly submitted a proposal for a boat of 285–300 tonnes capable of 25/26 knots, modifying his *grand torpilleur* design to meet the requirements for speed, armament and operational

Table 1: **From Torpedo Boat to Torpedo Boat Destroyer**

	Torpilleurs de Haute Mer	RN 27-knotter TBD	*Durandal*-class *Contre-Torpilleur*
Displacement	115-130t	275–300t	300t
Dimensions	45m x 4.5m	62m x 5.8m	55m x 5.9m
Horsepower	2,000ihp	4,000ihp	4,800ihp
Speed	24kts	27kts	26kts
Armament	2 x 381mm TT	1 x 12pdr	1 x 65mm
	2 x 47mm	5 x 6pdr	6 x 47mm
		2 x 18in TT	2 x 381mm TT

Note: 12pdr = 76mm; 6pdr = 57mm; 18in = 457mm.

range. Two armament options were proposed: one for the THM, the other for the the leader, now referred to as a *torpilleur divisionnaire/conducteur de flotille*.

The Navy asked for various changes, and Normand submitted revised proposals on 18 February 1896. The torpedo boat would mount three 381mm (15in) torpedo tubes with three reloads, and two single 47mm (3pdr) QF guns; speed was to be 26 knots with 4,800cv, endurance 2,800nm at 10kts with 50 tonnes of coal (3,900nm with 74 tonnes). The flotilla leader would mount a single 65mm 50-cal Mle 1891 gun, seven 47mm QF, and a single 381mm torpedo tube with two reloads; endurance would be 2,400nm at 10 knots with 37.6 tonnes of coal, but speed was reduced to 25 knots. The proposal was accepted by the *Conseil des Travaux*, which nevertheless wanted a uniform armament of one 65mm plus six 47mm, with the remainder of the armament to be decided following trials – it would eventually be decided to fit two 381mm tubes.

The *300-tonnes* as a Type

The *300-tonnes* was to become the standard French destroyer, with no fewer than 55 units completed for the *Marine Nationale* between 1899 and 1910. Ordered initially in batches of two/four, construction peaked during the early 1900s, ten boats of the *Arquebuse* class being ordered under the 1901 Budget and a further ten under 1902. They are normally divided into two groups: the 32 boats of the *Durandal* type, armed with the 381mm (15in) torpedo, and the 23 boats of the *Claymore* type, which had the 381mm replaced by the longer-range 450mm (17in) torpedo. However, within these broader groupings there were numerous sub-groups, all of which incorporated incremental 'improvements', most of which related to the hull-form (particularly above the waterline) and the superstructures.

The original Normand prototypes were intended to be fitted as *avisos-torpilleurs* (AT) or *torpilleurs d'escadre* (TE) as required. The first of the AT (for coast defence) was to be ordered under the 1896 Budget (submitted May 1895), and the contract with Augustin-Normand was signed on 20 May 1896. Normand offered to build a second boat at a 70,000FF discount if ordered straight away. Emile Bertin, currently *Directeur du matériel*, proposed accepting this offer using a supplement to the budget, and Minister Bernard agreed to fund a second unit, to be named *Hallebarde*. The budget for the following year, 1897, shows *Durandal* as an AT with the budgetary number M1, and *Hallebarde* as a TE with the number M'1, but by 1898 both were being designated TE and the boats as completed were identical.

At the same time it was proposed to employ the two recently-ordered *aviso-torpilleurs* of the *Dunois* class as 'torpedo catchers'. They would subsequently serve as division leaders for the *défenses mobiles*. Displacing

The 'torpedo catcher' (ex-*aviso-torpilleur*) *Dunois* dressed overall. During the early 1900s *Dunois* and her sister *La Hire* served as leaders for the coast defence torpedo boats. Armed with six 65mm (one fore and aft, four abreast the funnels) and six 47mm QF guns, they were not fitted with torpedo tubes. (Bougault, courtesy of Philippe Caresse).

900 tonnes and with light protection for the machinery spaces (25mm sides plus a 14mm vaulted deck), they had a relatively heavy gun armament of six 65mm 50-cal Mle 1891 guns and four 47mm Mle 1885 QF but no torpedo tubes. They were relatively slow (22.5 knots designed, less on trials) and only *Dunois* served in this role during the Great War, *La Hire* having been reassigned as a gunnery training ship in 1911.

Durandal/Fauconneau 1896–97

The hull lines of *Durandal* were based on Agustin Normand's sea-going torpedo boat *Flibustier*, but with a fuller stern and more robust scantlings. Hull weight was just under one third of displacement, and the length to beam ratio was a modest 9.0.

The bronze, four-bladed propellers and the rudder were submerged as deep as possible: their tops were almost one metre below the waterline with a bottom support for the rudder that doubled as a propeller guard. This arrangement increased efficiency when the ship was pitching, but because the rudder was well forward of the propellers it did not benefit from the wash; the consequence was a large turning circle (450m at 17 knots – twice the desired figure). All the *300-tonnes* had this configuration, which was likewise derived from *Flibustier* and was standard on all the torpedo craft built by Agustin-Normand, but in later types of *contre-torpilleur* the rudder would be moved to the extreme stern, as in the smaller torpedo boats.

The engines were in individual engine rooms amidships, between two outer boiler rooms (see GA Plans). This arrangement reduced the vibrations common in light craft and, more importantly, trim was unaffected by fuel consumption.[2] It also accounted for the wide spacing of the two funnels.

The hull was divided into nine watertight compartments separated by eight full-height transverse bulkheads. The layout from forward to aft was as follows:

1 a *compartiment de choc* (3.6m)
2 a seamen's mess (13.1m), divided into two by a partial transverse bulkhead above the waterline with magazines, provisions, stores and the cable locker beneath
3 the forward boiler room (6m), with a transverse coal bunker (1.2m) at its forward end and additional coal bunkers to the sides

The first French *contre-torpilleur d'escadre, Durandal.* Note the rounded configuration of the hull and the wooden walking deck almost one metre above it, on which most of the crew are standing. The structure topped by a stove pipe abaft the second funnel housed the galley. (NHHC, NH-64459)

4 the forward engine room (4.8m), with the reciprocating engine to starboard and the main condenser to port flanked by further coal bunkerage
5 the after engine room (4.8m), which was a mirror image of the forward room
6 the after boiler room (6m), which was likewise a mirror image of the forward BR with the tranverse coal bunker at its after end
7 accommodation for the CO and three officers (7.15m)
8 a mess for seven Petty Officers, including the chief engineer, who had his own cabin (4.2m)
9 the provision issue room (2.8m).

Above the waterline Normand limited the height of the hull by adding an elevated manoeuvring/walking deck (*pont surélevé à caillebotis*) comprising a wooden grating 0.8m above the upper deck from the stern to the twin breakwaters at the after end of the bow, which had a

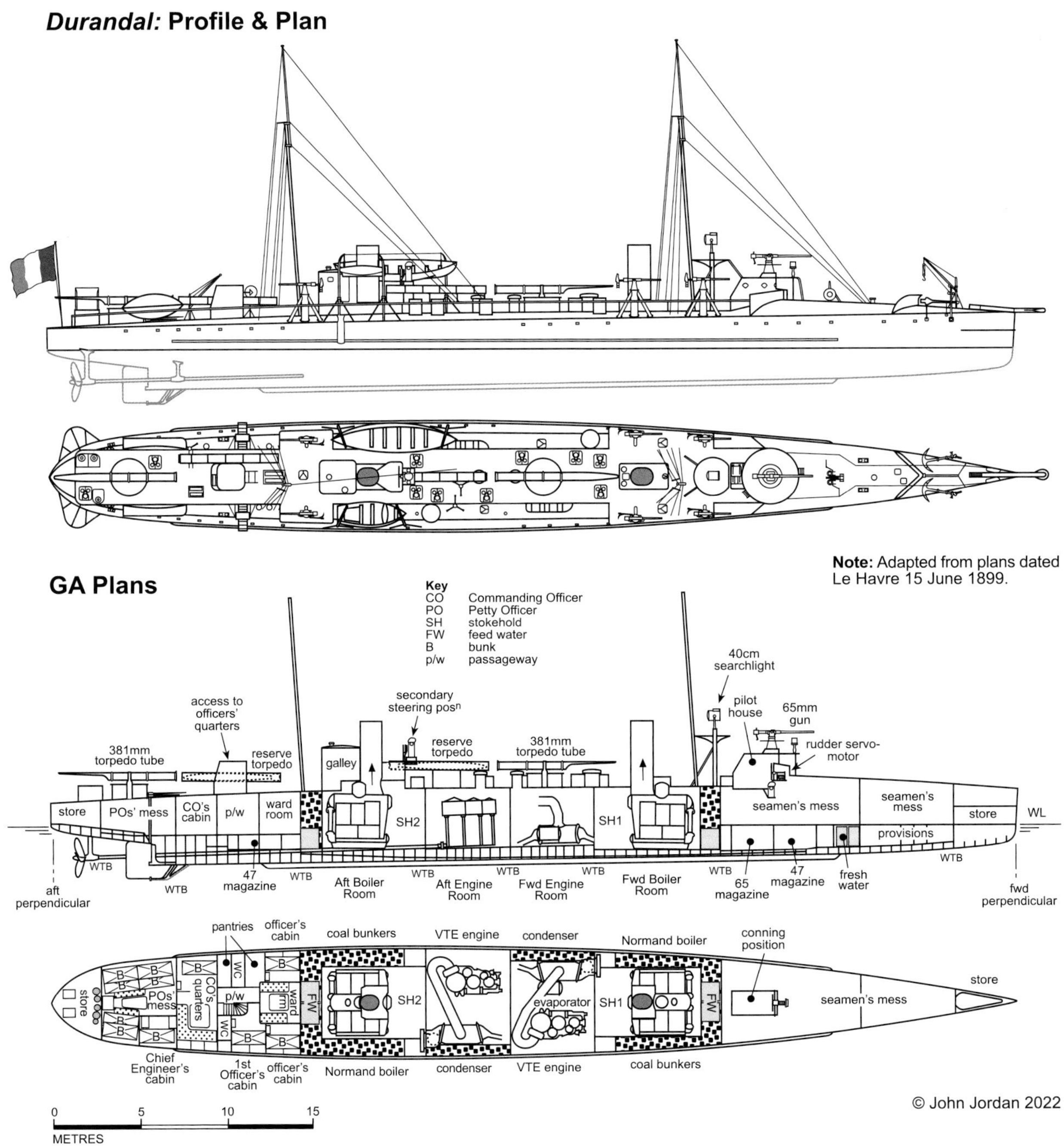

The Augustin Normand plans of *Durandal*, the lead boat of the 300-*tonnes* series. Note, in the inboard profile, the raised coamings for the numerous hatchways giving access to the compartments below decks. These were necessitated by the elevated walking deck (not depicted in this view in the interests of clarity), which was 0.8 metres above the upper deck.

prominent turtleback deck. The sides of the upper hull were inclined inboard so that seas passed between the hull and the walking deck instead of breaking on the hull sides and crashing on board (see Section drawing). This resulted in better sea-keeping and reduced the centre of gravity, so the metacentric height (GM) was higher. The walking deck also gave a better command for the guns in the waist, which were raised on centre pivots with supports; the crew was kept drier and the boats could maintain speed better in heavy seas. The downside was that the raised deck acted as encouragement for the crew to stow extra gear (including spares) on the upper deck beneath it to the detriment of stability. Also, the higher command of the guns was of limited benefit because of the way these tiny boats were tossed about.

***Durandal:* Midship Section**

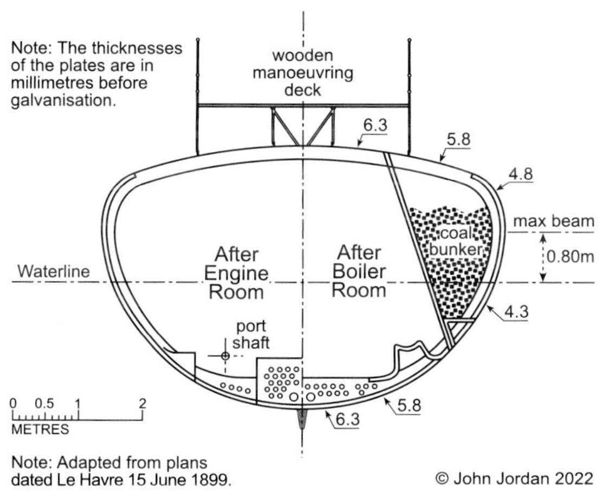

The two Normand return-flame boilers, which were rated at 16kg/cm^2, were scaled up from those of the fast THM *Aquilon* (completed 1896). These had good fuel efficiency, despite the long furnaces that made it difficult for stokers to maintain an even layer of burning coal. The reciprocating engines were scaled up by 15 per cent from those fitted in torpedo boat *No 201* and the fast THM *Cyclone*. They retained the robust structure of the latter with steam jackets on the cylinders and feed water heaters – these had been omitted from the early British TBD engines in order to reduce machinery weight and achieve very high speeds. The two engines were offset to port and to starboard, and each had its own condenser on the opposite side of the ship; the condensers were cooled by natural circulation using hull intakes, thereby dispensing with powered pumps. The engines were designed for 4,800cv with natural draught and 5,200cv with forced draught for 26 knots. The dimensions of the cylinders were: 47.2cm (HP), 67.8cm (IP) and 102.4cm (LP) with a 57.4cm stroke.[3] The engines of *Durandal*'s sister *Hallebarde* were uprated to 5,600cv forced draught by Minister Lanessan on 5 September 1899.

The machinery proved more robust and reliable than that of their British counterparts, despite its comparatively low weight (112 tonnes). Many destroyers of the British 27/30-knot types had their trials interrupted for long periods due to engine breakdowns. Normand thought they were too lightly built and too ambitious; many achieved 30 knots on trials, but this was not replicated in service. Normand was critical of the propellers

Table 2: *Durandal* Class

	Builder	Laid down	Launch	Trials	Speed (trials)
1896 Budget					
M1 *Durandal*	Augustin-Normand, Le Havre	25 Aug 1896	11 Feb 1899	Apr/Jul 1899	27.42 knots
M'1 *Hallebarde*	Augustin-Normand, Le Havre	25 Aug 1896	8 Jun 1899	Jul/Aug 1899	27.20 knots
1897 Budget					
M'2 *Fauconneau*	Augustin-Normand, Le Havre	29 Apr 1897	2 Apr 1900	May/Jul 1900	27.14 knots
M'3 *Espingole*	Augustin-Normand, Le Havre	29 Apr 1897	28 Jun 1900	Aug/Sep 1900	27.25 knots

Characteristics (*Hallebarde*)

Displacement:	
normal	308.50 tonnes
full load	364.59 tonnes
Dimensions:	
Length	55.00m pp/wl, 56.80m oa
Beam	5.96m wl
Draught	2.62m aft; 3.00m incl propellers
Propulsion:	
Boilers	Two Normand small-tube boilers, 16kg/cm^2
Engines	Two-shaft 3-cylinder VTE
Horsepower	4,800ihp (5,200ihp max)
Speed	26 knots (designed)
Coal	37.6 tonnes normal
Endurance	2,400nm at 10 knots
Armament:	
Guns	1 x 65mm/50 Mle 1891 QF (340 rounds)
	6 x 47mm/40 Mle 1885 QF (2,700 rounds)
Torpedoes	2 single tubes for 381mm torpedoes
	4 torpedoes Mle 1887 (2 reserve)
Boats:	1 x 7-metre whaler
	1 x 4-metre dinghy
	2 x 3.5-metre Berthon folding boats
Complement:	
As private ship	4 officers, 58 men

Durandal steaming at 25 knots during her trials. Note the effect on trim when these boats were running at high speed; in calm conditions the bow lifted out of the water and the propellers dug deep under the stern. (Philippe Caresse collection)

Close-up of the single 65mm QF gun of *Hallebarde*, which was mounted on a bandstand atop the pilot house. Two of the six 47mm Mle 1885 guns can be seen abreast the first funnel. (Philippe Caresse collection)

adopted for the British boats, which were designed by the builders to deliver high speed on trials at light displacement, but which were prone to cavitation at normal loading. The French destroyers also proved much more economical: coal consumption for *Durandal* was only 0.8kg/hp per hour.

The single 65mm 50-cal Mle 1891 gun was mounted on a bandstand forward directly above the pilot house, which was countersunk with its floor beneath the upper deck; height of command was 5.2 metres. The six 47mm Mle 1885 QF guns were disposed in the waist, three on either side; command was 4.15 metres for the forward pair of mountings and 3.75m for the after pair. The magazine for the 65mm gun, which was directly beneath the mounting, held 340 combat rounds (evenly distributed between steel and cast iron projectiles) stowed in cases of three. The magazines for the 47mm QF guns were located forward of the 65mm magazine and beneath the officer's accommodation aft (see GA Plans), and held a total of 2,700 steel and cast iron combat rounds (= 450rpg). Ready-use provision was eight rounds for the 65mm gun, sixteen for the forward 47mm guns and twelve for the after guns.

The 65mm and the 47mm guns were handworked and fired using local cross-hair sights; the 65mm gun had a crew of six and each of the 47mm guns a crew of three, making for a total of 24 out of a crew of only 59. Order transmission was by voice, whistle or bugle. For night fighting there was a single 40cm searchlight projector on a raised, circular platform directly abaft the 65mm mounting.

The armament was completed by two 381mm torpedo tubes for Mle 1887 torpedoes on single mountings located between the widely-spaced funnels and above the stern. Reloads for the torpedo tubes were stowed in lockers (*valises*) close to the tubes. The torpedoes in the lockers were not primed: they had been emptied of compressed air and needed three hours to be readied for launch. Reloading was laborious and could not have been carried out while the boat was underway. The Mle 1887 torpedo had a length of 5.68 metres and weighed 423kg with a 42kg dry cotton warhead; built by Whitehead and (under licence) at Toulon Naval Dockyard, it had a range of 400m at 28.5 knots and 600m at 27 knots.

There were two 380kg stock anchors plus a reserve anchor of 160kg. Handling on the early boats was manual; on later units a winch was provided. Electrical power for the searchlight projector, signalling and lighting was supplied by a single generator feeding an 80-volt circuit (50–75A). For damage control seven pumps with a capacity of 100 tonnes per hour were provided for the main compartments, while the two end compartments had smaller 20t/h pumps. The main drain was divided into two under the machinery spaces.

Accommodation for the designed complement of four officers and 59 petty officers and men was cramped. There was a large seamen's mess forward, divided into two sections by a partial transverse bulkhead. The officers' quarters were directly abaft the machinery spaces, while the petty officers were accommodated in an open mess in the stern fitted with three two-tiered bunks; the chief engineer had his own cabin in the same area. Habitability was poor. The seamen's and POs' messes were considered too small and ventilation was inadequate, especially in the forward part of the boat, resulting in excessive condensation in winter.

Durandal and her sister *Hallebarde* were funded under the 1896 Budget. Two further boats, *Fauconneau* (M'2) and *Espingole* (M'3), were funded under the 1897 Budget. As recommended by Jacques-Augustin Normand (July 1897), these two boats had slightly thicker scantlings, and freeboard was raised at the bow by 30cm and

Table 3: **French Destroyer Guns [*Durandal* to *Arquebuse*]**

	65/50 Mle 1891	47/40 Mle 1885
Total weight:	540kg	237kg
Total length:	3,450mm	2,048mm
Bore length:	3,247mm	1,878mm
Chamber length:	474mm	336.25mm
Chamber volume:	2.01dm^3	0.76dm^3
Breech pressure:	1,950kg/cm^2	2,000kg/cm^2
Rifling:	26 grooves at 6°	20 grooves at 7°10'
Shell:	4.17kg steel	1.50kg steel
	4.00kg CI	1.50kg CI
Bursting charge:	0.10kg powder/*mélinite*	0.022kg powder/*mélinite*
	0.20kg powder (CI)	0.055kg powder (CI)
Propellant:	0.88kg BM3 (cartridge)	0.29kg BM2 (cartridge)
Muzzle velocity (MV):	715m/s	650m/s
Remaining V 2000m:	402m/s	300m/s
Drift at 2000m:	1.5m	2.1m
Mean range error:	17.1m	16.8m
Mean line error:	0.63m	1.3m

Source: *Renseignements sur les Bouches à Feu de l'Artillerie Navale Tous Calibres et Modèles.*

at the stern by 10–15cm. They were otherwise identical to the first two boats.

Pique 1897

These four destroyers were ordered under a supplement to the 1897 Budget. Two were contracted to the private shipbuilder Ateliers & Chantiers de la Loire (ACL) at Nantes, the other pair to Forges & Chantiers de la Méditerranée at Graville (Le Havre). They were designed by the builders to hull plans from Augustin-Normand and incorporated a number of changes, the most significant of which was a move from two to four boilers, each served by its own funnel; they therefore had a distinctive appearance with four funnels in two widely-spaced pairs. The boilers were of the Normand type, not Normand-Guyot. The reduced length of the boiler (grating 1.7m vs 2.5m in *Durandal*) made it easier for the stokers, and the adoption of four smaller boilers led to greater flexibility and a more rapid increase in speed. However, accommodating the four boilers added length and weight and complicated the steam pipework. The length of the two boiler rooms had to be increased by 2.5m; 0.50m was recouped from the engine rooms, 0.50m from the seamen's mess and 0.55m from the officer accommodation. Waterline length increased by 1 metre and displacement by 10 tonnes; the distribution of weights was different and there was a higher centre of gravity, with a consequent loss of stability.

Trials were laborious, and *Pique* was the only one of the 75 *300-tonnes* completed that failed to make her contract speed; the engines were rated 5,700cv with forced draught in 1901, but were subsquently downrated by Minister of Marine Pelletan to 5,600cv on 23 July 1902. The two boats built by ACL were more successful than the two built at Le Havre, *Yatagan* achieving 27.07 knots on trials. *Yatagan* (and possibly her sister *Framée*) were fitted with three-bladed propellers, an experiment later repeated with *Dard* (*Arquebuse* class).

The hull steel of the *Pique* class incorporated 3% nickel. Together with the other two boats funded under the 1897 Budget (*Fauconneau* and *Espingole*) they were to have had specially fabricated ram bows, but it was decided on 20 September 1898 that all would retain the straight bow of *Durandal*. However, at the same time the Russians ordered five boats of the *Pique* type (*Vnimatelni* class: two from Normand, Le Havre, and three from FCM Graville) and purchased ram bows. They subse-

A starboard bow view of *Hallebarde* shortly after completion. Note the rounded hull with its prominent 35-metre rubbing strake, which was a feature of all the 300-tonne boats and which prevented damage to the thin hull plating when coming alongside. The searchlight on the raised platform forward of the foremast is a 40cm model. (NHHC, NH-64368)

quently placed a follow-up order for a further eleven boats, all with ram bows, during the Russo-Japanese War (*Lovki* class: three from Normand and eight from FCM, to be built at their Le Havre-Graville and La Seyne yards). The French Navy, on the other hand, did not repeat the 4-boiler type and did not fit ram bows in any subsequent destroyers.

Pertuisane 1899

Four copies of *Fauconneau* were ordered from Rochefort Naval Dockyard on 8 June 1899 under the 1899 Budget. Rochefort, on the Atlantic coast of France, was the smallest of the French naval dockyards and the only one engaged in the building of torpedo

Fauconneau steaming in company with one of the battle squadrons in the early 1900s. In the background: an armoured cruiser of the *Amiral Charner* class. (Philippe Caresse collection)

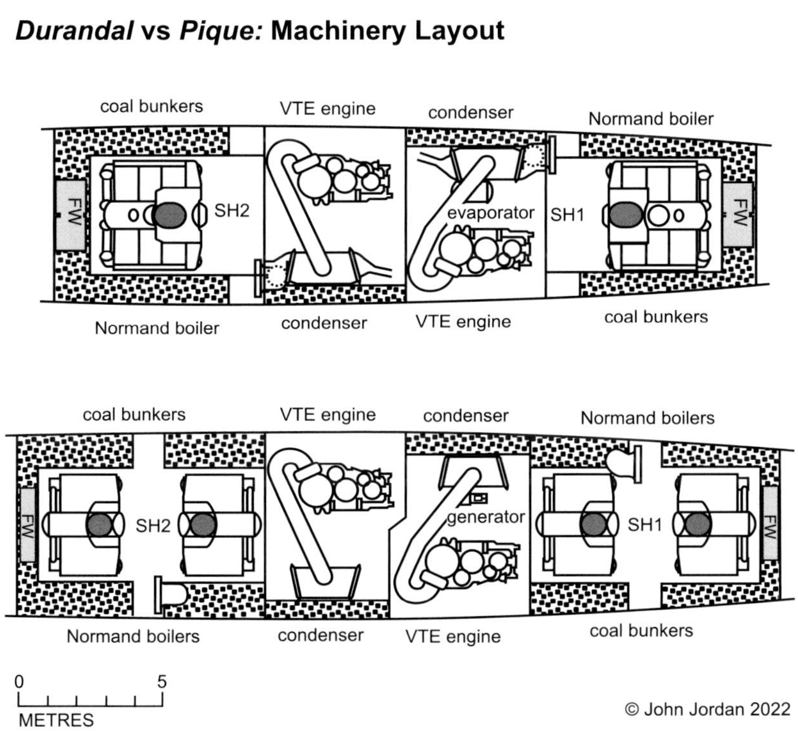

Schematic drawing of the machinery spaces of the two-funnelled *Durandal* and the four-funnelled *Pique*. The adoption of four smaller boilers in the *Pique* class was only possible by increasing the total length of the two boiler rooms from 14.3 metres to 16.7 metres, making economies necessary elsewhere to keep overall hull dimensions broadly the same. A reduction of 0.5 metres in the length of the two engine rooms was achieved by employing a staggered transverse bulkhead, a measure repeated in the later *Branlebas* class (see text and drawing). Other savings involved shaving more than one metre off the accommodation spaces, so these boats were even more cramped than *Durandal*. In the absence of published plans of *Pique* and her three French sisters, the drawing of the machinery spaces is based on the Russian *Forel*, one of five similar boats ordered from Normand and FCM.

Table 4: *Pique* Class

	Builder	Laid down	Launch	Trials	Speed (trials)
1898 Budget					
M'4 *Pique*	FC Méditerranée, Graville	Oct 1897	31 Mar 1900	Jul 00/May 01	25.88 knots
M'5 *Epée*	FC Méditerranée, Graville	16 May 1898	27 Jul 1900	Jan/Jul 1901	26.19 knots
M'6 *Framée*	AC Loire, Nantes	1897	21 Oct 1899	Feb/Jun 1900	26.91 knots
M'7 *Yatagan*	AC Loire, Nantes	1897	20 Mar 1900	May/Oct 1900	27.07 knots

Characteristics (*Pique*)

Displacement:	
normal	319.10 tonnes
full load	347.82 tonnes full load
Dimensions:	
Length	56.62m pp/wl, 58.21m oa
Beam	5.96m wl
Draught	2.79m aft; 3.02m incl props
Horsepower/speed	5,200ihp/26 knots

Remarks: Ordered 1897. Hull-form to Normand plans but had four boilers (16kg/cm^2) and four funnels. Five similar ships ordered by Russia (*Vinimatelni* class).

Yatagan, one of the four units of the *Pique* class, which had four smaller boilers and were easily distinguished from the other boats of the *300-tonnes* series by their four funnels, arranged as two pairs. The design was sold to Imperial Russia, which followed an initial order for four boats (*Vnimatelni* class) with one for eleven similar boats from French shipyards during the Russo-Japanese War. Note the distinctive *tampon de choc* protruding from the bow which was a feature of the *300-tonnes* boats and their 450-tonne successors; it was intended to protect the bow from a minor collision. (Philippe Caresse collection)

boats; it would remain the lead yard for destroyers even after Toulon became involved. Two of the four boats, M'8–9, were originally to have been contracted out to private shipyards but were reassigned, whereas M'10–11 were ordered from Rochefort from the outset.

The four boats of the *Pertuisane* class were laid down to Normand plans and had a similar layout and silhouette to *Fauconneau*, the only difference being that the two funnels were inclined aft. According to the *Feuille signalétique* produced for the *300-tonnes* series, *Flamberge* was fitted with two 3-bladed propellers on each shaft, but the experiment does not seem to have been repeated. The same boat, together with three later destroyers, would be converted for minelaying in 1909–11 in response to the lessons of the Russo-Japanese War; other *contre-torpilleurs* would be given light minesweeping gear from 1910 onwards.

Table 5: *Pertuisane* Class

	Builder	Laid down	Launch	Trials	Speed (trials)
1899 Budget					
M'8 *Pertuisanee*	Arsenal de Rochefort	9 Apr 1900	5 Dec 1900	Feb/Sep 1902	26.91 knots
M'9 *Escopette*	Arsenal de Rochefort	9 Apr 1900	20 Dec 1900	Jun 02/Jan 03	27.07 knots
M'10 *Flamberge*	Arsenal de Rochefort	2 Jan 1901	28 Oct 1901	Jan/Jul 1901	25.88 knots
M'11 *Rapière*	Arsenal de Rochefort	2 Jan 1901	16 Jul 1901	Oct 02/May 03	26.19 knots

Characteristics (*Fauconneau*)

Displacement:	305.77 tonnes
Dimensions:	
Length	56.00m pp/wl, 56.12m oa
Beam	5.94m wl
Draught	2.65m aft; 3.25m incl props

Remarks: Ordered June 1897; first from state dockyard. Laid down to Normand plans and similar silhouette to last two *Durandal* but with slightly inclined funnels.

Arquebuse 1901–1902

In late 1900 Augustin-Normand submitted two new proposals, both with a length of 56.30m pp/wl and with essentially the same hull and machinery. One was protected against the British 12pdr and 9pdr guns and featured an armoured box (*caisson blindé*) with 50mm plating for the hull and end bulkheads around the machinery spaces and a 17mm deck, the weight of protection being compensated by an increase in beam (6.75m max vs 6.3m) and deeper draught; speed was reduced to 25 knots. The other variant was a fast destroyer with a maximum designed speed of 28 knots but no protection. Horsepower was increased to 6,300cv in both types by adopting higher-pressure boilers (18kg/cm^2 vs 16kg/cm^2) and by improved combustion: the grating area was increased from the 6.2m^2 of *Durandal* to 7.15m^2, and shaft revolutions from 300rpm to 310rpm. Weight was increased by 3 per cent.

At around the same time the *Service Technique* presented a different proposal for an enlarged *Durandal*. Dimensions were length 66m, beam 6.5m and draught 2.05/3.30m; displacement was 400.8 tonnes. There were to be three boilers forward of the engine rooms, delivering 8,000cv for a speed of 30 knots; coal bunkerage was to be 45 tonnes. Possible alternative armaments were:

- 2x65mm + 8x24mm
- 10x47mm
- 2x65mm + 8x47mm

The 24mm was a new calibre that emerged from studies for lightweight destroyer guns capable of maximum penetration against light-skinned flotilla craft; it was to fire a 0.28kg projectile with a muzzle velocity of 1,050m/s. All three variants were to mount two torpedo tubes.

The above proposals were discussed by the *Conseil des Travaux* on 24 April 1900, and both the STCN and the Augustin-Normand protected variant were rejected. *Durandal* and *Hallebarde* had both by now run their

***Arquebuse:* Midship Section**

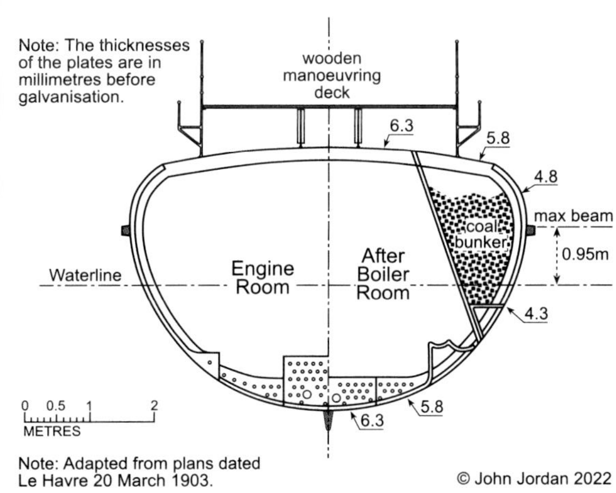

trials, and the Navy was very happy with them. The *Conseil* was, however, supportive of the Augustin-Normand proposal for a fast destroyer, despite its own reservations about the value of high speed, as the higher-powered machinery would have little impact on the basic design.

The new destroyers would be ordered in two batches of ten, the first under the 1900 Budget, the second under the 1901 Budget, which also featured the six battleships of the *Patrie* class. On 4 May 1900 Minister of Marine Lanessan ordered the Machinery Commission to contract with Augustin-Normand for two boats (M'14–15), and two further units (M'12–13) were ordered to the same plans from Rochefort. The remaining boats of the 1900 series were ordered in pairs from AC Loire (Nantes), FCM Graville and FC Gironde of Bordeaux. The 1901 boats had slightly modified boilers (area of grating: 7.72m^2). Normal displacement when completed was around 308/320 tonnes, and the average unit cost was 1,450,000FF (equivalent to £58,000 sterling).

The performance of the *Arquebuse* type would prove

Carabine & *Sarbacane*: GA Plans

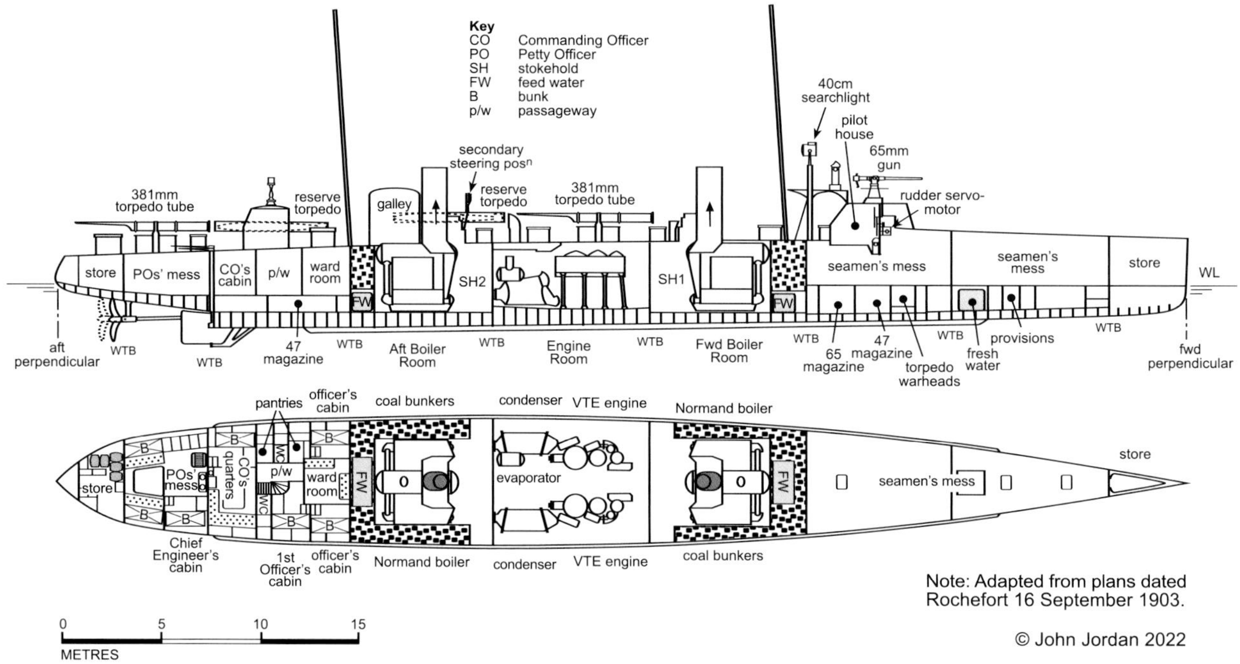

The GA plans of *Carabine* have been selected as those of the name-ship of the class, *Arquebuse*, are damaged and unclear. Note the single engine room with the reciprocating engines side by side and the main condensers in line. This saved 1.75 metres in length over *Durandal* which could then be invested in longer boiler rooms and accommodation for the crew. The *Feuille signalétique* produced for the *300-tonnes* series suggests that in this and the following *Claymore* class there was one fewer watertight transverse bulkead (seven vs eight), but this did not include the 'partial' bulkhead that divided the seamen's mess into two, despite this bulkhead being marked on the plans.

The destroyer *Carabine* of the *Arquebuse* class at Toulon. Ordered as two groups of ten in 1901/02, these boats were a development of the *Durandal/Fauconneau* type with more powerful engines and two knots greater speed. The engines were located side by side in a single engine room, so the funnels were slightly closer together than in the early boats. Initial stability problems were resolved by the suppression of the original mainmast. (Philippe Caresse collection)

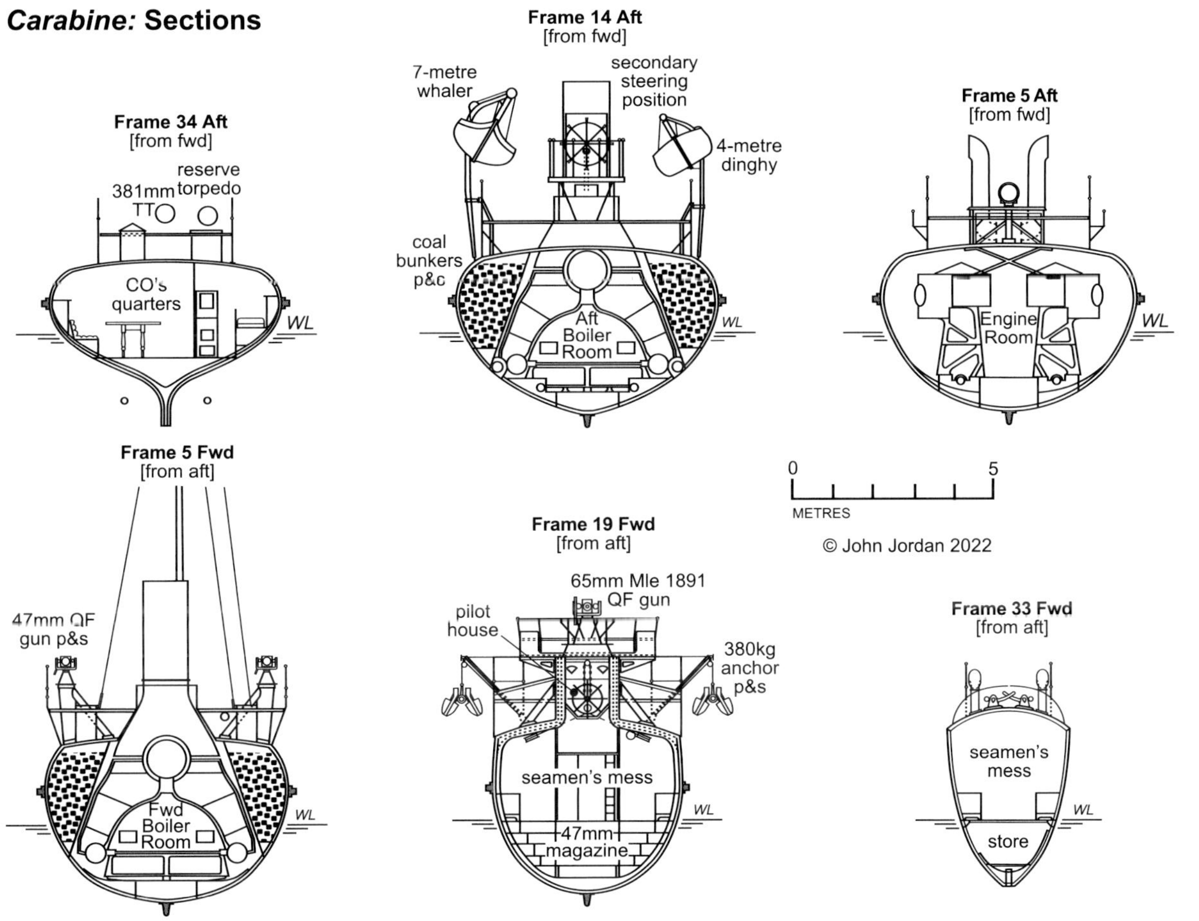

In the Sections plan, note the unusual frame numbering. For the destroyers up to and including the *Arquebuse* class the longitudinal mid-point of the hull was marked with a vertical broken line and the frames fore and aft numbered 1–46 to the bow and 1–48 to the stern. From the *Claymore* class that followed the frames were numbered in the conventional way: 1–98 from bow to stern.

to be even better than the early ships. Designed for 28 knots, the Normand-built *Arquebuse*, which would give her name to the class, attained 30.76 knots with 7,245cv (320rpm) on a displacement of 303 tonnes on trials 10 March 1903; three weeks later her sister *Arbalète* made 31.37kts with 7,136cv (311.13rpm) on a displacement of 304 tonnes.

The hull division of the destroyers of the *Arquebuse* class was broadly similar to that of *Durandal* but dimensions differed:

	Durandal	*Framée*	*Arquebuse*
Length BR	14.3m	16.7m	15m
Length ER	9.5m	9m	7.8m
Seamen's mess	13.1m	12.6m	15m
Officer accom^{n}	7.15m	6.6m	7.2m
POs' mess	4.2m	4.2m	4.2m

Source: Le Masson, *op cit*.

The 1.75m reduction in the length of the engine rooms was achieved by locating the engines side by side with the condensers abaft them in a single compartment and dispensing with the lateral coal bunkers (see GA Plans of *Carabine*). The bulkier superstructures housed enlarged command spaces, and there were concerns about stability following an incident during the trials of *Mousquet* (April 1903). These stability issues were remedied by landing the 16-metre mainmast.

Arbalète and *Mousqueton* were employed for the first trials of wireless telegraphy (W/T). *Harpon* was the first to have the letter of her name painted in black on either side of the bow as a recognition feature (from April 1903).

Claymore 1903

On 20 November 1901 Augustin-Normand submitted a new project for a modified *Arquebuse*. The principal changes were as follows:

- all ship control and navigation equipment to be consolidated in a new enclosed bridge
- the crew's galley to be moved forward of the foremast, with the bridge above and the searchlight projector atop it.

The proposals were examined by the *Conseil des Travaux* on 7 January 1902 and approved. However, following

Table 6: *Arquebuse* Class

	Builder	Laid down	Launch	Trials	Speed (trials)
1900 Budget					
M'12 *Carabine*	Arsenal de Rochefort	15 Jul 1901	21 Jul 1902	May/Sep 1903	30.16 knots
M'13 *Sarbacane*	Arsenal de Rochefort	1 Oct 1901	12 Mar 1903	Oct/Dec 1903	29.50 knots
M'14 *Arquebuse*	Augustin-Normand, Le Havre	1901	15 Nov 1902	Jan/May 1903	30.76 knots
M'15 *Arbalète*	Augustin-Normand, Le Havre	23 Nov 1900	28 Apr 1903	May/Aug 1903	31.37 knots
M'16 *Mousquet*	AC Loire, Nantes	Nov 1900	7 Aug 1902	Feb/May 1903	29.12 knots
M'17 *Javeline*	AC Loire, Nantes	Nov 1900	15 Oct 1902	Mar/Jun 1903	29.14 knots
M'18 *Sagaïe*	FC Méditerranée, Le Havre	1901	15 Nov 1902	Oct 02/Apr 03	30.62 knots
M'19 *Epieu*	FC Méditerranée, Le Havre	1901	17 Jan 1903	Mar/Jun 1903	31.21 knots
M'20 *Harpon*	FC Gironde	Nov 1900	20 Oct 1902	Nov 02/Mar 03	30.70 knots
M'21 *Fronde*	FC Gironde	Jan 1901	17 Dec 1902	Jan/Mar 1903	30.71 knots
1901 Budget					
M'22 *Francisque*	Arsenal de Rochefort	5 Feb 1903	2 Mar 1904	Mar/Apr 1904	30.09 knots
M'23 *Sabre*	Arsenal de Rochefort	1903	15 Apr 1904	May/Jun 1904	29.73 knots
M'24 *Dard*	Penhoët, Rouen	15 Oct 1902	10 Sep 1903	Dec 03/Apr 04	29.55 knots
M'25 *Baliste*	Penhoët, Rouen	20 Nov 1902	22 Oct 1903	Dec 03/Jun 04	29.90 knots
M'26 *Mousqueton*	Schneider	1901	4 Nov 1904	Sep 03/Aug 04	28.78 knots
M'27 *Arc*	Schneider	1901	24 Dec 1902	Jul 03/Jul 04	28.89 knots
M'28 *Pistolet*	AC Loire, Nantes	Sep 1901	29 May 1903	Jun/Aug 1904	28.94 knots
M'29 *Bélier*	AC Loire, Nantes	Sep 1901	29 May 1903	Jan/Mar 1904	29.93 knots
M'30 *Catapulte*	FC Méditerranée, Le Havre	1901	1 Apr 1903	May/Sep 1903	30.90 knots
M'31 *Bombarde*	FC Méditerranée, Le Havre	6 Dec 1901	26 Jun 1903	Aug/Nov 1903	30.48 knots

Characteristics (*Pistolet*)

Displacement:	
normal	307.21 tonnes
full load	357.16 tonnes
Dimensions:	
Length	56.30m pp/wl, 56.58m oa
Beam	5.99m wl
Draught	2.85m aft; 3.20m incl props
Horsepower/speed	6,300ihp/28 knots

Remarks: Modified *Durandal* with slightly increased beam. Boilers with higher steam pressure; designed for 28kts and generally exceeded. Turned out slightly overweight and mainmast suppressed.

the near-capsize of the destroyer *Pique* in good weather due to poor stability in March, the design was again reviewed (24 June 1902) and the *Conseil* requested modifications to the hull as follows:

- centre of gravity (CG) to be as low as possible
- metacentric height (GM) to be 0.6m as *Arquebuse*
- hull to be enlarged amidships but reduced aft
- raised bridge to be compatible with the stability recommendations
- turning circle to be improved – it was currently 400m at 13 knots, 500m at 18 knots – by reducing the after extension of the axial keel which was a feature of the earlier boats and by fitting of bilge keels.

The contract for *Claymore* was signed with Augustin-Normand on 2 September 1903. Rochefort was still to lay down five boats of the *Arquebuse* class ordered on 12 May 1902 (M'32/33) and 5 Aug 1903 (M'34/35/36), and it was decided that these too would be of the *Claymore* type. M'38/39 would subsequently be ordered from Rochefort on 5 July 1904 and M'40/41[4] on 5 May 1905. On the same date M'44/45/46 were ordered from Toulon Naval Dockyard. The latter would share with Rochefort all destroyer orders not placed with private shipyards from this date to 1914. However, lack of experience with this type of craft meant that the Toulon boats were the least successful of the series.

The boats of the *Claymore* class had a very different hull form from the earlier types with a broader upper hull (see Section drawing). The scuttles were at the top of the vertical section before the hull turned sharply inwards beneath the walking deck. The fore part was raised for better sea-keeping and these boats could otherwise be distinguished by their high bridge with the seamen's galley beneath (see GA Plans). The funnels were raised so that the exhaust gases cleared the bridge, and the forward funnel was closer to the foremast; the funnels of *Tromblon* were slightly tapered towards the top. The two main anchors were uprated to 450kg to match the increase in hull size and displacement.

These boats were the first of their type to mount the

Table 7: *Claymore* Class

	Builder	Laid down	Launch	Trials	Speed (trials)
1903 Budget					
M'32 *Stylet*	Arsenal de Rochefort	21 Mar 1904	18 May 1905	Jan/Apr 1907	27.12 knots
M'33 *Tromblon*	Arsenal de Rochefort	8 July 1904	17 Jun 1905	Dec 06/Apr 07	28.48 knots
M'34 *Pierrier*	Arsenal de Rochefort	6 Oct 1904	28 Feb 1907	Jul/Nov 1908	27.12 knots
M'35 *Obusier*	Arsenal de Rochefort	10 May 1904	9 Mar 1906	Jun/Sep 1907	28.30 knots
M'36 *Mortier*	Arsenal de Rochefort	12 Sep 1904	23 Mar 1906	Oct 07/Jan 08	29.50 knots
M'37 *Claymore*	Augustin-Normand, Le Havre	1904	14 Mar 1906	Apr/May 1906	30.35 knots
M'38 *Carquois*	Arsenal de Rochefort	10 Jul 1905	26 Jun 1907	Apr/Aug 1908	30.10 knots
M'39 *Trident*	Arsenal de Rochefort	10 Jul 1905	5 Dec 1907	Oct 08/Jan 09	28.15 knots
M'40 *Fleuret*	Arsenal de Rochefort	1906	14 Dec 1906	Aug 07/Mar 08	28.98 knots
M'41 *Coutelas*	Arsenal de Rochefort	3 Feb 1906	12 Jan 1907	Sep 07/Mar 08	29.57 knots
M44 *Cognée*	Arsenal de Toulon	1 May 1905	26 Nov 1907	Aug/Dec 1908	27.04 knots
M45 *Hache*	Arsenal de Toulon	1 Aug 1906	15 Feb 1908	Dec 08/Jun 09	28.48 knots
M46 *Massue*	Arsenal de Toulon	1 Nov 1906	19 Sep 1908	Feb/Jun 1909	28.47 knots

Characteristics (*Obusier*)

Displacement:	337.36 tonnes
Dimensions:	
Length	58.00m pp/wl; 58.28m oa
Beam	6.20m wl
Draught	2.86m aft; 3.39m incl props
Propulsion:	
Boilers	Two Normand small-tube boilers, 18kg/cm^2
Engines	Two-shaft 3/4-cylinder VTE
Horsepower	6,800ihp
Speed	28 knots (designed)
Coal	30 tonnes normal
Endurance	2,400nm at 10 knots
Armament:	
Guns	1 x 65mm/50 Mle 1902 QF 6 x 47mm/50 Mle 1902 QF
Torpedoes	2 single tubes for 450mm torpedoes 4 torpedoes Mle 1892/1906 (2 reserve)
Boats & Complement:	[as *Durandal*]

The destroyer *Sarbacane* (*Arquebuse* class) undergoes an inspection; note the white dress uniforms of the officers. There is a 381mm Mle 1887 torpedo loaded into the after tube; these early models of torpedo had an ogival 'front end' to minimise resistance when running through the water. Until 1908–10 French destroyers had the *toile mouillée claire* ('light wet canvas') livery that was employed for the upperworks of the battleships of the period. They were subsequently repainted in a uniform blue-grey. (Philippe Caresse collection)

65mm Mle 1902, a 50-calibre gun firing the same ammunition as the Mle 1891 but with superior range and penetration; muzzle velocity was 870m/s (vs 715m/s for the older gun). The 47mm/40 Mle 1885 QF was likewise superseded by a 50-calibre gun of the same Mle 1902 series with improved penetration qualities (see Table 8). Following an instruction from Minister of Marine Pelletan dated 2 March 1904, the 381mm torpedo tubes were uprated to 450mm. The torpedoes carried were Mle 1892 for M'32–37 and Mle 1906 for the later boats of the class. Mle 1892 was a Whitehead torpedo that was also built under licence at Toulon; it had a length of 5.05m and weighed 530kg with a 75kg warhead; range was 400 metres at 31 knots or 800 metres at 27.5 knots. Mle 1906, likewise built at the Whitehead factory and by Toulon, was slightly longer and significantly heavier:

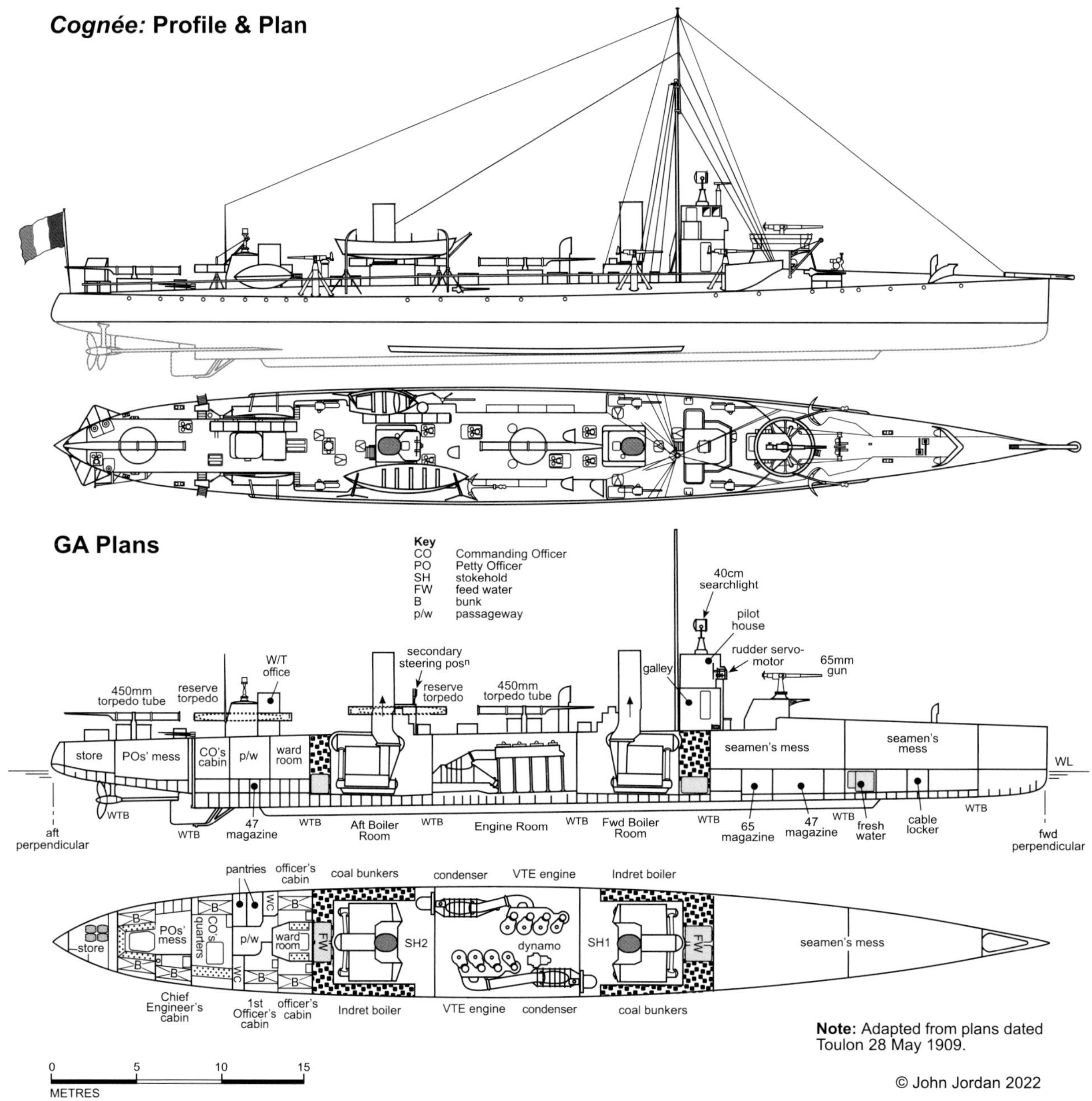

Cognée was one of three boats of the *Claymore* class allocated to Toulon Naval Dockyard. This class and the ten boats of the *Branlebas* class had more modern weaponry than the earlier boats of the series, and could be distinguished from them by the raised pilot house, which was located atop the galley (moved from abaft the second funnel). This had an adverse impact on stability, and the mainmast was suppressed in favour of a short 'L'-shaped pole mast to carry the W/T aerials. The plans of the last two boats built at Rochefort (*Fleuret* and *Coutelas*) and the three boats assigned to Toulon show engines with the LP cylinder divided into two, the two functioning as a single cylinder; the result was three cylinders of about the same diameter plus a smaller high pressure cylinder. The engines were built by FCM at their Marseille and Le Havre factories. The standard three-cylinder model was built by Schneider and Indret.

5.07m with a weight of 640kg which included a 108kg warhead; range was 600 metres at 36.5 knots or 2,000 metres at 24 knots.

The decision to build all but the lead-ship, *Claymore*, from the *arsenaux* backfired, and trials were not particularly successful: those of *Stylet* and *Tromblon* took more than a year and their engines were downrated to 6,000cv. The Normand-built *Claymore* was the fastest, achieving 30.35 knots with 7,200cv at 311rpm. The boilers for the units built at Toulon were a Du Temple model from the state establishment at Indret, and the engines were from Schneider – a regular sub-contractor for the state-owned dockyards. The machinery trials of the Rochefort-built *Pierrier* were prolonged due to the adoption of a prototype 4-cylinder 6,800cv engine with forced lubrication from Delaunay-Belleville, the twin aims being to secure reduced manning – always an issue with these small ships – and more efficient lubrication. The engines were delivered 25 months late. Serious oil leaks were then experienced during trials. *Pierrier* was accepted into service only in November 1910, six years after the keel had been laid. There were new leaks during the guarantee period, resulting in further extensive modifications. The engines were finally judged insufficiently robust for a destroyer, especially at high speeds, and in 1911 it was decided that they should revert to classical lubrication.[5]

On entry into service the *Claymore* class was found to have excessive topweight – these boats emerged from the building yards 20 tonnes heavier than *Arquebuse*. The bridge wings were removed and some navigation equipment, including the rudder servo-motor and the compass, was lowered to the deck below.

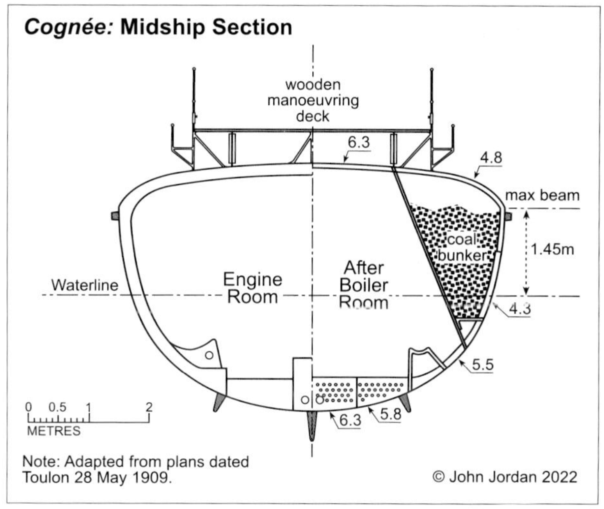

Branlebas 1904

These were the last boats of the *300-tonnes* series to be completed for the French Navy. Built to Augustin-Normand plans, they had a slightly modified hull with

Mousqueton moored at Toulon prior to the Great War. The mainmast, which was landed to improve stability, has been replaced by spreaders fore and aft for the W/T aerials. Note the prominent reload locker for the after 381mm torpedo tube. (Philippe Caresse collection)

Table 8: **French Destroyer Guns [*Claymore* Onwards]**

	65/50 Mle 1891	47/40 Mle 1885
Total weight:	594kg (loaded)	338kg (loaded)
Total length:	3,450mm	2,538mm
Bore length:	3,250mm	2,530mm
Chamber length:	674mm	336mm
Chamber volume:	2.25dm^3	0.99dm^3
Breech pressure:	2,800kg/cm^2	2,200kg/cm^2
Rifling:	26 grooves at 5°	20 grooves at 6°
Shell:	4.17kg steel	2.00kg steel
Bursting charge:	0.10kg *mélinite*	0.025kg *mélinite*
Propellant:	1.43kg BM5 (cartridge)	0.46kg BM3 (cartridge)
Muzzle velocity (MV):	870m/s	690m/s
Remaining V 2000m:	490m/s	356m/s
Drift at 2000m:	0.6m	1.2m
Mean range error:	35.4m	20.9m
Mean line error:	1.2m	1.21m

Source: *Renseignements sur les Bouches à Feu de l'Artillerie Navale Tous Calibres et Modèles.*

greater freeboard (2.9m vs 2.6m). The curve inboard of the upper hull began 1.8m above the waterline compared to 1.5m for the *Claymore* type. Beam was increased in compensation to improve stability – GM was 0.64m vs 0.6m for *Claymore* in normal condition, 0.57m vs 0.54m in light condition.

This class reverted to two separate in-line engine rooms, the overall length of which was reduced by adopting a staggered transverse bulkhead between them (see drawing), as in the earlier *Pique* class. The machinery compartments were protected by light 20mm plating, which cost 12 tonnes. Compartmentation compared to *Claymore* was as follows:

	Claymore	*Branlebas*
Length BR	15m	14.5m
Length ER	8.5m	10m
Seamen's mess	15m	14.5m
Officer accom^{n}	7.15m	6.75m
POs' mess	4.2m	4.8m

Claymore, which gave her name to a class of thirteen slightly-enlarged boats ordered between 1903 and 1905. They could be distinguished externally from the earlier boats by their high bridge, which was located atop the galley – the latter being moved from abaft the second funnel. More modern weaponry, in the form of new models of 65mm and 47mm gun and 450mm torpedoes, added topweight, and stability was marginal despite the suppression of the mainmast. (Philippe Caresse collection)

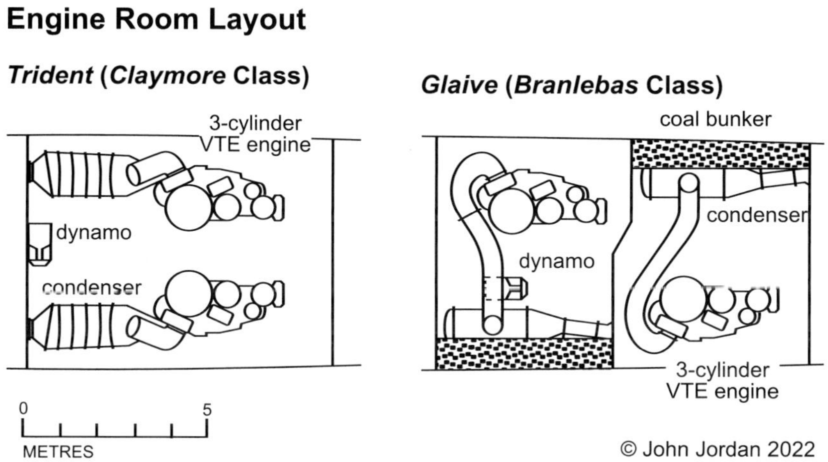

Schematic showing the engine room layouts of *Trident* (early *Claymore* class) and *Glaive* (*Branlebas* class). Both have the standard 3-cylinder reciprocating engine, but whereas these are arranged side-by-side in *Trident* with the main condensers in line abaft them, in *Glaive* they are *en echelon* in two separate compartments, as in the early boats of the *300-tonnes* series, with the lateral coal bunkers reinstated. Note, however, the staggered transverse bulkhead separating the two engine rooms, adopted in order to accommodate the larger, more powerful 6,800ihp engines without a significant increase in hull length.

There were to have been eight boats in the class, but when the first two were ordered from Normand on 4 November 1905 it was realised that two units of the *Claymore* class assigned to Rochefort (M'42–43) had yet to be laid down, and these were therefore built to the new plans. Three further pairs of boats were contracted from private shipbuilders on 8 November.

They were designed for 28 knots, but given the increase in displacement resulting from the larger hull, a ministerial directive of 8 October 1907 revised this downwards to 27.5 knots. *Branlebas* attained 28.76 knots with 6,700cv at 303rpm, and *Fanfare* 28.91 knots with 6,560cv at 298rpm; the fastest was the Penhoët-built *Sape*, which made 29.82 knots at 300rpm.

The *Branlebas* class had the same raised bridge with the searchlight projector atop the pilot house. The S/L projector was a larger model: 60cm vs the 40cm of earlier ships. The raised forecastle gave them good sea-keeping qualities, although the increase in topweight resulted in a prominent roll with a beam sea. The class had an axial keel 35m long and 0.5m deep, as in the earlier boats, but no bilge keels. With the wind three quarters on the bow they could easily sustain 12 knots in heavy seas even during wartime (as noted in a 1915 report); however by this time displacement was around 400 tonnes, with adverse consequences for stability.

The *300-tonnes* in the Great War

Two of the boats were early losses. On the night of 10/11 August 1900 *Framée* (*Pique* class) was in collision with the battleship *Brennus* off Cape St Vincent and promptly

Table 9: *Branlebas* Class

	Builder	Laid down	Launch	Trials	Speed (trials)
1904 Budget					
M'42 *Glaive*	Arsenal de Rochefort	3 Feb 1908	10 Sep 1908	Apr/Sep 1910	27.90 knots
M'43 *Poignard*	Arsenal de Rochefort	1908	3 Jul 1909	Apr/Nov 1910	28.65 knots
M47 *Sabretache*	De la Brosse & Fouché	Jun 1906	5 Feb 1908	Apr/Sep 1908	29.65 knots
M48 *Oriflamme*	De la Brosse & Fouché	Jun 1906	4 Apr 1908	May/Sep 1908	29.49 knots
M49 *Etendard*	Dyle & Bacalan, Bordeaux	4 Dec 1905	20 Mar 1908	May 08/Feb 09	28.53 knots
M50 *Fanion*	Dyle & Bacalan, Bordeaux	Dec 1906	4 May 1908	Jul 08/Feb 09	29.06 knots
M51 *Sape*	Penhoët, Rouen	25 May 1906	23 Sep 1907	Mar/Nov 1908	29.82 knots
M52 *Gabion*	Penhoët, Rouen	2 Jul 1906	21 Dec 1907	Feb/Nov 1908	29.77 knots
M53 *Branlebas*	Augustin-Normand, Le Havre	Nov 1905	8 Oct 1907	Nov 07/Jul 08	28.76 knots
M54 *Fanfare*	Augustin-Normand, Le Havre	Nov 1905	19 Dec 1907	Jan/Sep 1908	28.91 knots

Characteristics (*Glaive*)

Displacement:	345.60 tonnes
Dimensions:	
Length	58.00m pp/wl, 60.12m oa
Beam	6.28m wl
Draught	2.87m aft; 3.39 incl props

Remarks: Built to Normand plans. Similar to *Claymore* with raised pilot house, but slightly heavier and with greater freeboard. Topweight problems. Designed speed reduced to 27.5 knots due to increase in weight.

sank. And on 4 February 1902 *Espingole* (*Fauconneau* class) struck a rock at Cape Lardier and sank in the Hyères roadstead in 29 metres of water; salvage attempts failed (see the article by Philippe Caresse in *Warship 2013*). However, the remaining 53 boats remained in service when war was declared in August 1914. Of these the 23 boats of the *Claymore* and *Branlebas* classes, which had modern guns and mounted the 450mm torpedo, were still regarded as first-line units. Although superseded by larger, more modern destroyers of the *450-tonnes* and *800-tonnes* types, most of these boats were little more than six years old, and two had been completed as recently as 1910. The older *300-tonnes* served in a variety of roles, including as leaders for the submarine squadrons.

The standard unit of organisation for French 'squadron' destroyers[6] was the *escadrille* (= flotilla), which comprised two divisions each of three boats. Submarines were likewise assigned to particular *escadrilles*, and these generally comprised two or three three-boat divisions, each of which had a small destroyer as leader. The disposition and composition of the destroyer and submarine flotillas in August 1914 is detailed in Table 10.

The 23 units of the *Claymore* and *Branlebas* classes were distributed equally between the *Armée Navale*, based at Toulon in the Mediterranean, and what was now the *2e Escadre Légère*, a force of cruisers and flotilla craft based at Cherbourg and intended to operate in the Channel as far as the Straits of Dover. Because there were only 23 boats of this type, the *4e Escadrille* had the 450-tonne *Hussard* as leader to ensure uniformity of machinery and armament. The *2e Escadre Légère*, which as a second-line force had been allocated only two of the latest destroyers (the 800-tonne *Capitaine Mehl* and *Francis Garnier*), had three *escadrilles* of older 300-tonne boats, all but one of which (*Rapière*) belonged to the *Arquebuse* class. There were also two 300-tonne boats that had been converted as minelayers. The leaders of the submarine flotillas and divisions were all of the earlier types, armed with the older-model guns and mounting 381mm torpedoes.

The *Armée Navale* would be tasked first with the escort of convoys bringing troops from North Africa to metropolitan France, then with the blockading of the Austro-Hungarian Fleet in the Adriatic. The principal mission of the *2e Escadre Légère* during the first months of the war would be to ensure the safe transit of the British Expeditionary Force from the south coast of England to northern France, while assisting the Royal Navy in sealing off the Straits of Dover against any attempted incursion by German surface forces or submarines. Three boats of the *Arquebuse* class operated out of Saigon.

The *300-tonnes* were very active during the Great War. They still constituted two thirds of the fleet torpedo boats: 29 in the North, 21 in the Mediterranean and three in the Far East. However, they had gained some 50 tonnes in weight since their entry into service and were no longer capable of the speeds attained on trials; most could sustain only 22/23 knots in formation, although some were still capable of 25 knots. They proved eminently well-suited to convoy work in the

A port quarter view of *Gabion* taken between 1912 and 1914. The multiple aerial spreaders have been replaced by a single spar forward and a short 'L'-shaped pole mast aft. *Gabion* has a broad white band on the first funnel, marking her out as belonging to the 1st Division of the 2nd *Escadrille* based at Cherbourg, and the recognition letters 'GB' are painted in black on either side of the bow. (NHHC, NH-64468)

Table 10: **French Destroyer Formations 1914**

Armée Navale: Toulon	
Flotille des torpilleurs d'escadre: *Bouclier* (CA Lejay)	
1re escadrille [800t]:	*Casque, Dague, Boutefeu, Fourche, Faulx, Mangini*
2e escadrille [450t]:	*Carabinier, Lansquenet, Asp Herber, EV Henry, Mameluck, Spahi*
3e escadrille [450t]:	*Fantassin, Janissaire*[2], *Cavalier, Tirailleur, Voltigeur*[2], *Chasseur*[1]
4e escadrille [300t]:	*Hussard, Sape, Pierrier, Mortier, Hache*[2], *Massue*[2]
5e escadrille [300t]:	*Poignard, Fanfare, Sabretache, Coutelas, Trident, Cognée*
6e escadrille [800t]:	*Cdt Rivière, Cimeterre, Magon, Bisson*[1], *Renaudin*[1], *Cdt Bory*[3], *Protet*[3]
Flotille des sous-marins:	*Dehorter* (CV Moullé)
1re escadrille:	[300t] *Arbalète, Hallebarde* [S/M] *Monge, Gay-Lussac, Ampère, Papin, Messidor, Cugnot, Fresnel*
2e escadrille (Bizerte):	[300t] *Mousqueton, Sarbacane, Arc* [S/M] *Joule, Faraday, Coulomb, Bemoulli, Arago, Le Verrier, Curie, Circé*
2e Escadre Légère: Cherbourg	
Flotille des torpilleurs d'escadre: *Dunois* (CA ??)	
Unattached [800t]:	*Capitaine Mehl, Francis Garnier*
1re escadrille [300t]:	*Obusier* (fl), *Branlebas, Oriflamme*; *Tromblon, Etendard, Carquois*
2e escadrille [300t]:	*Glaive* (fl), *Gabion, Fanion*; *Stylet, Fleuret, Claymore*
3e escadrille [300t]:	*Catapulte* (fl), *Rapière, Epieu*; *Bélier, Bombarde, Arquebuse*
Minelayers [300t]:	*Baliste, Flamberge*
Refit/Reserve	
[300t]	*Sagaïe, Harpon, Dard, Carabine, Pertuisane, Pique*
Flotille des sous-marins:	
1re escadrille:	[300t] *Francisque, Fauconneau, Sabre* [S/M] *Prairial, Pluviôse, Floréal, Fructidor, Watt, Germinal, Berthelot, Ventôse, Giffard*
2e escadrille (Calais):	[300t] *Escopette, Durandal* [S/M] *Mariotte, Brumaire, Frimaire, Volta, Nivôse, Euler*
3e escadrille:	[300t] *Javeline, Epée*[1] [S/M] *Foucault, Montgolfier, Franklin, Am Bourgeois*[1], *Archimède*[1], *Thermidor*[1], *Gustave Zédé*[1]
Channel Station	
[300t]	*Yatagan*
Saigon	
[300t]	*Pistolet, Mousquet, Fronde*

Notes:
1 Under repair
2 Minelayers
3 Undergoing trials/work-up

Atlantic, at the Dardanelles, off the Syrian coast, and off Dunkirk and Flanders.

At the Dardanelles the 5th Flotilla (*Poignard, Fanfare, Sabretache, Cognée, Coutelas*) was attached to Guépratte's division and bombarded Turkish positions during the spring of 1915. These boats were later joined by *Trident, Mortier, Arc, Massue, Arbalète, Sarbacan, Pique* and *Sape* for the landings that took place in April.

Off Dunkirk and Flanders German aircraft posed a threat, and some boats had the two mountings for the after 47mm guns modified, raising maximum elevation from +20° to +83°. In 1917 many also had a single 8mm MG mounted on the after torpedo tube. When the tube was later disembarked (see below), the machine gun was relocated atop the after steering position. By late 1918 many boats carried two MG capable of being mounted on two of four posts.

Several of the boats had the 65mm replaced by a 75mm G (G=*Guerre* – an Army artillery piece) to enable them to engage submarines on the surface more effectively. And during the autumn of 1917 some had their after torpedo tube replaced by a Thornycroft anti-submarine mortar, with two rails on the quarterdeck for six or twelve 75kg depth charges. The after torpedo tube was initially paired with the midships torpedo tube, but was replaced in late 1918 when the mortar was disembarked.

The anti-submarine war also resulted in changes to the composition of formations. The 3rd Flotilla based at Cherbourg was transferred to the Mediterranean in the late summer of 1915; it now comprised *Catapulte, Rapière, Epieu, Bélier, Arquebuse* and *Sagaïe* (replacing *Bombarde*), and was renumbered 9th Flotilla. On 20 September 1915 the *Flotille mixte anti-sous-marine de l'Armée navale* was formed with these six boats plus ten trawlers (*chalutiers-patrouilleurs*). In October the flotilla was reinforced by 21 more trawlers. The flotilla was based on Bizerte – the focal point of the trans-Mediterranean convoy routes – and divided into two formations: one for the Western, the other for the Eastern Mediterranean. The 9th Flotilla was then attached to the *Patrouilleurs d'Algérie/Tunisie*, a grouping of 57 units which included submarines and the new 300/350-tonne A/S gunboats. The *300-tonnes* still constituted half of the fleet destroyer strength, and they were faster than the other boats in the A/S formations.

Considering the level of activity of these boats, war losses were comparatively light. Two were lost from navigation accidents: *Yatagan* collided with the British cargo ship *Teviot* off Dieppe on 3 December 1916, and *Catapulte* sank following a similar collision with the

The destroyers *Branlebas* and *Fanion* moored alongside, probably at Cherbourg, and dressed overall. The structure with the open doorway forward of the torpedo tube, and the cylindrical coaming with the canvas roof aft are the *descentes* for the officers' and petty officers' quarters respectively. The photograph gives an excellent view of the elevated wooden grating that served as a 'walking deck'. (Philippe Caresse collection)

British steamer *Warrimoo* off Bizerte on 18 May 1918. *Mousquet* was sunk off Penang (Malaya) on 28 October 1914 while pursuing the German raiding cruiser *Emden*; *Branlebas* was lost after striking a German mine off Nieuwpoort, Flanders, on 30 September 1915; and *Etendard* was sunk during an engagement with German destroyers tasked with bombarding Dunkirk on the night of 25 April 1917. In an earlier action, on 22/23 August 1915 *Oriflamme*, in company with *Branlebas*, sank the German coastal torpedo boat *A-15* off Ostend. During the 13-minute engagement *Oriflamme* fired off four rounds of 65mm and 140 of 47mm, sustaining only a single hit in return that passed through her whaler.

Most of the surviving *300-tonnes*, worn out by wartime service and with their maximum speed now reduced to 20 knots, were stricken during 1919–21. Some continued to serve with the training schools; the last two (*Carquois* and *Trident*) were stricken in 1930 and 1931 respectively.

Acknowledgement:

The author wishes to thank Philippe Caresse for his assistance with photos from his extensive collection.

Principal sources:

Henri Le Masson, *Histoire du Torpilleur en France*, Académie de la Marine (Paris, 1963).

Stephen S Roberts, *French Warships in the Age of Steam 1859–1914*, Seaforth Publishing (Barnsley, 1921).

Centre des Archives de l'Armement, Châtellerault: Plans.

Endnotes:

1. Torpedo Boats Nos 65–74, built by Augustin-Normand of Le Havre during the mid-1880s, were 33 metres long; they had a displacement of 50 tonnes and a designed speed of 20 knots; armament was two 356mm (14in) bow-launched torpedoes.
2. A similar layout was adopted by Bertin for his contemporary armoured cruiser designs.
3. In later boats it proved possible to expand the basic design further to obtain designed ratings of 6,300cv (*Arquebuse* class) and even 6,800cv (*Claymore* class), with a maximum of 7,200cv being attained by the lead boat of the latter class on trials.
4. The M' (formerly D') was originally used to distinguish the TE *Hallebarde* from her sister *Durandal*, initially typed *aviso-torpilleur* (D, then M); the system was retained until 1905, but for the 1906 budget the designation was changed to MM, then from 1907 to simply M.
5. Forced lubrication would be perfected only with the later boats of the 450-tonne *Spahi* type after further teething troubles.
6. From 14 October 1913 the *contre-torpilleurs d'escadre* were reclassified as *torpilleurs d'escadre* (TE). The latter term would remain in use throughout the interwar period, while the *contre-torpilleur* designation would be revived for the large, fast 'scouts' built under the 1922 Programme.

ITALIAN DESIGNS FOR MONITORS AND BATTLECRUISERS DURING THE INTERWAR YEARS

Michele Cosentino looks at some of the lesser-known interwar projects for capital ships proposed for the *Regia Marina*.

In 1920, then-Commander Romeo Bernotti wrote a 553-page book entitled 'Maritime Power during the Great War'.[1] At the time, Bernotti was a lecturer at the Italian Navy's Naval Academy and delivered lectures to lieutenants attending senior courses.

Romeo Bernotti, photographed here in the uniform of a Rear Admiral, influenced Italian naval thinking during the interwar years. However, his support for aircraft carriers and embarked aviation was ignored. (Italian Navy Historical Office, INHO)

The conflict that had just ended had upset old certainties in the field of naval strategy and tactics and created contradictions, making it difficult to determine new ways of using fleets and naval bases and developing new naval weapons. Since 1917 this situation had inspired Bernotti to reflect on these problems and document them in the above-mentioned volume.

Showing remarkable poise, Bernotti emphasised the role played by battleships in implementing a total blockade of supplies for the Central Powers, but advocated the construction of battleships that were faster and better protected than the battlecruisers of the day. He warned against believing that submarine warfare would be decisive in future conflicts, noting that the submarine threat could be repulsed by escorted convoys, while unrestricted submarine warfare could anger neutral countries and prompt them to intervene. Additionally, he understood the value of embarking aircraft in warships and exploiting their potential in future naval combats. As for the *Regia Marina*, Bernotti called for the construction of a large new naval base in the southwest corner of Sardinia as the most suitable location for protecting Italy's maritime traffic. From a general political point of view, he pointed out that Italy would still seek alliances with nations able to play decisive roles in maritime warfare, including guaranteeing communication routes and supply lines.

The book heralded the busiest and most significant years for Bernotti both in his naval career and as a publicist, years in which his strategic thinking reached the highest level of maturity and originality. He gave further lectures to the Italian Naval Academy's midshipmen, and was ordered to plan a new Naval War College for the *Regia Marina* that aimed to improve Italian naval officers' grasp of doctrine and technology.

In his book Bernotti analysed the employment of British and German battleships and battlecruisers at Dogger Bank and Jutland, the pursuit of *Goeben* and *Breslau*, the battles of Coronel and the Falklands, and the

The pre-dreadnought *Regina Elena* underway in the Gulf of La Spezia during the first decade of the 20th century. (Author's Collection)

wartime naval programmes of the Triple Entente and the Triple Alliance. In the last chapter of the book, Bernotti addressed the Italian maritime 'problem', the word referring to an Italian geostrategic scenario calling for a new military posture to be adopted by the Italian military authorities. Bernotti emphasised that the problem could not be resolved by a strategy based only on passive coastal defence, but that it was necessary to adapt the overall construct of the Italian fleet to include the development of new active tactics and procedures focused on a dynamic combination of surface, subsurface and air assets. Bernotti did not call for the construction of new battleships because he was aware that funding constraints did not allow Italy to do so. However, he stressed the need to scrap those warships that did not add any military value to the Italian fleet and to convert the four battleships of the *Regina Elena* class[2] into battlecruisers, which Bernotti judged a 'vital necessity' for the *Regia Marina*.

The pre-dreadnought *Emanuele Filiberto* moored at Fiume (now Croatia) after the Great War. (Courtesy of Maurizio Brescia)

It is worth noting that several aspects of Bernotti's books were taken into account in the report related to the 1920 budget for the *Regia Marina* which the then-Finance Minister, Carlo Schanzer, presented to the Italian parliament for approval. They included the accompanying table illustrating the main types of warship then in commission or being built in the United Kingdom, the United States, France, Japan and Italy.[3]

Although the proposed conversion did not take place and the four battleships in question were decommissioned in 1923–26, Bernotti's thoughts on battlecruisers were taken on board by numerous Italian naval officers and engineers, kick-starting a period of concepts and designs that spanned the interwar period up to a few years before the outbreak of the Second World War. In short, Bernotti influenced Italian naval thinking for at least fifteen years, leading to several studies and designs for monitors and battlecruisers.

Nabor Soliani

Nabor Soliani served as an officer of the *Regia Marina*'s Naval Engineering Corps from 1874 to 1897 and retired as a General of that Corps.[4] After three years spent in the Italian Parliament, Soliani joined the Genoa-based ship-

Table 1: **Bernotti's Table of Naval Strength 1920**

	United Kingdom		United States		France		Italy		Japan	
	C	B	C	B	C	B	C	B	C	B
Dreadnought	33	–	17	5	7	5	5	–	5	2
Semi-dreadnought	2	–	–	–	5	–	–	–	2	–
Pre-dreadnought	6	–	11	–	5	–	–	–	6	–
Battlecruisers	9	1	–	–	–	–	–	–	4	–
Armoured cruisers	19	–	–	–	13	–	7	–	10	–

C = Commissioned
B = Building

building company Giovanni Ansaldo & Co and worked there from 1900 to 1925. During this period, Soliani supervised the reconstruction of *Mesudiye*, a former Turkish sail-rigged armoured frigate that Ansaldo converted into a pre-dreadnought battleship 1898–1903. In the early 1920s, Soliani studied an alternative to battleships and battlecruisers, bearing in mind the potential threats to them that had emerged from the Great War, namely submarines and aircraft.

The first aircraft carriers then in service were still in their operational infancy. Thus it appeared that the performances achieved especially by the German U-boats during the Great War could make the submarine the most influential candidate to replace the battleship as the new capital ship of the fleets. In this context, it is worth noting that during the Great War the Royal Navy commissioned a number of submarines of the 'K' and 'M' classes to cope with the German submarine cruisers. Armed only with a single 305mm gun, the M-class boats demonstrated the limits of such a solution, the others being the submarine's low underwater speed and manoeuvrability. Bearing in mind these aspects, in 1920 Soliani proposed a new type of 'innovative' warship: a semi-submersible 'battle monitor' armed with large-calibre guns. In an article published in June 1921 in *Rivista Marittima*, the official monthly magazine of the *Regia Marina*, Soliani addressed this new concept firstly by criticising both the British battlecruiser *Hood* and the battlecruisers of the *Lexington* class currently being built for the US Navy.[5] Soliani pointed out that these ships were costly and poorly protected, especially against underwater threats such as the mine and the torpedo. He therefore proposed that an ideal fleet of the future would include a class of semi-submersible, well-protected but relatively slow 'battle monitors' and a class of fast, well-armed but poorly protected battlecruisers.[6]

Having studied their performance during the Great War, Soliani was aware of the inherent limitations of monitors and devised a proposal specifically aimed at satisfying the *Regia Marina*'s main requirement: a cheap warship to be used for defensive purposes, well protected and well armed. The principal innovation of Soliani's proposal was the concept of a partial submergibility for this monitor, a condition to be achieved by flooding specific ballast tanks. This would allow the monitor to submerge up to the weather deck, leaving only the gun turrets, a boat deck, a conning tower (termed 'blockhaus'), a large funnel and a short forecastle above the surface.

After serving in the *Regia Marina*, Nabor Soliani joined the Genoa-based shipbuilding company Giovanni Ansaldo & Co, and proposed a number of designs for monitors and battlecruisers. (www.academiabrig.org)

Displacing around 20,000 tonnes, the Soliani semi-submersible battle monitor was a slow warship, able to reach 18–20 knots with horsepower totalling on average 22,000–26,000hp on four shafts. Manoeuvrability was provided by three rudders. A double hull, similar to that of *Hood* and the British monitors, and tight internal compartmentation would ensure protection against shells, bombs, torpedoes and mines. Horizontal protection was to be provided by a 150mm armoured deck with

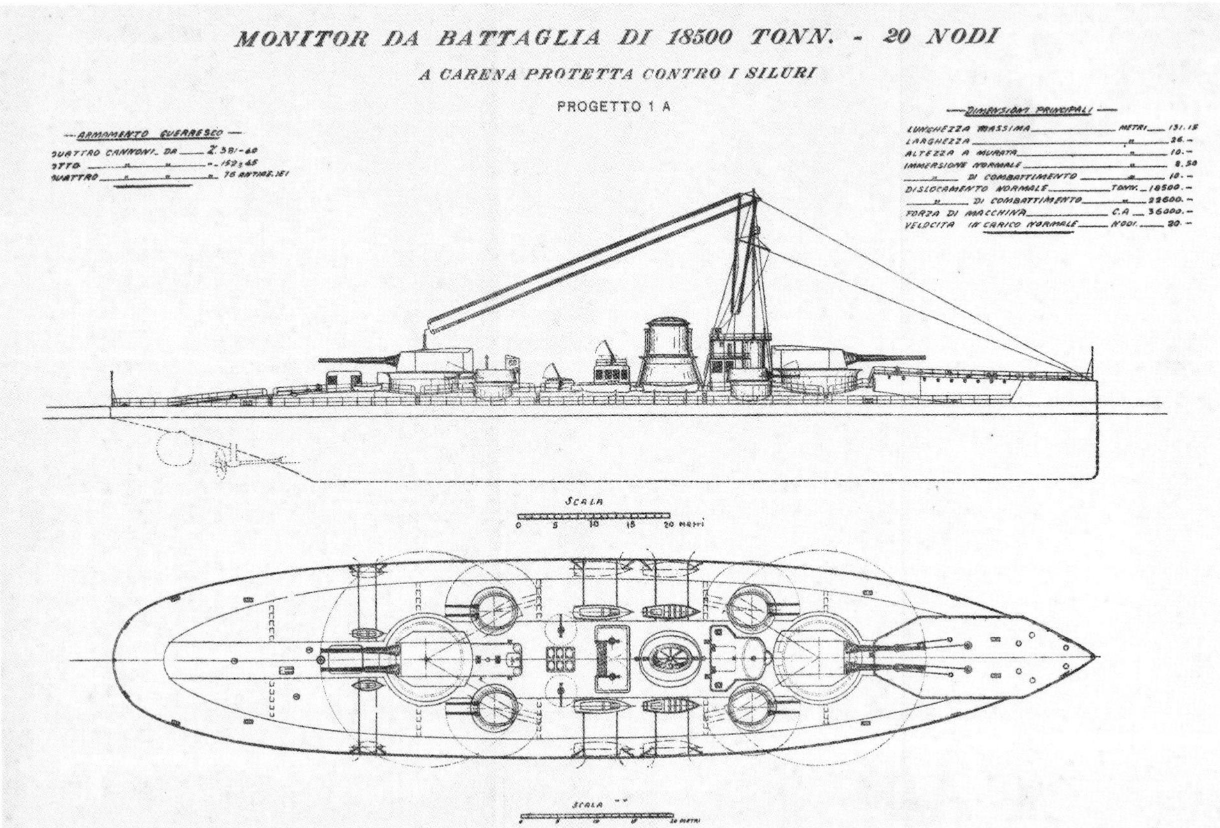

Profile and plan of the monitor variant 1A proposed by Soliani. The double waterline shows the extent of submergibility. (From the June 1921 *Rivista Marittima* article quoted in Fn 6)

Table 2: **Semi-Submersible Battle Monitors**

Characteristics	1 A	2 A	2 B	3 A	3 B
Length	130m	150m	150m	150m	150m
Beam	26m	29m	29m	29m	29m
Draught (max)	10m	10m	10m	11m	11m
Displacement (fl)	22,600t	24,500t	25,000t	26,000t	26,000t
Machinery	steam turbines	diesel engines	diesel engines	diesel engines	diesel engines
Maximum power	36,000shp	24,000bhp	24,000bhp	24,000bhp	24,000bhp
Funnels	3	1	1	1	1
Propellers	2	4	4	4	4
Maximum speed	20kts	18kts	17kts	18kts	16.5kts
Solid ballast	–	6,000t	6,000t	6,900t	7,200t
Liquid ballast	4,100t	3,000t	3,000t	3,000t	3,600t
Armour:					
deck	150mm	150mm	150mm	–	150mm
belt	–	150mm	150mm	150mm	150mm
Armament:					
381mm	4	4	6	6	6
152mm	8	–	–	–	–
120mm	–	8	8	8	–
76/100mm	4	4	4	4	2
Reserve of buoyancy:					
on surface	7,100t	5,200t	5,700t	4,500t	4,500t
submerged	3,000t	2,200t	2,200t	1,500t	900t

Note: All Soliani's proposals: no indications are provided regarding complement.

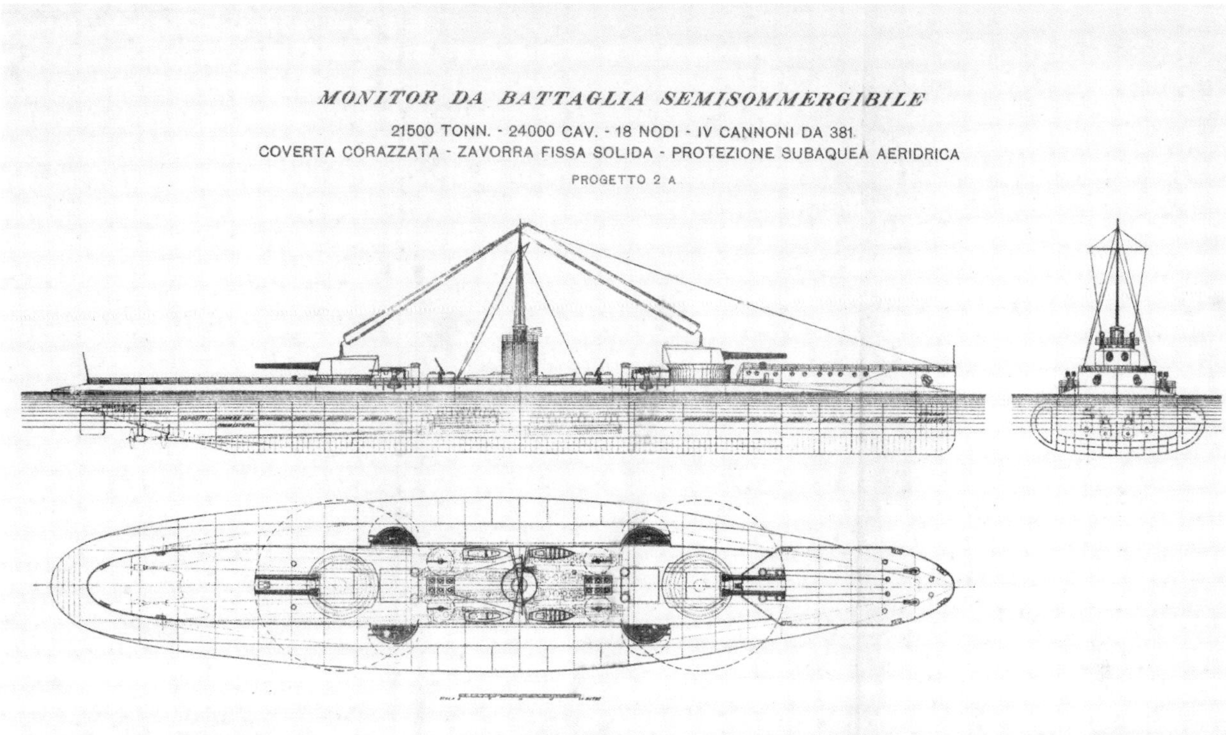

Profile and plan of the Soliani monitor variant 2A. It features an upper armoured deck, fixed solid ballast and underwater protection comprising bulges filled with water. There were to be four diesels each rated at 6,000bhp driving four shafts. (*Rivista Marittima*, June 1921)

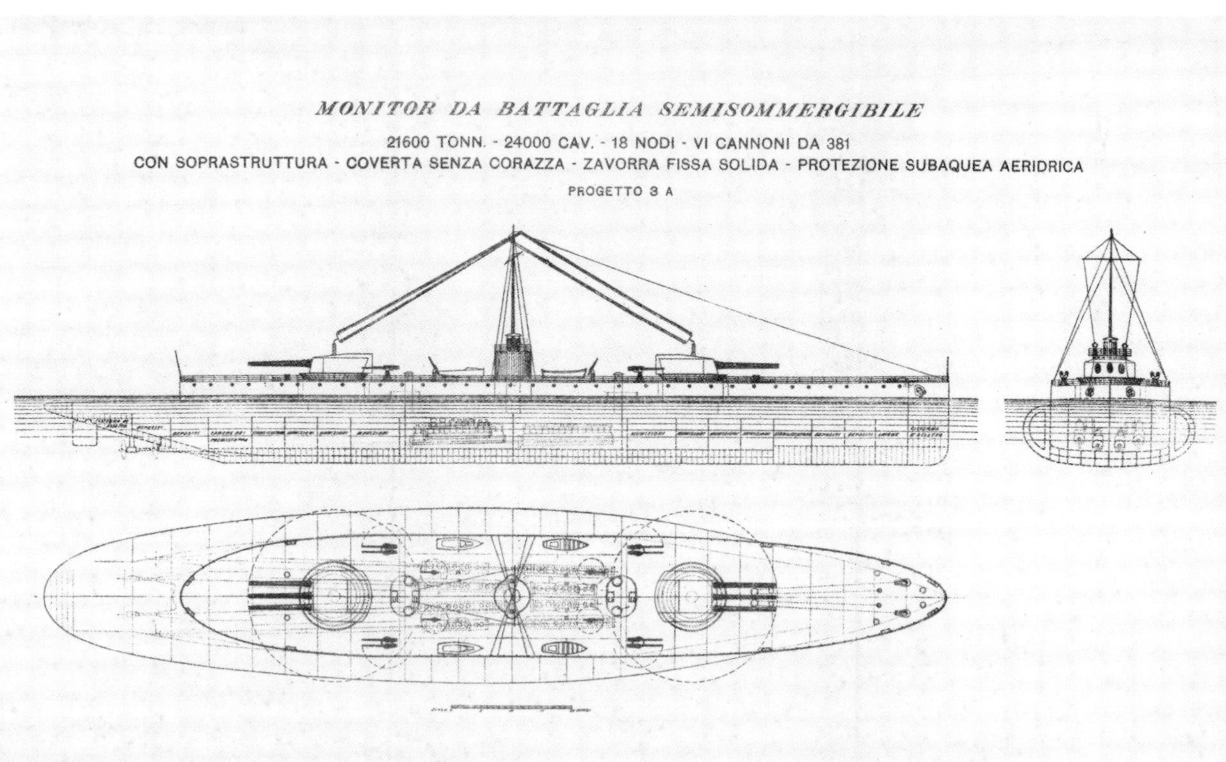

Profile and plan of the Soliani monitor variant 3A. The upper deck is not armoured, while the submergibility and propulsion systems are as in variant 2A. Main armament includes four 381mm (15in) guns in twin turrets. (*Rivista Marittima*, June 1921)

a layer of water-excluding cork beneath. Armament included two main turrets, each equipped with two or three 381mm (15in) guns, while four or six 120mm (4.7in) guns were devoted to antiaircraft protection.

Soliani put forward five different possible variants, with differences in length, beam, draught, displacement, machinery, armament and ballast configuration. The principal characteristics of these preliminary designs – identified as 1A, 2A, 2B, 3A and 3B – are listed in the accompanying table. An interesting aspect of Soliani's concept was cost. He stated that the money spent on *Hood* (about £6,000,000) would be sufficient to build three 20-knot monitors each armed with four 381mm guns or four 9,000-tonne, 35-knot cruisers (three knots more than *Hood*), each armed with four 381mm guns.

As far as battlecruisers were concerned Soliani, in the afore-mentioned article, also proposed a preliminary study for a 'very fast battlecruiser', equipped with a powerful armament but poorly protected against surface and underwater threats. Soliani proposed two versions of his battlecruiser, one displacing 9,000 tonnes with a maximum speed of 35 knots, the other displacing 12,500 tonnes and capable of 33.5 knots. For both versions, the main armament would have comprised 381mm twin turrets fore and aft, each having particularly wide arcs of fire: the secondary battery was to be located on the middle line and was to comprise either four 152mm (6in) twin turrets or six 120mm twin mountings. Armament was completed by four triple torpedo mountings. These battlecruisers had a flush-deck hull with an overall length of 175 metres, while beam was 15 metres. There were small superstructures fore and aft, each supporting a pole mast; a long, narrow one-level deckhouse ran along the middle line and supported the secondary battery and other equipment. Although other information about these designs was sketchy, machinery would probably have comprised geared turbines with steam supplied by an undisclosed number of boilers exhausting through three funnels driving four propeller shafts. The lower part of the hull was fitted with bulges.

Although superficially striking, Soliani's concept of a semi-submersible battle monitor had several features open to criticism, mostly due to the type of warship taken as a reference: the submergibility of the monitor would undoubtedly have created major difficulties in manoeuvring, let alone a very poor efficiency of guns and their crews, even with moderate seas, when the hull was partially submerged. Additionally, habitability and ventilation would be inadequate. The 120mm guns selected for antiaircraft purposes were poorly suited to the role because they were incapable of ensuring a high rate of fire, this being an indispensable capability for creating a barrier against air strikes on warships. Soliani was also aware that his proposals regarding semi-submersible battle monitors were opposed by numerous naval authorities in Europe, United States and Japan. In a meeting organised by the Royal Institute of Naval Architects on 9 April 1921 Sir Eustace d'Eyncourt[7] stated:

> A good deal has been written and talked about the surface capital ship being dead and the necessity for a submersible. But with our present knowledge it would be quite impossible to design a submersible ship which on the same displacement and cost had anything like the fighting qualities on the surface which are now possessed by *Hood*. Every ship is a compromise, and if in addition to the ordinary qualities of a battleship she is required to submerge, or even partially submerge, a very considerable percentage of weight has to be added to give this additional capability of submergence. She becomes still more of a compromise, and the added weight must detract from the fighting qualities of the ship when on the surface, so that whatever is

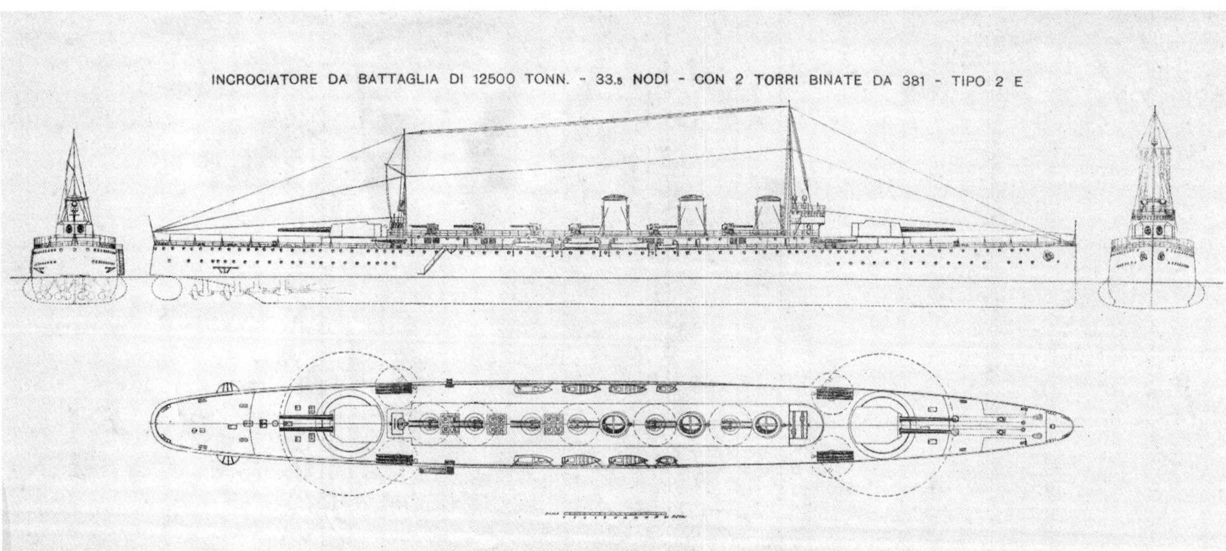

Profile, plan and bow/stern views of the battlecruiser proposed by Soliani. It features 4 shafts, while all guns are mounted on the middle line; the drawing depicts the variant with six twin 120mm mountings. Maximum speed was 33.5 knots. (*Rivista Marittima*, June 1921)

> done, other things being equal, the submersible ship must be inferior to a surface ship in an ordinary action. There are many difficulties of details in the design of a submersible battleship ... a little thought will make it clear to anybody that if naval warfare is to continue, the surface ship of the line must still hold the field as the principal fighting unit of any great navy. This view is apparently shared by other countries that are developing their navies, and both Japan and the United States are building large capital ships.

Although Soliani's proposals were probably little more than an academic exercise in the wake of the debate on the future of battleships, they were nevertheless widely publicised, including in the Journal of the American Society for Naval Engineers.[8] However, regardless of the criticisms directed at the concept of the semi-submersible battle monitor, Soliani's proposals for a fast battlecruiser did not attract the same degree of attention, suggesting that this concept remained popular within several European navies.

An Opposing View

In October 1921 *Rivista Marittima* published an article by Ferdinando Cassone, another officer of the *Regia Marina*'s Naval Engineering Corps, entitled 'The Current Evolution Status of Capital Ships: a Critical Essay'.[9] Cassone began by criticising Soliani's proposals for semi-submersible battle monitors and focused his attention on the correct balance between armament, speed and protection, stating that this balance impacted on an adequate distribution of weights and volumes on board. Assuming that a naval battle involving two fleets would, as occurred at Jutland, mean a number of engagements of battlecruiser versus battlecruiser and battleship versus battleship, Cassone favoured the conceptual merging of these two types of warship. (He considered the final phase of the battle of Jutland, when British battleships engaged German battlecruisers, to be an exception to the rule.)

He also stated that the evolution of the battlecruiser was a consequence of the Anglo-German race towards very fast battleships, with *Hood* and the US *Lexington* class epitomising the eventual merging of the two types. In order to achieve this objective, Cassone elaborated a concept aimed at overcoming the incompatibilities arising from the percentage ratios of weights and volumes. He defined maximum limits for the three technical factors of naval warfighting efficiency – propulsive power, armament and protection – taking into account the naval technologies available at the time.

With regard to horsepower, Cassone identified 200,000shp as the maximum figure and distributed it between four shafts, because it was neither desirable to have more than 50,000shp on each shaft nor feasible to increase the number of shafts beyond four.

As for main armament, Cassone considered the number of turrets, the number of guns in each turret (no more than three), the fire control systems and the maximum calibre of the main guns. He assumed an upper limit of eight 457mm (18in) guns in four twin turrets and a related number of secondary guns. Cassone also assumed that protection would not exceed 32 per cent of displacement. Extrapolating his assumptions from known information, Cassone defined a first iteration of his 'maximum ship-of-the line', with the following characteristics:

Displacement: 57,000 tonnes
Length: 280m
Beam: 34m
Draft: 9.5m
Horsepower: 200,000shp (boilers + steam turbines)
Main armament: eight 457mm/45cal guns in twin turrets
Protection: 18,240 tonnes (457mm max)

Cassone was well aware that these features would lead to huge and costly leviathans of the sea and would be unaffordable even for the major navies. He therefore aimed to find more affordable solutions by performing a detailed analysis of particular parameters: main, secondary and antiaircraft armament (including layout), fire control, ventilation, machinery and speed. In his analysis, Cassone used as a reference 29 classes of British, US, Italian and Japanese battleships and battlecruisers dating back from the late 19th century to the early 1920s.[10] Cassone then drew up two groups of comparative tables featuring particular British and US battlecruisers and battleships, notably *Hood, Courageous, Royal Sovereign, Queen Elizabeth, Lion, Renown* and *Lexington*. The first table was related to displacement, horsepower, speed, and type of boiler, while the second focused on percentages of hull, machinery, armour and armament weights relative to displacement. Cassone stated that the percentage of weight allocated to machinery in comparison with the percentage allocated to protection penalised battlecruiser designs. The exception to this conclusion was *Hood*, because she had lighter machinery than other battlecruisers, thus allowing a percentage of armour weight that was 33.5 per cent of displacement and therefore comparable to the German and British battleships. This meant that *Hood* could be considered as a warship incorporating all the characteristics of a dreadnought battleship, but with the speed of a battlecruiser.

Cassone then proposed a solution for the design of a ship-of-the line with these characteristics, but focusing on the disposition of the major components of the machinery: boilers and turbines. In order to improve both the flow of air in the engine rooms and the forced draught for the boiler rooms, Cassone proposed to locate the latter above the former: in addition to drawing air from the engine rooms, the fans of the boiler rooms could also draw air from outside. Cassone then addressed a preliminary design of a battlecruiser equipped with 18 boilers and four sets of steam turbines and a total horsepower of 216.000shp. The boilers were located in six boiler rooms, which were located above three compartments for the steam turbines. According to Cassone, such a machinery layout would allow the design of a battlecruiser featuring displacement and dimensions more limited than those identified for the 'maximum ship-of-the line'.

His battlecruiser had a length of 265 metres, a beam of 32 metres and a draught of 9.5 metres; displacement was 45,000 tonnes. Maximum designed speed was 35–40 knots – an extremely ambitious figure that would be difficult to verify without tank tests. The main battery comprised eight 457mm guns in four twin turrets, two forward and two aft; the secondary battery comprised six 6in/152mm DP guns in three twin centreline turrets and additional low-angle 152mm guns in casemates. From a sketch attached to Cassone's essay, it is possible to ascertain that his battlecruiser had a forecastle that extended as far as the after superimposed main turret and a high central deckhouse (in order to accommodate the boiler rooms above the engine rooms). As for vertical protection, Cassone envisaged an inclined 457mm (18in) belt over the lower hull, decreasing to 356mm (14in) and 152mm (6in) on the upper hull sides. Horizontal protection comprised 127mm (5in) plating above the battery deck. An armoured citadel some 160 metres long enclosed all the machinery spaces and magazines. As for underwater protection, Cassone proposed that the lower part of the citadel was to be 20 metres wide, leaving a depth of 6 metres on either side for a bulge divided into watertight compartments to absorb the explosion of torpedo warheads, a compartment filled with sealed steel tubes and a double longitudinal bulkhead. Cassone affirmed that such a scheme had never been attempted in any warship design and could not be achieved with the traditional machinery layout (*ie* boiler and engine rooms located at the same lower level).

Cassone closed his essay by stating that if Italy was compelled to have new ships of the line, these should be very large and not compromises because, in the words of von Tirpitz:

> A modern naval battle is reduced to a number of duels between one ship and another adversary, so that the decisive technical factor is the concentration of firepower in individual ships rather than a number of ships.

Interwar Italo-French Rivalry and the Race for Cruisers

Neither the proposals of Soliani nor those of Cassone materialised. At the time, Italy's financial situation did not permit the construction of new battleships or battlecruisers, while the international political situation favoured the proposal by then-US President Warren Harding for a global disarmament conference that would satisfy domestic demands and prevent other naval powers, notably the United Kingdom and Japan, from improving their fleets.

For Italy, the Washington Treaty signed on 6 February 1922 allowed the country to reach a level of parity with its historical competitor, France. Italy was permitted the same number of battleships, which was a result of considerable political and military significance. However, in 1923 the then-French Navy Minister, George Leygues, stated:

> We reject parity with Italy because France has coasts facing three oceans and seas and manages a vast empire with an 18-hour sailing time from Marseille and sea lines of communication that need a high level of protection.

While this statement was correct, it was also true that Italy had 8,000 kilometres of coastline vulnerable to invasion from the sea and had to protect sea lines of communication with its own colonies, namely the Dodecanese islands in the Eastern Mediterranean, Libya in North Africa and Somalia in East Africa. A rivalry between Rome and Paris therefore developed during the interwar years and impacted on the development of their respective military forces, including their navies.

As far as major surface warships were concerned, the *Regia Marina* and the *Marine Nationale* followed a similar pattern, both building battleships and heavy/light cruisers, with an eye on similar developments occurring in the United Kingdom and, from 1933, Germany. As the 1922 Washington Treaty froze the number of battleships allowed to the major naval powers and instituted a 'battleship holiday' (initially to last ten years) that prohibited the construction of capital ships, attention turned to a new type of 10,000-ton cruiser armed with 8in (203mm) guns, known as the 'Washington type'. In this category of naval armaments, France and Italy established a competition in construction that had both political (the need for parity) and industrial objectives. During the 1920s, the new type of cruiser became the 'capital ship' of Italy and France, and was also important for demonstrating Italian and French shipbuilding prowess abroad and for 'showing the flag', a new expression of the gunboat diplomacy that had been developed over the previous 50 years. In 1924–25, less than three years after the entry into force of the

Table 3: **The First Generation of French & Italian 'Treaty' Cruisers**

Class	Names	Displacement (fl)	Main armament	Max speed
Trento	*Trento* *Trieste*	13,500 tonnes	8 x 203mm (4 x II) 16 x 100mm (8 x II)	35.0 knots
Duquesne	*Duquesne* *Tourville*	12,200 tonnes	8 x 203mm (4 x II) 8 x 75mm (8 x I)	33.7 knots

The heavy cruiser *Pola* underway at full speed. The general layout of the second group of Italian heavy cruisers (comprising four ships) inspired all the battlecruiser studies and preliminary sketches drawn up within and outside the Italian Navy. (Author's Collection)

Washington Treaty, the Italian and French shipyards laid down their first classes of 10,000-ton cruiser, the principal characteristics of which are shown in the accompanying table.

The Italo-French naval arms race continued into the 1930s, in the wake of changes in the geostrategic situation that had occurred in the meantime in Europe which included the new conferences on naval armament limitations held in London in 1930 and 1936. At the time, a debate was underway in Italy and within the *Regia Marina* concerning whether it would be better to build new capital ships or to modernise the old battleships commissioned during the Great War. The former choice implied drawing up new designs, taking into account constraints stemming from the Washington Treaty and confirmed by the London Treaty. It also implied the need to find affordable solutions, as the *Regia Marina* needed also to improve the capabilities of other warship types such as destroyers and submarines, leaving aside decisions concerning aircraft carriers and embarked aviation. Given this context, a decision was to develop a balanced battle fleet that would include either battleships or battlecruisers or both types.[11]

The Italian naval shipbuilding industry was called to support the *Regia Marina* in order to reach an informed and appropriate solution to a problem that was complicated by the emergence of a new type of capital ship, the German *Panzerschiff* which, in turn, influenced the decisions made by the *Marine Nationale*. Within this overall context, Italian studies for new or converted battleships ran side by side with preliminary designs related to battlecruisers. Several projects of this nature were conceived during the period 1928 to March 1940, either to meet the requirements of the Italian Navy or for foreign customers, with a key role played by the Italian shipbuilding industry.

Battlecruiser Proposals for the *Regia Marina*

The very first preliminary study for an Italian battlecruiser was conceived in 1928, and stemmed from an investigation into the best way to utilise the 70,000 tons of new construction allotted to Italy under the Washington Treaty. The proposal was for a small battlecruiser with protection sacrificed for heavy armament and high speed. A design displacement of 23,000 tons would allow three units to be built within the 70,000-ton limit. The study envisaged an armament comprising six 381mm guns in twin turrets (two forward, one aft), eight 152mm guns in twin turrets and twelve 100mm guns in twin HA mountings. With a length of 194.5 metres and a beam of 28.9 metres, the hull featured a long forecastle and a superstructure block placed amidships, with two trainable catapults for IMAM Ro.43 seaplanes and a massive single funnel. The propulsion machinery comprised four sets of geared turbines driving four propellers, and total horsepower was 150,000shp for a maximum speed of 29 knots; endurance was 2,000nm at 20 knots. Probably drawn up by the Technical Department of the *Regia Marina*, this effort was abandoned because it was deemed too costly and of limited effectiveness.

At the beginning of the 1930s, and still in the context of naval competition with France, the debate in the *Regia Marina* concerning renewal of the battle fleet centred on whether to modernise/reconstruct the two old battleships of the *Cavour* class or to build new capital ships from the keel up. Two preliminary studies for a battlecruiser were drawn up around 1930 and dubbed 'Design 45367' and 'Design 770', their numbering indicating the involvement of the Technical Department of the *Regia Marina*. However, the results of these two studies and others that

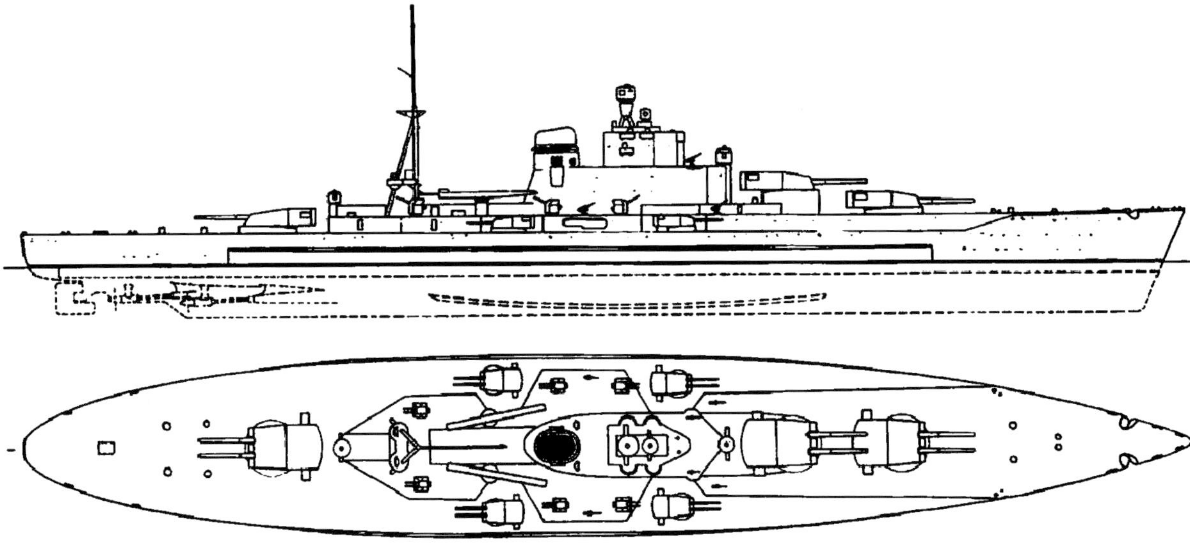

Profile and plan of a 23,000-ton battlecruiser armed with six 381mm guns in twin turrets, probably drawn up by the Technical Department of the *Regia Marina* around 1928. (Courtesy of Maurizio Brescia)

were worked on in the following years are not entirely clear. The only certainty is that the Leghorn-based company Odero Terni Orlando (often known as OTO) was involved in the development of at least one of them. It is likely that the work on one or more of these projects went on for some time, up to the production of a scale model later exhibited at the Technical Naval Museum at La Spezia and still there today. The plaque that accompanies the scale model designates the ship as a 'battlecruiser' (*incrociatore da battaglia* – see photo); it states that 'this ship was never built' and lists the key characteristics.

Two photos of the model of a battlecruiser proposed by OTO exhibited in the Naval Technical Museum of La Spezia. (Author)

INCROCIATORE DA BATTAGLIA

PROGETTO O.T.O. – LIVORNO

Questa unita' non fu costruita.

Il modello e' relativo ad un progetto elaborato verso il 1930-35 per la trasformazione delle unità delle Classi Cavour e Doria o per la costruzione ex novo di unità conformi agli accordi internazionali allora vigenti

Le caratteristiche approssimative di progetto di queste navi sarebbero dovuto essere:

DISLOCAMENTO STANDARD	26.670 TON.
LUNGHEZZA	215 mt. circa
LARGHEZZA	32 mt. circa
VELOCITA'	30 nodi circa
ARMAMENTO	9-381 = 12-152 = 12-100 38 MITRAGL. 8 TUBI L.S. 533

A close-up of the plaque that accompanies the battlecruiser model exhibited at La Spezia. It states clearly that the ship was never built and that the design was drawn up during the early 1930s. (Author)

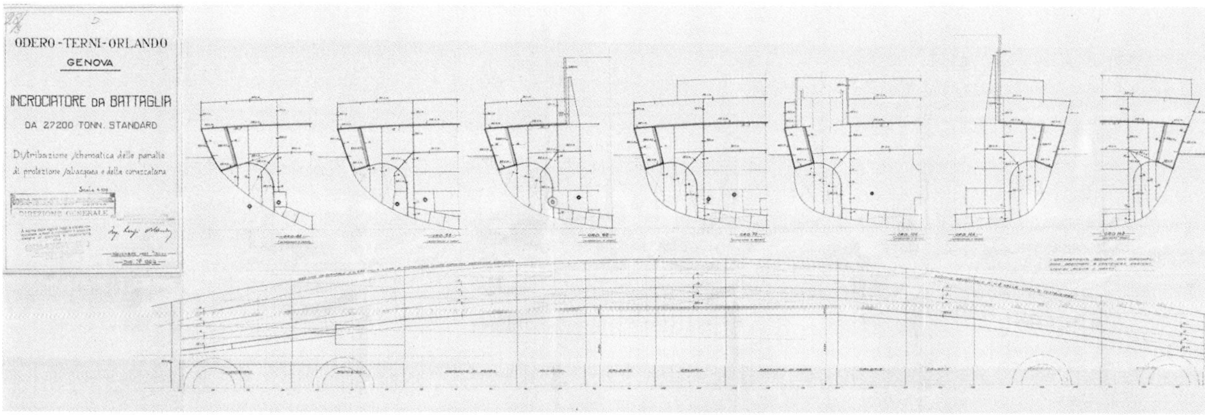

Plan of the protection scheme of the 27,200-ton OTO battlecruiser design dating from November 1932. They show that the armour on the barbettes of the main turrets was 250mm KC above the upper deck and 220mm KC below; that on the barbettes for the secondary turrets was 150mm KC above and 120mm KC below. The plans show four main turrets, and the diameter suggests that these were twins rather than triples. Calibre was either 381mm or, more likely, 343mm as in the 'Type F' – see below. (INHO)

Searches at both the Italian Navy Historical Office (USMM), the La Spezia-based Technical Naval Museum and the OTO company for additional details and information about the model and its design have proved disappointingly fruitless. However, it is important to highlight that this scale model differs from what has been reported elsewhere, with claims that this project was armed with six 381mm guns in twin turrets. From the photos of the scale model it is evident that the main armament comprised nine 381mm guns in triple turrets, two forward and one aft. The upperworks of the model show the influence of the heavy cruiser *Pola* on the design adopted by the *Regia Marina*, although the battlecruiser had only a single funnel. Other information available from the plaque includes a secondary battery comprising twelve 152mm guns in triple turrets and twelve 100mm HA guns in twin shielded mountings. There were also no fewer than 38 machine guns and eight 533mm torpedo tubes. A pair of catapults is visible abaft the funnel, suggesting that two seaplanes were to be embarked. Maximum speed was to be 30 knots, with steam turbine machinery driving four shafts. Length is given as 215 metres and beam 32 metres, and the ship would have had a standard displacement of 26,670 tons.

Although none of the studies and concepts for battlecruisers carried out in the early 1930s bore fruit, the *Regia Marina* and the Italian shipbuilding industry did not hesitate to carry forward the idea of developing proposals for novel designs of capital ship as alternatives to the construction of new battleships or the reconstruction of ships dating from the Great War.

In this context, it is worth mentioning a project named 'Design 1933' for a battlecruiser which aimed to counterbalance the laying down of the French fast battleship *Dunkerque*.[12] Developed for naval operations within the Mediterranean basin and involving OTO, Design 1933 was meant to oppose the *Dunkerque*s using the remaining capital ship tonnage assigned to Italy by the Naval Treaties. Up until a few years ago, information about Design 1933 was scarce. The discovery and analysis of some general plans in the archives of the Directorate of Naval Armaments of the Italian Ministry of Defence testify that the project, which probably had the status of a preliminary design, was worked on for at least a couple of years, with variations mainly relating to the armament. It is therefore possible to attribute the designation of 'OTO Project' to the 1933 Design, comprising at least two versions, one of which was dated November 1932 and characterised by a standard displacement of 27,200 tons, the other – designated 'Project Type F' – with a standard displacement of 33,610 tons and drawn up in February 1933.

The plans of the November 1932 version show only the layout of the underwater protection and the midships section. The underwater protection extended along the lower part of the hull and encompassed the magazines and machinery spaces. It comprised a curved torpedo bulkhead 35mm thick in the form of an arc, this 'internal bulge' being divided by three longitudinal bulkheads: of the four compartments three were filled with fuel or water while the the second was void. The aim of this layout was a gradual absorption of the underwater explosion to prevent water ingress into the magazines and machinery spaces. Atop the curved torpedo bulkhead was an inclined box that formed a spaced armour belt, with the outer wall intended to decap AP shell – a common feature of Italian warships of the period. The outer wall, roof and floor of the box had plates of 60mm nickel-chrome (CN) steel, while the face of the inner wall comprised 250mm plates of Krupp Cemented (KC) armour. The main armoured deck had 110mm CN plating, and there was a 20mm CN splinter deck beneath. There were four boiler rooms housing eight boilers and four sets of geared turbines in a unit arrangement, with a magazine for the secondary guns between the second pair of boiler rooms and the after engine room (see GA plans).

The most interesting documentation of OTO's activities is that related to the larger, 33,160-ton 'Project Type F'. The survival of such detailed drawings indicates that the preliminary design was probably at an advanced stage,

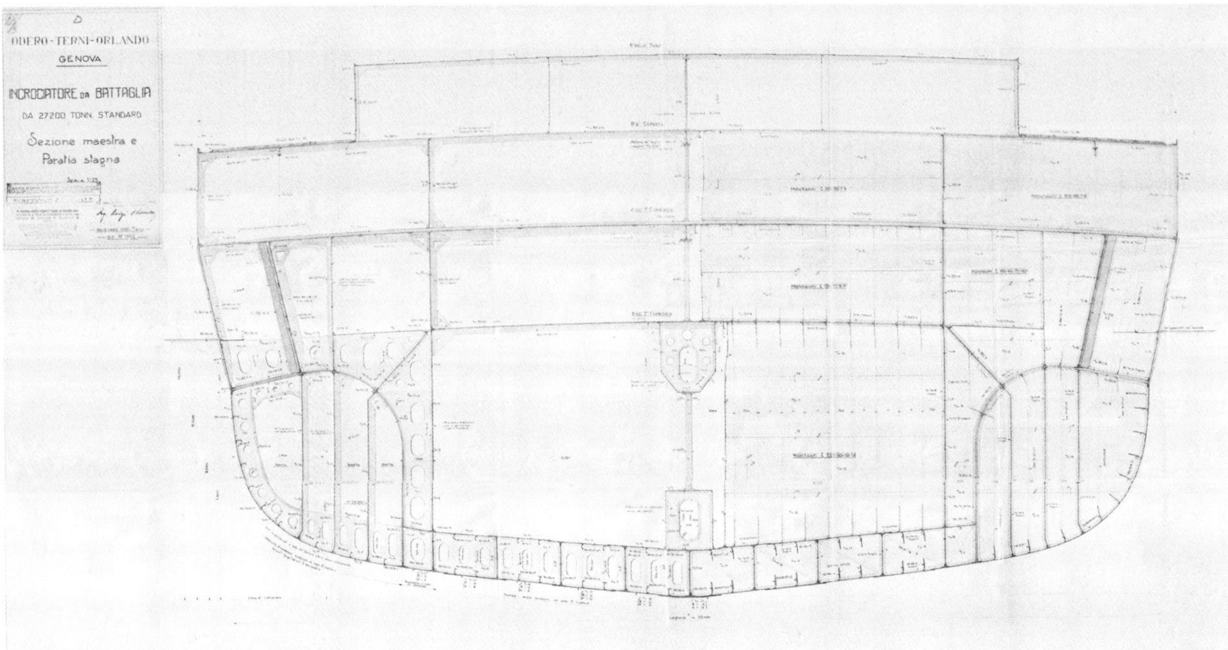

Midship section of the 27,200-ton OTO battlecruiser. (INHO)

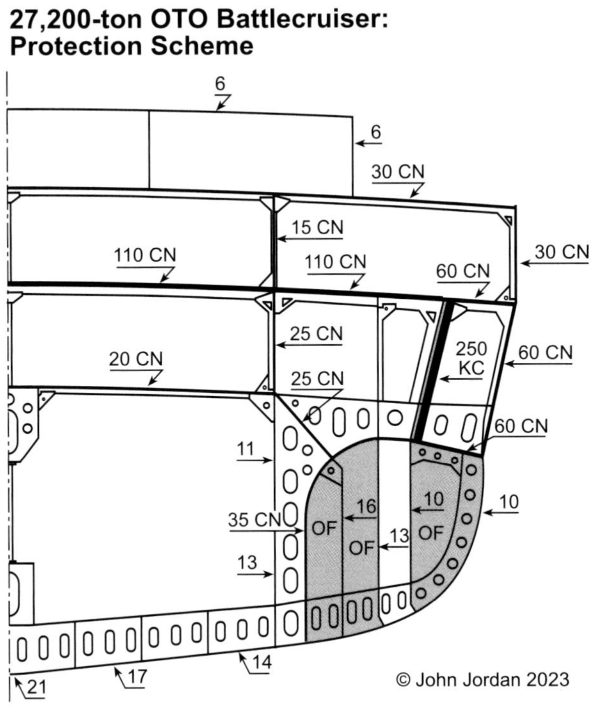

Midship half-section of the 27,200-ton battlecruiser based on the OTO plans. Note the curved torpedo bulkhead with the armoured 'box' of the inclined belt above it. The plans of the 'Type F' battlecruiser show a similar arrangement, but with the multiple longitudinal bulkheads of the underwater protection system replaced by a Pugliese-type cylinder surrounded by liquids (oil fuel or feed water). There was also a small increase in some armour thicknesses: 70mm (vs 60mm) CN for the outer 'decapping' plate of the belt, and 25mm (vs 20mm) CN for the splinter deck beneath the main armoured deck.

making them a useful tool to better describe the ship.

The principal characteristics of the 'OTO Type F' battlecruiser are listed in the accompanying table. The increase in dimensions and displacement was dictated by the decision to adopt a main armament comprising twelve 343mm/53 (13.5in) guns in triple turrets, which were positioned at either end of an amidships superstructure block. It is difficult to understand the rationale for the 343mm gun because at this time such a weapon was neither designed nor manufactured by Italian companies. A possible hypothesis is that 'Type F' was a preliminary design intended for export, as requested by a foreign Nation. If this was the case, the 13.5in guns would be procured from a British company. Furthermore, it cannot be ruled out that these guns would be procured from a British company and then installed in a battlecruiser built for the Italian Navy.

Apart from these considerations, the Type F' battlecruiser had a flush-deck hull, with nineteen watertight compartments below the armoured deck, five decks and a double bottom. The superstructures featured two tripod masts and a single central funnel. The secondary battery comprised twelve 152mm/53 guns in four triple turrets located at the four corners of the upper deck; anti-aircraft protection was to be provided by 100mm/47 HA guns, and 37mm and 13.2mm light AA and machine guns in twin mountings, all located on the upper deck and the deckhouses. There were three main fire control positions, two atop the bridge structure with a third aft, equipped with 5-metre rangefinders, two for the secondary guns (3.5-metre RF) at the base of the tripod foremast, and two for the HA guns (3-metre RF) at the foot of the mainmast. There were also four 120cm searchlight projectors, which because of the somewhat cramped arrangements amidships were mounted in pairs on platforms on the foremast and mainmast respectively.

Two trainable catapults were mounted at the sides of the funnel, with space provided to stow two seaplanes with wings folded inboard of their after end. As for machinery, the plans feature four shafts driven by geared turbines, with steam supplied by boilers (probably sixteen) housed in four separate rooms in a classic unit arrangement (BR1&2+ER1, BR3&4+ER2). This layout implied a large volume of exhaust trunking below decks before the uptakes converged into the single funnel. It is possible that this layout was selected in order to limit hull length, which would increase if separate funnels for the forward and after boiler groups were to be adopted.

The protection scheme for 'Type F' was inspired by that developed by General Umberto Pugliese, Naval Engineering Corps, Italian Navy, in which a cylindrical structure inside a 50mm CN curved torpedo bulkhead was integrated into the lower part of the hull to absorb the force of a torpedo warhead explosion.[13] Vertical protection called for an inclined main multilayered belt extended in length for 180 metres to include magazines and machinery spaces. There was a 70mm CN external plate and a KC 250mm internal plate with a 60mm wooden backing. The armoured citadel was closed by 200mm KC transverse bulkheads and was topped by an armoured deck of 110mm CN plating. There were 50mm CN steel plates on the upper part of the hull sides. Forward and aft of the citadel, the thickness of the belt was reduced to 150mm KC with a 50mm wooden backing, applied directly to the shell (*ie* without an external decapping plate). There were 150mm KC transverse bulkheads at the outer ends of these secondary armoured boxes, which were topped by a 50mm CN deck. The lower edge of the main multilayered armoured belt was secured to the curved torpedo bulkhead of the Pugliese system.

The upper decks had 30mm CN plating. The barbettes of the main gun turrets had 250mm KC plates with a 60mm CN decapping plate, increasing to 280mm (wihout the decapping plate) above decks and decreasing to 200mm + 25mm beneath the forecastle deck. The forward fire control post had KC plates whch varied in thickness from 120mm (roof) to 280mm (walls), while the communications tube was formed by 150mm hoops with an outer ring of 60mm CN plates. The after fire control post had a similar arrangement, but the walls were reduced to 150–200mm plates of KC, the roof to 30mm CN, and the communications tube to 120mm KC + 60mm CN. The highest of the three main control positions had relatively light protection, with only 25mm CN plates over the director but a lower cylinder (serving as a conning tower) with 90mm KC walls. The funnel had a 25mm CN cap and there was 50–100mm CN plating on the boiler uptakes. Design 1933 – and possibly Project 'F' – were probably the last Italian battlecruiser designs the *Regia Marina* considered before deciding on a reconstruction of the older battleships of the *Cavour* and *Duilio* classes and the development of the new fast battleships of the *Littorio* class.

Table 4: **OTO Type F Battlecruiser**

Displacement (trial)	35,900 tonnes
Length:	
overall	228.3m
waterline	220.0m
Beam (max)	30.5m
Draught:	
maximum	15.0m
trials	9.5m
Machinery	16 *Regia Marina*-type boilers 4 Ansaldo geared turbines
Horsepower (max)	150,000shp
Speed (max)	30 knots
Armament	12 – 343mm/53 (4 x III) 12 – 152mm/53 (4 x III) 16 – 100mm/47 HA (6 x II) 16 –37mm AA (8 x II) 16 – 13.2mm MG (8 x II)
Fire controL	three 5.0-metre RF two 3.5-metre RF two 3.0-metre RF four 120cm S/L
Protection:	
deck	110mm CN
belt	250mm KC + 70mm CN

Export Designs

For completeness, it is worth noting that in 1937 Ansaldo developed a project designated UP 102 for the Romanian Navy. This was for a 13,000-ton armoured cruiser with six 253mm (10in) main guns in two triple turrets, one forward and the other aft. The layout was similar to that of the German *Panzerschiffe*, albeit with a dual-purpose secondary armament comprising twelve 127mm guns in twin mountings. There were two triple 533mm torpedo tubes and a single catapult for a total of three aircraft. The propulsion machinery was intended to develop 60,000shp (75,000shp max) for a design speed of 30–31 knots, this being significantly higher than the diesel-powered German *Panzerschiffe*. In the event, this UP 102 project was not taken up.

The final design effort made by Ansaldo before Italy entered the Second World War was a preliminary study dated March 1940 for an armoured cruiser which strongly resembled a battlecruiser because of its speed (37 knots) and firepower – see the accompanying profile and plan. The study was probably carried out for Spain or a Spanish-speaking country in South America, possibly Argentina, as it is referred to as a '*crucero acorazado*', and the characteristics on the plans are likewise in Spanish. The overall layout of this preliminary study resembled that of the Italian *Zara*-class heavy cruiser but with more firepower and speed. The Ansaldo *crucero* featured a trial displacement of 16,000 tonnes, an overall length of 213.9 metres and a maximum beam of 22 metres. Armament was based on nine 203mm/55 guns in three triple turrets, twelve 90mm antiaircraft guns in six twin mountings, sixteen 37mm light guns in eight

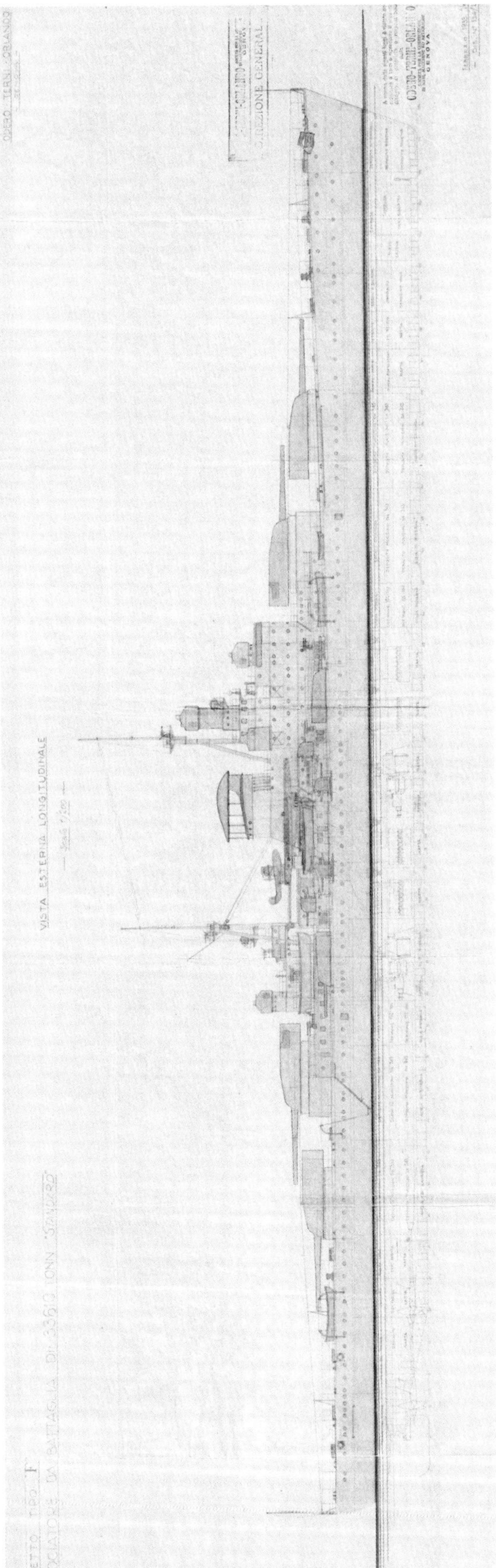

Above: Profile and plan of the battlecruiser dubbed 'Progetto Tipo F' by OTO dating from February 1933. The lower part of the profile shows the arrangement of the magazines and the machinery compartments. (INHO)

Below: Plan of the protection scheme for the 'Type F' battlecruiser project of February 1933 proposed by OTO. Note that the cylinder in the Pugliese underwater scheme is elliptical. (INHO)

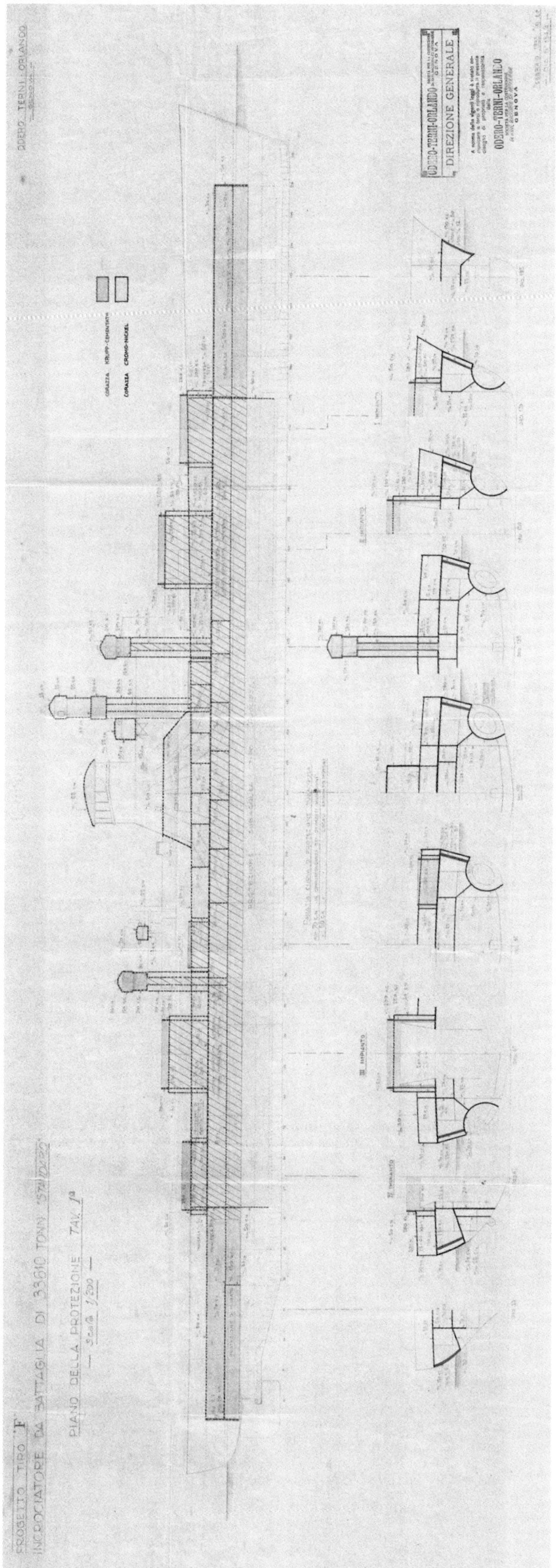

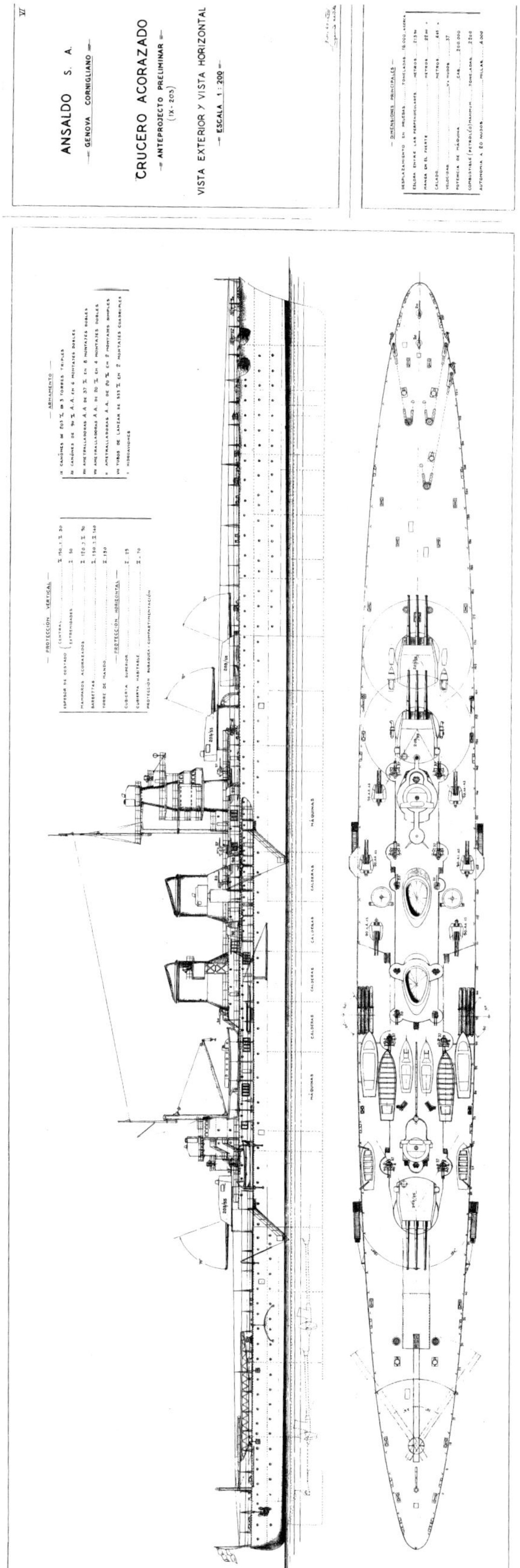

The final design effort made by Ansaldo before Italy entered the Seond World War was a preliminary study dated March 1940 for an armoured cruiser which strongly resembled a battlecruiser because of its speed (37 knots) and firepower (nine 203mm guns). (INHO)

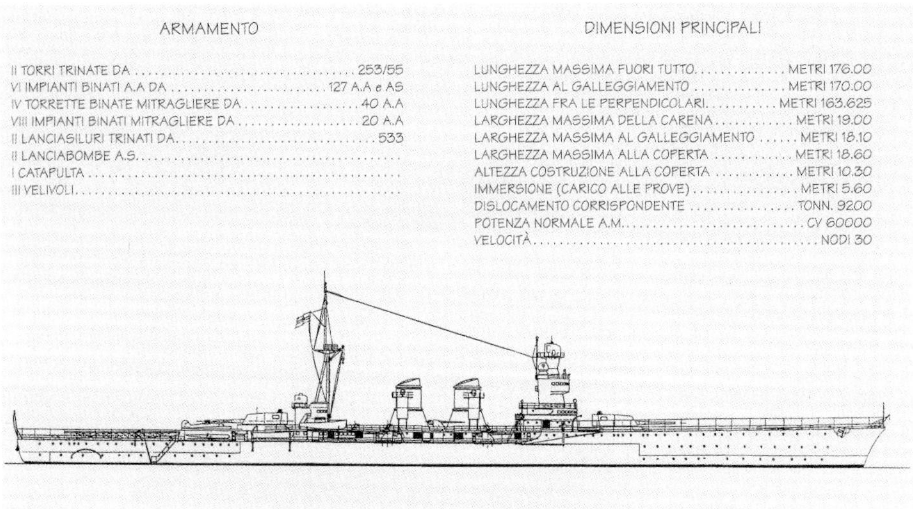

Profile and characteristics of UP 102, a preliminary design for an armoured cruiser to be built by Ansaldo for the Romanian Navy. It featured six 253mm main guns, arranged in two triple turrets and, and was to have a maximum speed of 31 knots. (INHO)

twin mounts and ten 20mm light guns in four twin and two single mounts. The main guns could elevate to 70 degrees. There were eight 533mm torpedo tubes in two quad mountings. Maximum designed speed was 37 knots, achievable through a four-shaft propulsion system delivering 200,000shp. The layout of the propulsion plant was similar to that of contemporary Italian Navy cruisers.

As usual in battlecruisers, speed and firepower prevailed over protection, which comprised a 150mm vertical belt, reducing to 50mm fore and aft, and a 70mm armoured deck.[14] The barbettes for the main guns were 150mm, reducing to 140mm between decks. Underwater protection was provided only by tight subdivision, while a trainable stern catapult was used to launch a seaplane – a solution similar to that adopted in the Italian battleships of the *Littorio* class. The plans suggest that two seaplanes were stowed in a hangar beneath the quarterdeck, served by a lift.

Conclusion

Although several naval architects and designers put forward proposals, studies and projects for battlecruisers, the *Regia Marina* in the interwar years first decided in favour of heavy and light cruisers – which would not always meet expectations – and later would opt to use the available resources for the modernisation of the older battleships and for the construction of the new *Littorios*.

Endnotes:

1 Romeo Bernotti, *Il Potere Marittimo nella Grande Guerra*, Raffaello Giusti (Leghorn, February 1920).

2 These four pre-dreadnoughts – *Regina Elena*, *Vittorio Emanuele*, *Roma* and *Napoli* – were commissioned in 1907–08. Their main armament comprised two 305mm guns in single turrets and twelve 203mm guns in twin turrets. Their top speed was 21 knots.

3 Italian Navy Historical Office, *Raccolta di base*, *Busta* 1504, *Foglio* 18.

4 During his career in the Regia Marina, Soliani was charged, together with Giuseppe Rota, with the design and construction of the model basin at La Spezia. The first of its kind in Italy, the basin was 146 metres long, 6 metres wide and 8 metres deep. Soliani had also designed the scout cruisers of the *Agordat* class and the water tanker *Tevere*.

5 The construction of three more battlecruisers of the *Hood* class (*Anson*, *Howe* and *Rodney*) was halted for economic reasons, while the US Navy decided to convert two of the six battlecruisers laid down – *Lexington* and *Saratoga* – as aircraft carriers and scrap the others.

6 Nabor Soliani, '*L'avvenire delle navi da guerra*' ('The Future of Warships'), *Rivista Marittima*, June 1921 (Rome).

7 In his capacity as Director of Naval Constructions of the Royal Navy, d'Eyncourt had been in charge of the *Hood* design.

8 'The Future of Warships', *Journal of the American Society for Naval Engineers*, August 1921.

9 Ferdinando Cassone, '*La fase attuale dell'evoluzione delle navi da battaglia (studio critico)*', *Rivista Marittima*, October 1921 (Rome).

10 They were: *Dreadnought*, *Temeraire*, *Collingwood*, *Neptune*, *Hercules*, *Orion*, *King George V*, *Iron Duke*, *Queen Elizabeth*, *Royal Sovereign*, *Tiger*, *Renown*, *Courageous*, *Hood*, *Doria*, *Caracciolo* (building), *Florida*, *Arkansas*, *Nevada*, *New York*, *New Mexico*, *Tennessee*, *Maryland* (building), *Indiana* (building), *Lexington* (building), *Fuso*, *Nagato*, *Kongo* and *Akagi* (building).

11 These efforts also included the development of a so-called 'limited displacement battleship': see Michele Cosentino, 'The Italian Navy and The Battleship in the 1930s: Theory and Practice', *Warship 2023*, 81–93.

12 *Dunkerque* was laid down on 24 December 1932, *Strasbourg* would follow on 25 November 1934.

13 Michele Cosentino, 'The Bonfiglietti Project: An Aircraft Carrier for the Regia Marina', *Warship 2015*, 44–61. The Pugliese system was later used for the battleships of the *Littorio* class, and would also be installed in the modernised *Cavour* and *Doria* classes.

14 For comparison, the *Zara*-class cruisers had eight 203mm guns and a maximum speed of 32 knots. The guns in the Ansaldo Model 1927 turrets were 53 calibres long and could elevate to 45 degrees. Protection was similar: a 150mm belt and a 70mm deck.

THE BEGINNINGS OF SOVIET NAVAL POWER:

THE 'S'-CLASS SUBMARINES

The 'S' class, designed by the Germans and built with German assistance, is generally regarded as the best of the early Soviet submarine types. **Przemysław Budzbon** tells the story of the fraught negotiations between the Soviets and the Germans that laid the basis for the class, and details its further development prior to and during the Second World War.

After the First World War numerous navies displayed an interest in German submarine technology. On the other side the *Marineleitung* (German Naval Command) and the German shipyards Germaniawerft, AG Weser, and Vulcan AG were anxious to continue submarine development, the first with aim of further developing submarine capabilities, the latter in order to turn the expertise gained during the war into profits. Leading these endeavours was *Korvettenkapitän* (Ret) Karl Bartenbach.[1]

The first foreign nation to seek German assistance was Japan in 1920, followed by Argentina, Italy and Sweden. Restrictions imposed by the Treaty of Versailles made it impossible to conduct such activity in Germany, so the nearby Netherlands was chosen: in July 1922 a company named NV Ingenieurskantor voor Scheepsbouw (IvS) was formed as a common undertaking of the three major German submarine building yards. The company, with Hans Techel[2] as technical director and *Korvettenkapitän* (Ret) Ulrich Blum[3] as sales director, initially operated from the offices of the Germaniawerft at Kiel until it was moved, together with its eleven-man staff, to The Hague in the summer of 1925. The company was secretly backed financially by the *Seetransportabteilung* (Maritime Transport Department) of the *Reichsmarine*. The department, which was headed by *Kapitän zur See* (Ret) Walter Lohmann[4] between 1923 and 1927, financed nearly all the clandestine and semi-clandestine naval projects of IvS via a shadow accounting company Mentor Bilanz.

In 1925, on the advice of *Korvettenkapitän* Wilhelm Canaris,[5] the secret U-boat department (codenamed 'Au' – Anti-Uboat Defence Questions) of the *Marineleitung* was formed under retired U-boat commander *Konteradmiral* Arno Spindler with the objective of developing, in collaboration with IvS, U-boat types for mobilisation contingency plans. The first to be selected was the proven UBIII coastal type, but in 1926 it was decided to replace it with the UG design (Project 51) of the *Kaiserliche Marine*. There was not a complete set of plans or documentation available, as the design had not been finalised before the end of the war. IvS was therefore charged with completion of the design work (Project 51a), which was performed by naval architects Friedrich

UB-148 of the UBIII type in a heavy sea, demonstrating the seakeeping qualities of her bow lines. These were used as a model solution for Project 51, to be repeated in the Pu111 and Pr224 designs. The photograph was taken after the First World War, during trials conducted by the US Navy. (National Archives, 512979)

UB-126 (centre) of the UBIII type (incorrectly identified in the NHHC caption as *UB-133*) photographed after surrender at Cherbourg, showing off her fine bow lines. She is accompanied by *UB-24* of the UBII type (right) and the French submarine *Montgolfier* (*Brumaire* class, at left). A comparison of the configuration of the three boats testifies to the progress made by German designers during the war. *UB-126*, which was never commissioned into the *Marine Nationale*, was employed for explosives trials and broken up at Toulon in 1921. (Naval History and Heritage Command, NH 2448)

Schotte[6] and Albert Ehrenberg. Due to a scarcity of documentation, the reconstruction had to be based on the memories of former employees of the technical office of the *Uboot-Inspektion*. This partial double-hull boat had a displacement 100 tonnes greater than the UBIII and had a layout modelled on the Ms ('Mobilisation') type: the control room was located, conventionally, between the after accommodation/battery compartment and the engine room. The bow section of the outer hull was modelled on the sharp lines of the UBIII, which proved to be particularly seaworthy, while the pressure hull aft was spindle-shaped to ensure the best diving characteristics and a reduction in drag when submerged. The lightweight, fast-running geared diesels of Project 51a were not used on the original UG design, which featured UBIII-type machinery but with supercharging.

The intention was that Project 51a would undergo trials using personnel from the *Reichsmarine* after the prototype boat had been built in a friendly country under German supervision. Spain seemed particularly appropriate to implement the plan because of earlier contacts. In August 1926 Lohmann and Canaris went to Spain with the Project 51a plans to explore the possibilities of building the submarine. The design had to be re-worked several times to match the requirements of the Spanish Navy, the most important being the selection of MAN diesels. The resulting Project Pu94 of November 1927 marked a serious deviation from the original design, with

Table 1: **Project 51a characteristics**

Date	Summer 1926
Surface displacement	655 tonnes
Length, wl	59.00m
Beam, max	6.42m
Draught max	4.0m
Engines	2 diesels / 2 electric motors
Total power	2,400bhp / 1,000hp
Speed	15.5 / 8 knots
Oil fuel	70 tonnes*
Battery	2 x 62 cells, 7,200Ah
Range	5,000nm (8,400) at 6 knots surfaced 60nm at 4 knots submerged
50cm TTs	4 bow, 2 stern
Reloads	6
Main gun	1 x 10.5cm
Diving depth	100m

* Could be increased by 35 tonnes using MBTs II and V.

the control room and conning tower moved forward and placed between the two accommodation/battery rooms in order to accommodate the new two-stroke MAN D5Z 30/42 diesel (offered in February 1928), which was rated at 1,200bhp and 500rpm and used direct drive. The diameter of the pressure hull had to be increased to 4.30m and displacement to 601.3 tons (665.9 tonnes), just on the edge of the 600-ton limits discussed at the Geneva Conference.[7]

The Spanish Navy considered the boat to be underpowered, so MAN proposed the D5Z 30/44 engine of 1,350bhp; this imposed a further extension of the engine room aft, but the diameter of the pressure hull could be reduced by 10cm. The electrical motors and the after compartments had to be likewise extended in order to accommodate the larger compressor requested by the Spanish. This completely changed the initial UG layout, putting the control room forward of the longitudinal centre of gravity (LCG), which had moved farther aft due to the increased weight of the diesels. In May 1928 MAN withdrew the offer of the D5Z 30/44 and proposed a supercharged single-acting 8-cylinder, four-stroke M8V 39/45 engine, delivering 1,350 hp at 500rpm – later changed to the M8V 40/46 (1,400 hp at 480rpm). This implied an increase in the pressure hull diameter to 4.28m, while an additional supply of lubricating oil and a second large air compressor in the electric motor room further inflated the size of the boat, which in the November 1928 version had a displacement of 751.5 tonnes (681.9 tons). However, the Geneva Conference failed, so the project was saved. The design received the IvS designation Pu111; the torpedo tubes were changed from the original German 50cm diameter to the international 21in (533mm) and fitted to fire electric torpedoes.

The submarine, christened *E-1*, was laid down on 22 March 1929 at the Echevarrieta y Larrinaga shipyard in Cádiz (Yard Number Ech-21) as a speculative collaboration between the Spanish industrial tycoon Horacio Echevarietta and IvS, with the Spanish Navy having the first option of purchase. The boat, the cost of which was estimated at 4.6 million Reichsmarks (RM), was owned nominally by Echevarietta but in reality by IvS, backed by a 4 million RM loan from the secret fund of the *Marineleitung*. The Spanish yard's only experience was in mercantile construction, and the hull was built in sections by the Fijenoord Shipyard[8] in the Netherlands, while the engines and fittings came from Germany. Assembly was supervised by a group of IvS engineers (including Schotte, Hugo Seligmann, Fritz Ebschner and Seidel), supported by a group of naval architecture students.

E-1 was launched on 22 October 1930 and was visited by the Spanish King, Alfonso XIII, and by government officials who gave verbal assurances that Spain would purchase the boat on completion; this convinced Raeder of the desirability of continuation of funding. The engines were started for the first time in January 1931, while static (dock) trials began in the spring. For the sea trials, which took place from May 1931, the German crew was initially commanded by *Kapitänleutnant* Robert Groom with Heinrich Papenberg[9] as chief engineer.

Table 2: **Pu111 characteristics**

Date	June 1928
Displacement	755 / 965 tonnes
Length, oa	72.38m
Breadth, max	6.20m
Draught max	4.00m
Engines	2 MAN M8V 40/46 diesels / 2 electric motors
Total power	2,800bhp / 1,000hp
Speed	17 / 8.5 knots
Oil fuel	116.74 tonnes max
Range	7,000nm at 10 knots surfaced
	160nm at 4 knots submerged
21in TTs	4 bow, 2 stern
Reloads	4
Main gun	1 x 10.5cm L/45
AA gun	1 x 2cm
Diving depth	100m

There were several innovations as compared with First World War technology, so the Germans had a first opportunity to test:

- the reduction in air bubbles when firing torpedoes
- blowing out the main ballast tanks (MBT) using diesel exhaust gases
- pneumatic control of the Kingston valves and venting flaps
- silencing
- sonic communication
- the application of welding to the fuel tanks
- the arrangement of the fuel tanks.

The results of the trials were sufficiently satisfactory to officially accept *E-1* as the template for the mobilisation contingency plan by a decree of the *Reichswehrministerium* dated 15 November 1932, despite the relatively high displacement and the numerous changes made at the request of the Spanish Navy.

The abolition of the monarchy in Spain and proclamation of the Republic on 14 April 1931 prevented the purchase of *E-1* by the Spanish Navy. She was offered to France, Poland, the Soviet Union and Yugoslavia, and again to the Spanish Republic. As a part of this sales campaign, in January 1933 the boat was inspected by Russian naval architects. Finally purchased by Turkey, *E-1* was taken over in Valencia on 27 December 1934 and renamed *Gür*. She left for Istanbul in July 1935 under the command of four Turkish naval officers, with a mixed crew of Germans and Spaniards.

Standstill

The issues encountered during dock trials of the first Soviet-built submarines (*Dekabrist* class) were the direct cause of the arrest of leading submarine constructors Boris Malinin, Ernest Kriuger and Sergej Bazilevskij[10] on

E-1 in dock, displaying the features which would be inherited by the Pr224 designs: hull configuration, sliding bridge fairing and the flatiron-shaped fairing around the gun platform. (Author's collection)

27 July 1930 (see *Warship 2020*). This brought Soviet submarine development to a standstill. As the detainees constituted the elite of Soviet submarine designers, the rest – even the outstanding theoreticians or engineers – were overcome by an overwhelming desire to cut themselves off from a project viewed by the authorities as 'wrecked' (*vreditel'skij*). The mood of despondency was only heightened by the head of the Shipbuilding Section of the Naval Scientific-Technical Committee[11] (acronym NTKM), Iulian Shimanskij, who recommended scrapping 16 hulls of the Series I (*Dekabrist* class), Series II (*Leninets* class) and Series III (*Shchuka* class) submarines, *ie* practically the entire Soviet submarine programme.

Some months later Malinin and Kriuger were sentenced to death by firing squad. As the Soviet naval armaments programme was at that time based on submarine construction, the irreversible liquidation of the experience accumulated in the heads of convicts was considered dangerous for Stalin's close associates. The court was therefore instructed accordingly, and the death sentences changed to ten years' imprisonment. The naval architects were assigned to an elite labour camp, informally known as a *sharashka*, which effectively became the design and construction bureau of OGPU. The expertise of these men was retained and utilised, but they lost influence on the general design process, being involved in secondary works on projects (*Pravda*-class fleet and *M*-class coastal submarines) run under the auspices of OGPU by Aleksej Asafov, the jailed submarine constructor and amateur designer.

The efforts to obtain the assistance of the *Reichsmarine* in submarine construction dated from 1926, but any co-operation with the Soviets was rejected on the ground of the limitations of the Treaty of Versailles. Instead the Germans proposed the services of IvS, which was gladly accepted as the Russians were already informed about IvS negotiations with Finland for the design and supervision of construction of submarines (later the *Vetehinen* class). The first visit took place on 2 April, and the Russians left with sketch plans of a 600-ton boat, which allowed them to spread rumours in Helsinki that they were co-operating with IvS and contemplating purchase of the same design as the Finnish Navy (*Suomen Merivoimat*). The objective was to hamper the latter project (the Finns had a justified distrust of the Russians), but when this attempt failed the Soviets, through the *Profintern*,[12] initiated and financially supported a 9-month strike at the Crichton-Vulkan Yard in 1927 that delayed the Finnish submarine construction programme. The truth was that because of financial shortcomings, the Soviets were not able to acquire the services of IvS at that time, so they did not persist with these overtures. Finally, they did receive from the *Marineleitung* the general arrangement plans of the First World War U-boats. However, without calculations or an understanding of theoretical principles, machinery and weapon and production technology, these were of little practical value.

The reluctance of the *Marineleitung* to engage with the Soviets was weakened under pressure from the German defence minister General Wilhelm Groener, and on 24 February 1930 a Soviet clandestine naval delegation headed by Vladimir Orlov, then C-in-C of the Naval Forces of the Black Sea, was received in Berlin.

During a stay which lasted a few weeks the Soviet

An IvS marketing brochure from 1936 featuring the Turkish Navy's *Gür* (ex-*E-1*). (Courtesy of Mr Burak Reis)

naval commanders visited the most modern ships of the *Reichsmarine*, the shipyards and the naval dockyards (see Table 3). The Germans refused to share progress in areas forbidden by the Treaty of Versailles: submarines, MTBs and chemical weaponry. However, the most important objective of Orlov was to obtain the agreement of Admiral Erich Raeder, the *Chef der Marineleitung*, to the creation on Soviet territory of a joint submarine construction and training centre similar to the Lipetsk fighter-pilot school (*Kampffliegerschule Lipezk*) active from 1925. This idea was, however, rejected.

The Soviets were in turn visited on 26 July 1930 by a German delegation headed by Vice Admiral Friedrich Brutzer, the head of the Naval Command Office in the *Reichswehrministerium*. The Red Navy was anxious to show itself under the best light, inviting the German officers to the naval facilities of the Baltic and the Black Sea Fleets and offering cruises aboard the training cruiser *Komintern* and the destroyer *Uritskij*. The impressions of both sides were quite different. In his report to Voroshilov, Orlov recommended the extensive use of German naval technology and the involvement of German officers in the combat training of the Red Navy. Brutzer, on the other hand, reported to Raeder that due to the inexperience of the crews and the lack of modern technology, the potential of the Soviet Navy could not be considered sufficiently high to bring to the Germans any benefit. The German naval authorities therefore became convinced of the futility of cooperation, and

Table 3: **Shipyards, Factories and Naval Establishments visited by the Soviet delegation 1930**

Place	Specification	Remarks
Kiel	*Panzerschiff A*	armoured ship, on the slipway
	Königsberg	light cruiser, in commission
	Naval arsenal	
	Naval radio establishment	
	Naval training facility	artillery and mechanicians
	Deutsche Werke shipyard	
Wilhelmshaven	*Leipzig*	light cruiser, fitting out
	Naval training facility	coastal and AA artillery
	Naval dockyard	
Düsseldorf	Rheinmetall factory	

firmly adhered to a policy of non-involvement with the Red Navy.

Breakthrough

At that time the initial batch of twelve submarines of the 1926 Programme had already grown to 36 hulls, and following the Mukden incident of September 1931 Stalin's appetite was increased: a total of 55 hulls was approved by the *Revvoensovet* on 22 February 1932. The Soviets were in a hurry, and therefore decided to take up the opportunity offered by IvS, who approached Orlov during his stay in Berlin in February 1930. IvS had already acquired a favourable reputation and, thanks to its network of spies and the complacency of the Western elites, the Soviets were already in possession of the plans of the Finnish mine-laying submarines (*Vetehinen* class), copies of which they purchased from France. This only served to enhance the perception of IvS's expertise in Soviet eyes.

The Soviet delegation arrived at the offices of IvS in The Hague early in 1932, identified by the Germans under the codename 'Flensburgers'. They sought to purchase plans of the Pu110 design (the future Finnish *Saukko*), but were satisfied instead with sketch plans for eight types of submarine for which they paid 3,000 guilders. The submarine *E-1* was offered to them by Blum at a price of US $1.5m. Recognising Soviet interest and their undisguised admiration for German technology, in March 1932 IvS offered the Soviet commission the complete technical documentation for a replica of *E-1* and additional plans for two submarines of 600 tons, one of them for minelaying.

As the influx of modern technical information was found to exceed the technical abilities of their own submarine designers, in April 1932 the Soviets released Malinin and some of his colleagues, and a month later appointed him as chief engineer's assistant of the design

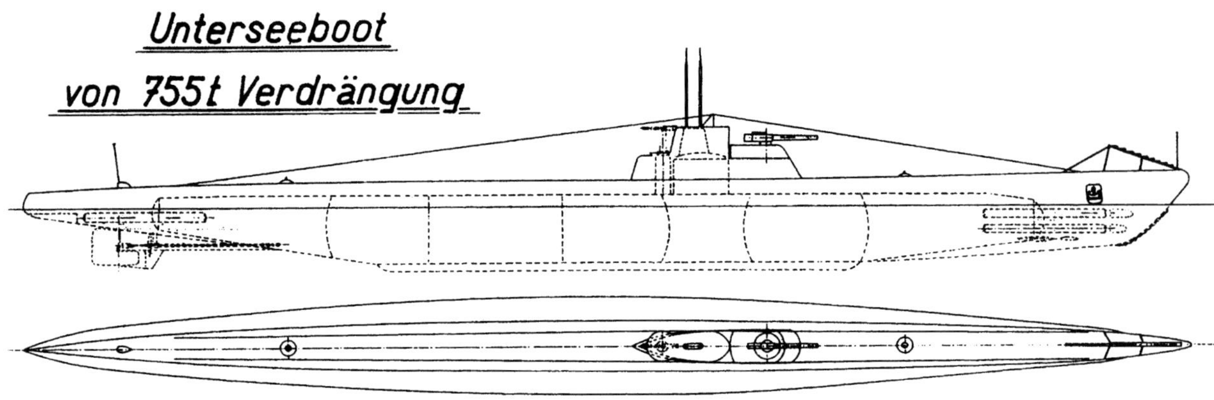

Draft of general arrangement of a 755-tonne boat (*E-1*) submitted by IvS to Sivkov in December 1932. (Author's collection)

bureau TsKBS No 2.[13] However, Malinin and his colleagues were not trusted by OGPU, and were therefore not allowed to become directly involved in the negotiations with IvS, which constituted a simple task for the Germans against relatively inexperienced Soviet staff.

Striking while the iron was hot, Techel, accompanied by a former U-boat commander, visited Moscow in September to ease the negotiations and to prepare the next visit of the Soviets. The Soviet delegation arrived on 9 November headed by Sivkov, with Paul Oras representing the Navy, Vladimir Nikitin (Chief Engineer of the TsKBS design bureau, specialising in surface warships) representing *Soiuzverf*',[14] and novice submarine designers including Vladimir Peregudov, future chief constructor of the *Proekt 627* (NATO 'November' class), the USSR's first nuclear-powered submarine. The objectives of the team were:

- investigation of the possibility and conditions of the direct purchase of submarine designs of 600–700 and 1,000–1,200 tonnes, and support for the launch of series production in the Soviet Union
- investigation of the possibility of purchasing samples and obtaining assistance for production in the Soviet Union of the following components:
- diesel engines with a weight-to-power ratio of 12kg/hp or less
- high-pressure bilge pumps
- electrically-powered evaporators
- low-noise electric motors and gearing
- red lead batteries
- tubular batteries.

On 22 November Sivkov submitted to IvS the Soviet requirements for submarines of 700 tonnes and 1200 tonnes. The continuing lack of confidence in Malinin is evidenced by the fact that these requirements were prepared by Sivkov personally during the trip. In response, two preliminary designs were submitted on 7 December.

The Russians were busy in December as they completed visits to the factories of Siemens (electrics), MAN (diesels) and AFA (batteries), which proved unsuccessful due to distrust on both sides. The Germans feared that their sales efforts would end in failure and the production of 'Chinese copies' would be initiated in the Soviet Union instead, while the Russians feared prison

Table 4: **Preliminary Characteristics of the Medium Submarine**

	Sivkov's requirements	Improved Pu111	TsKBS No 2
Date	November 1932	December 1932	April 1933
Displacement	700 / – tonnes	755 / 965 tonnes	n/a
Diesel engines		2 MAN M8V40/46	2 MAN M6V 49/48
Total power		2,800bhp	4,000bhp
Maximum speed	20–21 / 9 knots	20 / 9–10 knots	20 / 9–10 knots
Range surfaced	2,000nm at 20–21kts	1,000nm at 20kts	n/a
	8,500nm at 8.5kts	10,000nm at 10kts	10,000nm at 10kts
Range submerged	9nm at 9 knots	10nm at 9 knots	n/a
	120nm at 3 knots	150nm at 3 knots	135nm at 3 knots
Crushing depth	90m	90m	*c*100m
21in TTs	4 bow, 2 stern	4 bow, 2 stern	4 bow, 2 stern
Reloads	6	6	n/a
Main gun	1 x 100mm	1 x 100mm	1 x 100mm/51 B-24*
AA guns	1 x 0.5in	1 x 20mm or 0.5in	1-45mm/46 21-K*

* Soviet types

(or a firing squad) should the components acquired prove to have poor performance.

The Soviet delegation remained at The Hague into the New Year of 1933, conducting a detailed review of documentation and calculations.[15] To convince the Russians of the feasibility of the project, Techel invited them to inspect *E-1*. Sivkov went on board on 8 January and the entire delegation a month later. The impression was positive, and a comparison between the German boats and current Soviet designs was enlightening (and dispiriting!).

This enthusiastic assessment was shared by the Kremlin, but the Russians decided not to buy *E-1*. Instead, consent was given to proceed immediately with the drafting of a contract for the design of a submarine based on the Pu111 project. The final contract, valued at US $400,000, was signed between *Soiuzverf'* and IvS (with the collaboration of Deschimag[16]) at The Hague on 25 April 1933. The contract was for:

- development of a preliminary project and detailed technical documentation of the submarine based on specifications supplied by TsKBS No 2
- technical assistance in the production of workshop drawings
- supervision of construction of the prototype by delegating three experienced German naval architects to the Russian Ordzhonikidze shipyard at Leningrad
- hosting of two Soviet naval architects at IvS to supervise the design work
- design methodology training at IvS for four Soviet naval architects
- assistance in placing orders in Germany for the machinery and fittings necessary for the construction of two submarines
- supply of technical information concerning submarine developments in other countries.

The contract was to enter into force following a final evaluation of feasibility by the special commission, so Techel went to Leningrad a second time to agree the details. The Soviet delegation, headed by Captain 2nd Rank Arnol'd Zel'ting[17] (CO of the 2nd Submarine Flotilla of the Black Sea Fleet) visited *E-1* in Cadiz in mid-July. The boat was manned by only twelve men, and during the presentation a fire broke out in the forward battery room and the electric control system of the hydroplanes failed. Neither incident prevented positive feedback by the Russians. The contract signed on 12 August 1933 was the culmination of eight years of Soviet efforts to gain access to German submarine technology. German/Soviet cooperation proved short-lived, as relations were adversely impacted by Hitler's swearing-in as Chancellor. *Reichswehr* access to three military training and testing sites (Lipetsk, Kama, and Tomka) was abruptly terminated by Stalin within the month. However, on the German side, Hitler did not attempt to terminate Deschimag's lucrative contract as he was reluctant to upset the giants of the German arms industry.

From Project Pu111 to the 'E-2' Type

The improved Pu111 was assigned the IvS internal number Pr224, while the drawings were marked 'E2'. The Internal layout of the pressure hull generally repeated that of Pu111, with the control room some 6 metres forward of the LCG. The plans delivered in November 1933 were returned to IvS because some requirements had not been met; of these the most important were a 25 per cent shortfall in maximum range when submerged, and the guns, which were German models. It was requested that the AA gun mounting on the upper deck (see drawing) be moved to the after platform of the fin.

The amended design was designated Pr224^{II}, and the drawings were duly delivered to Leningrad in March 1934. The submarine was of the partial double-hull type; it had a riveted pressure hull with a diameter of 4.28m based on the latest version of Pu111 (June 1928) while the diameter over the engine room was increased to 4.60m. The pressure hull was lengthened by 5 metres: 3 metres at the ends of the boat, 1 metre for the engine room and 1 metre for the after battery room. The internal layout generally repeated that of Pu111 with seven watertight compartments as follows: (i) forward torpedo compartment and bunks for the crew, (ii) officers' quarters with battery banks below, (iii) control station with conning tower above, (iv) crew space with battery banks below, (v) engine room, (vi) motor room, and (vii) aft torpedo compartment and bunks.

The welded outer hull covered the whole length of the boat and housed pipework, 14 compressed air cylinders and the MBT vents, while the upper deck had the mooring fittings, the fin and a prominent net cutter at the bow. The fin housed the conning tower with the bridge placed at its forward end and an AA gun platform aft. The after part of the fin housed the air intake trunking for the diesels and general ventilation trunks with the AA gun mounted in a recess between them. The fin tapered towards the stern, and the branched intake air lines protruded beyond the sides at deck level and were masked with prominent fairings which obstructed passage fore and aft. Nevertheless, the hydrodynamics of the 'tear drop' fin with its protruding fairings offered less resistance than one with straight sides rounded at both ends. The 100mm gun forward of the fin was enclosed in a prominent shaped housing that was aligned with the fin. The nine main ballast tanks, which were faired with the outer hull, included two bow and two dual-use midship tanks (ballast/fuel). Two of the fuel tanks were located within the pressure hull beneath the batteries. The emergency diving and trim tanks were aligned with the main ballast tanks outside the pressure hull.

In April 1934 a new design bureau, SKB, was set up in Leningrad as a department of TsKBS No 2 tasked with the production of working documentation for the *E-2* type. Headed by a Communist Party *apparatchik*, the bureau employed the top Soviet submarine designers: Malinin, Kriuger and Bazilevskij – all freshly released from prison.

Project works were supervised by German engineers from IvS: Leopold Bänder, Ebschner, and Mauritz.[18]

A clash of cultures was readily apparent. When IvS engineers refused to make changes or introduce technical innovations beyond the signed contract free of charge, the Russians found this difficult to accept. The Germans likewise declined to hand over theoretical calculations and results of the model tests, because the inexperienced Soviet negotiators had not provided for this in the contract. Instead, to the annoyance of the Russians, they assured them that the technical parameters outlined in the contract would be met, and that this was covered by the contractual guarantee terms.

On 4 August 1934 Voroshilov approved the construction of three boats to the *E-2* design, under the designation 'Series IX', and preparations began in the Baltic Works[19] in Leningrad. In late 1934 six Soviet constructors were sent to the IvS office at the AG Weser shipyard in Bremen,[20] where the *E-2* draft plans were listed under the designation Job No 1100. Due to the lack of trust in Malinin and his colleagues, novice constructors were delegated, thereby hampering the possibility of acquiring knowledge regarding design processes and calculations.

Table 5: **Project Pu224^{II} Characteristics**

Date	March 1934
Displacement	828 / 1,080 tonnes
Length, oa	77.8m
Beam	6.4m
Mean draught	4.0m
Machinery	2 MAN 6V49/48 diesels, 4,000shp = 20kts
	2 electric motors, 1,100shp = 9kts
Range [nm at kts]	10,200 at 10kts / 135nm at 3kts
Oil fuel	115 tonnes (incl 37t internal tanks)
21in TTs	4 bow, 2 stern
Reloads	6
Main gun	1 x 100mm/51 B-24
AA gun	1 x 45mm/46 21-K
Complement	40

From the 'N' Class to the 'S' Class

Hulls numbered 266 and 267 were laid down on 29 and 31 December 1934 respectively, to be named *N-1* and *N-2* (naming decided on 11 September 1934); the 'N' stood for *Nemetskaia* (German). The chief constructor was Malinin, supported by the IvS engineers Richard Wagner, Seligmann, and Ebschner.

One can only imagine the headache that NKVD (the new name for OGPU from 10 July 1934) officers had 'protecting' Soviet secrets and preventing Soviet citizens from having contact with the foreign engineers. A separate office was set up to limit access to the Germans and to protect the latter from accidental contact with the designers and drawings of the other Soviet projects. Appropriate security procedures had to be created in the shipyard, which only served to hamper the process. The Germans noted the growing distrust and the high level of secrecy. They could walk from the shipyard office to the hull only along a certain path and were accompanied by an 'interpreter' – in reality a secret agent of the NKVD. Access to the workshops was not permitted, with a few extraordinary exceptions. They were allowed to stay at the yard only until 4pm, while the work was carried out on board until 8pm and during night shifts.[21]

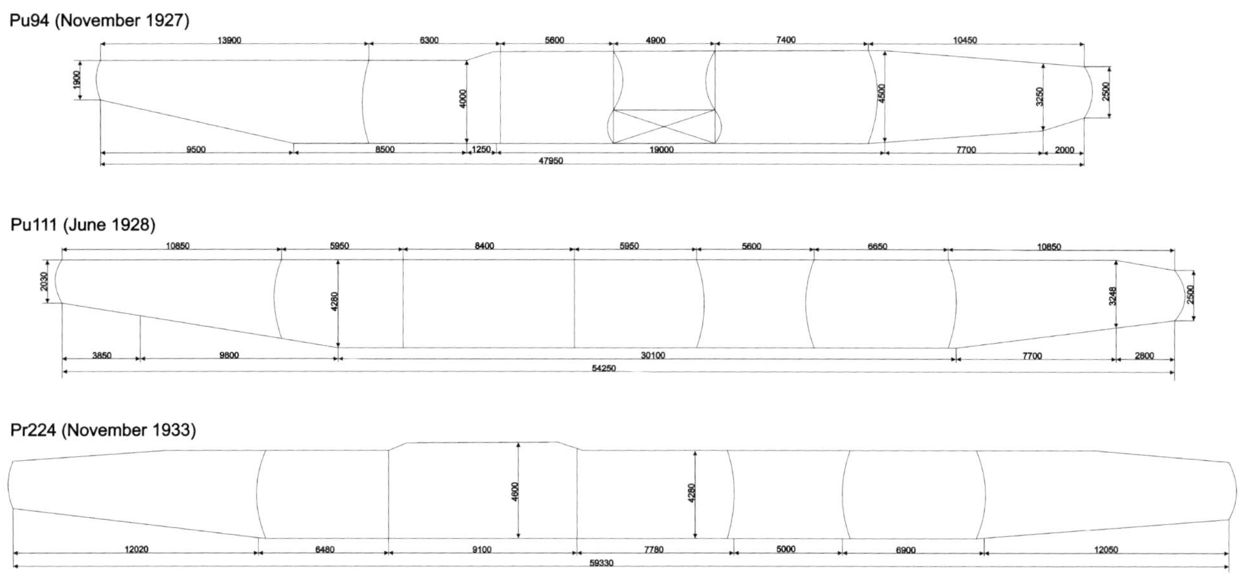

Evolution of the pressure hull design. The November 1927 version (Pu94) marked a significant deviation from the UG project (see drawing on page 161), with the control room and conning tower moved forward and placed between the accommodation/battery compartments, which changed the longitudinal balance. In the June 1928 Pu111 design (*E-1*) there was an increased imbalance, with the longitudinal centre of gravity (LCG) even farther aft of the control room due to the adoption of heavy diesel engines and additional equipment. The Pr224 design (*E-2*) of November 1933 further increased this imbalance as it featured a super-heavy diesel engine. Note the increase in the diameter of the hull over the engine room. (Drawn by Jerzy Lewandowski)

Complete sets of machinery and fittings were ordered in Germany for the first pair of boats only. Machinery and fittings for the third boat were to be copied and produced in the Soviet Union. The total value of the orders was 17 million RM. The main items were as follows:

- main diesels (German MAN)
- batteries (German AFA)
- friction clutches (German Bamag)
- thrust bearings (British Michell)
- centrifugal pumps (French Rateau and German Brown-Boveri)
- compressors (German Brown-Boveri)
- electric steering gear (German Siemens)
- changeover switches (German Brown-Boveri)
- high-pressure air tanks (German Mannesmann)
- air blowers (German Brown-Boveri and Buchi)
- periscopes (Italian Galileo and German Zeiss)
- hydrophones (German Atlas Werke)
- gyrocompasses (British Sperry and German Anschütz)
- echosounders (German Anschütz).

The working documentation was ready in March 1935 and the SKB bureau was closed, as the only reason for its existence was to keep the Germans away from TsKBS No 2. To facilitate construction of the 'N'-class submarines, the bureau was put in the hands of the Baltic Works in October, to be renumbered TsKB-18 in January 1937. The final task of SKB was the adaptation of the working documentation to incorporate Soviet-built diesel engines. The third boat (and the first to be fitted with them), *N-3* (hull no 268), was laid down on 25 April 1935.

The hull of *N-1* was launched on 8 August 1935 with a launch weight of 408 tonnes; with the exception of the Kingston valves nothing mechanical was on board. The Russians worked in a hurry: by 4 September the main electric machinery was brought on board and assembled within five days; the diesel engines were run by 22 September. The shipyard management aimed to complete the boat by 7 November – the 18th anniversary of the Bolshevist coup. Another motivation for this unaccustomed haste was to dispense with the Germans as soon as possible in order to secure the NKVD's grip on the yard and speed up construction. At the beginning of December *N-1* went to a sea for the first time, crossing the Gulf of Finland from Leningrad to Kronshtadt and back. On 15 December the first diving trials were conducted in the River Neva.

The prototype *N-1* during trials in late in December 1935 or early 1936, with the gun fairing but without the gun mounting. Note the Yard Number painted on the fin. (Boris Lemachko collection)

Dock and sea trials were completed by August 1936, and as the trials progressed *N-1* built a reputation for success. In April it was suggested that the boat be named *Voroshilovets*, while Molotov and Kalinin waited in a queue for *N-2* and *N-3* to be named *Molotovets* and *Kalininets* respectively. It was Stalin himself who apparently put a stop this, and the boats retained their numbers as originally planned. However, the pennant 'N' was a reminder of the German origin of the project, so the following year a new classification was introduced, officially designating these ships as 'medium' submarines. Medium translates as *Sredniaia* in Russian, so the 'N' class was renamed the 'S' class[22] on 20 October 1937. The change affected only ships referred to as the 'N' class (the *Shch* designation remained unchanged[23]), thereby confirming the political motivation behind this move. The German origins of the project were kept secret, and were finally revealed only after *Perestroika*. The respected Russian magazine *Sudostroenie* (= 'Shipbuilding'), in an article published in 1975 on the 'S' class, noted that the ship was co-designed by Malinin, and there was no mention of Techel, Blum or the IvS whatsoever.

N-1 ran acceptance trials at the turn of August and September 1936, during which her superiority over all other Soviet submarine classes was confirmed. The favourable opinion of Soviet users was a double-edged sword: the high quality of the design contrasted with the deplorable state of the remaining Soviet submarine fleet. Most striking was the practical arrangement of action stations, mechanisms, electrical and piping systems as well as fittings, with easy access for operational deployment and maintenance. The boat was considered remarkably seaworthy, able to run on the surface in heavy seas, and to dive and surface in adverse weather conditions, while her long periscopes meant that she was not in danger of being exposed. Diving was fast and the whole process simple to manage. There were no restricted shaft revolution zones for the diesels. At full ahead on the surface *N-1* was able to turn 180 degrees in 3 minutes (9 minutes at 6 knots when submerged). Firing torpedoes caused only an insignificant wake on the surface because of the relatively short length of the tubes and the clever design of the bow casing around them.

Later, Russian naval publications tended to downplay the design's merits, and focused on a laundry list of minor improvements compared to Soviet practice at the time. Neither did Bazilevski, who in his memoirs – written in 1977 and therefore parsimonious in its praise for this German project for reasons of personal safety – focused on numerous things. His list of advantages (the original order is preserved) comprised:

- elliptical CT cylinder with stiffeners only in the area of greatest curvature, which reduced drag and allowed for the housing of all periscopes and steering gear
- aft TT shutters in the form of a half-cylinder revolving around the horizontal axis, with a plated surface hidden in the outer hull when rotating and uncovering the TT wells
- prominent net cutter comprising a steel saw with large teeth above the deck supported on stand-up collars; thanks to the streamlined connection between the stem and the deck, the bolt ropes of nets and boom defences slipped off onto the cutter; it showed outstanding performance when, during trials, the moorings of a small dredger in Kronshtadt were cut
- hydraulic control of hydroplanes, periscopes, Kingston valves and MBT ventilation valves
- Kingston valves with two connected flaps, one flap opening inside the tank, the second to the sea, so that a force equal to the differential pressure was sufficient to open or close the flap
- thin-walled valve castings of the Kingston valves
- high-pressure valve boxes of alloyed brass
- the possibility of copying purchased mechanisms and devices to launch their production in Russia[24]
- outstanding body lines with low drag and high hull efficiency
- battery tanks without an inspection corridor; instead a small trolley was fitted which allowed an electrician lying on it to move along the tank, having access to inspect and maintain the batteries; the volume saved by eliminating the corridor allowed the height of the tanks to be lowered to the advantage of the accommodation spaces above.

A more objective assessment, free of censorship, was given in 1996 in the authoritative work *History of Domestic Shipbuilding*[25] by an editorial team headed by Igor Spasskij, general designer of 'almost 200' Soviet and Russian nuclear submarines: 'Development of the Series IX boats was of great importance for domestic submarine shipbuilding. Many design solutions and types of mechanisms that were first adopted for these boats were implemented in subsequent projects.'

On the other hand, the list of complaints reflected in

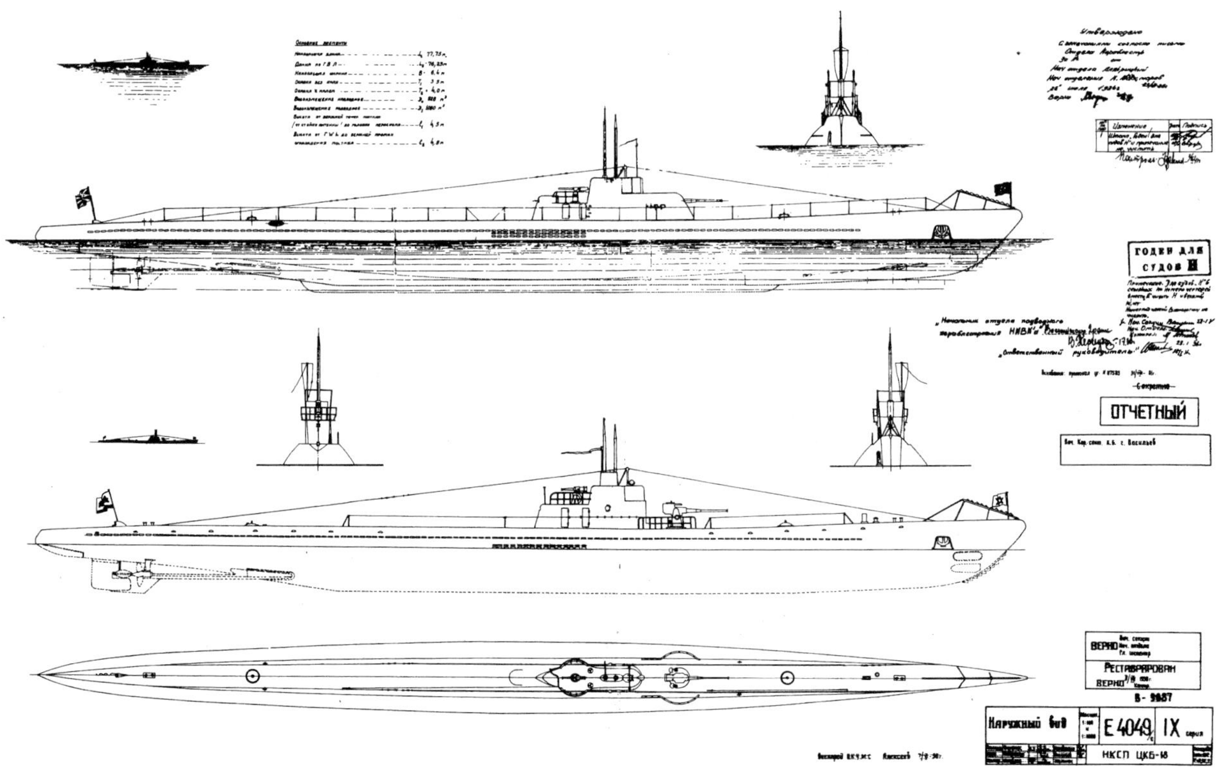

Official drawing No E4049/s from TsKB-18, an appendix to the acceptance protocol of the Series IX conversion, which featured the removal of the main gun fairing and a stiffening of the periscopes. The plans were declassified in 1961.

The upper drawing depicts the original design from March 1934. Note that the 45mm AA gun, already moved from the main deck to the fin platform, is mounted in the recess between the air intakes and ventilation ducts. This position was necessary because of its impact on stability of the heavy mounting, but was not convenient for operating the gun. This was changed during the final stage of construction, apparently at the request of the Navy, so the recess was plated up and the gun relocated to the higher position. The large stamp on the right has the text 'годен для судов Н' (fit for N type).

The lower drawing depicts the Series IX after conversion, with the main gun fairing removed and the periscopes stiffened. The 45mm AA gun is already in its the final position, but this was done before trials, so was not an aspect of the 1938/39 conversion. The final drawing was accepted in July 1939. (Author's collection)

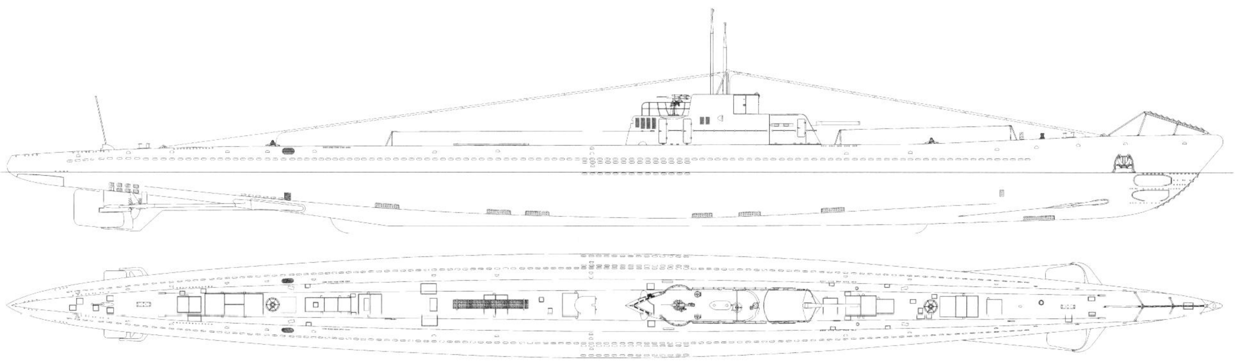

S-1 as completed in 1938 with her armament complete. The Series IX boats had the free-flood holes in two rows, the upper in a single uninterrupted grouping. (Drawn by Jerzy Lewandowski)

some respects the inadequacies of Soviet submarine practice. The one considered most significant was the location of the main gun on the upper deck, not on the bridge deck as it was in *L-55* and subsequent Soviet boats. The latter feature was intended to provide protection for the gun crew against waves washing over the casing when operating in the North Sea. For submarines operating in the Baltic and Black Seas this solution proved to be unnecessary, and the increased topweight affected stability. The second was the capacity of the fresh water tank, which was considered too small – it appears that Russian water consumption and usage was less disciplined than German practice. The third complaint related to battery charging practices, as the output of a single diesel was considered insufficient to recharge the batteries. This 'defect' became apparent in Arctic conditions, where only ships with Soviet-made batteries served, and may therefore relate to the mediocre quality of the batteries.

The list of more substantial flaws comprised:

- The boats failed to reach their designed surface speed by 0.5 knots.
- The fairing of the main gun was cramped and inconvenient in operation – the fairings were mounted only on *N-1* and *N-2*, and the idea was abandoned after trials; no other boats were so fitted, and the existing fairings were removed during the winter of 1938/39.
- The periscopes tended to vibrate when fully raised at high speeds – this defect was eliminated by extending the bases of the periscopes – introduced from *N-3* and retro-fitted in the earlier boats during removal of the gun fairings.
- If MBT No 2 was used for fuel stowage it made the bow heavy with a noticeable trim, so the fuel in this tank was to be used first, before reaching the open sea – the German origins of the project need to be borne in mind.
- The sliding fairing of the bridge was considered inconvenient in use and failed to protect against spray and wind – the first attempts to increase protection against the elements were small wave deflectors fixed (often at the initiative of the CO) to the upper rail of the bridge fairing.

Arguably the most serious deficiency, acknowledged by the Soviets, was that their most experienced designers were not allowed to participate in project selection and negotiations with IvS.[26] The second-rate Soviet designers were not able to talk as equals in a way that would enable them to be treated as partners by German engineers. Bazilevskij complained that it was only in 1935 that, while chatting with Mauritz, he learned about the use of elastic pads as a means to reduce vibration of machinery and resist shocks that were a feature of German submarine projects. The immediate request to employ elastic mounting of machinery on the *E-2* project was rejected, as it was not in the contract, and obtaining approval for a change that would require additional funding and which would therefore have to be put through the whole Soviet military bureaucratic machinery, was beyond the possibilities of the former prisoner. The latest Soviet submarines were, therefore, to remain as noisy as their predecessors, and could be detected from a distance five to six times greater than their foreign counterparts.

Other deficiencies could be traced to the experience of the *Kriegsmarine* with the trials of the Type I submarines, which were in fact sisters of the *E-2* type. Following completion of two boats (*U-25* and *U-26*), which took place five months before the completion of *N-1*, the Germans decided to discontinue the project and replace it with the Type IX. There were many reasons for this decision (including both insufficient range and the number of reloads), which was not influenced by Doenitz's preference for the smaller, more manoeuvrable Type VII boat.[27] In Type IA depth-keeping was a quite an arduous business, and the submarine had a tendency to instability during a crash dive and at periscope depth. These boats were quite demanding, and among the crews there was a belief that newly-trained personnel who mastered the Type IA would successfully cope with any other submarine. The primary reasons for these difficulties were considered to be:

- the forward position of the control room and the 6-metre separation from the LCG, which meant that the position of the CO and the first officer was less intuitive, especially in a situation requiring rapid evasion.

- the ventilation system of the fuel bunkers, which did not work reliably: air bubbles wandered forward or aft depending on the trim, and the volume changed depending on the depth kept.

The Germans resolved this by moving the control room and the CT one compartment farther aft in the Type IX design. This problem went largely unnoticed by the Soviets, as the boat was apparently much more controllable and manoeuvrable than their own designs. It is possible that more experienced COs quickly got to grips with it, whereas for the the Germans, who were training commanders *en masse* and quickly in wartime, it was significant.[28]

Series Production Boats

Series IX-bis

Following positive results of the first trials of *N-1*, the decision was taken to start series production of this type to replace the *Shch* class as the standard medium submarine. The key prerequisite was starting the production of copies of engines and equipment in Russia, based on the documentation and devices obtained under the contract with Deschimag. In the event the hull of *N-3*, renamed *S-3* from 20 October 1937, remained immobile after launch awaiting the availability of the Soviet-made fittings. The key element was the MAN M6V49/48 diesels, which were to be replaced by the 47LN8 type developed by the Kolomenskij Works;[29] this was a four-stroke, eight-cylinder, non-reversible type with mechanical supercharging rated at 2,000hp at 475rpm. As it had no built-in water pumps for circulation, autonomous electric cooling pumps were to be installed.

The diesels required an increase in the diameter of the outlet for the exhaust gases from 360 to 420mm, so the silencers had to be reworked. The engines revealed serious flaws during factory trials – the crankshaft was too light so cracks developed, and blower performance was unsatisfactory – so had to be reworked with turbine supercharging and received the designation 1D.[30] To avoid sea water getting onto the hot blades of the gas turbine of the supercharging unit when starting and stopping engines, a non-return flap was provided in the silencer. The changes required a redesign of the fuel and oil feed systems as well as water cooling and air intake piping, which could be seen externally; the exhaust was moved forward by six frames, and the angle of inclina-

Two photos of *S-2* taken by the Finns on 28 August 1938. The fairing for the main gun is prominent. (Sotamuseo, Helsinki)

Contemporary German submarine development by class. From top to bottom:

Type UG (1918)

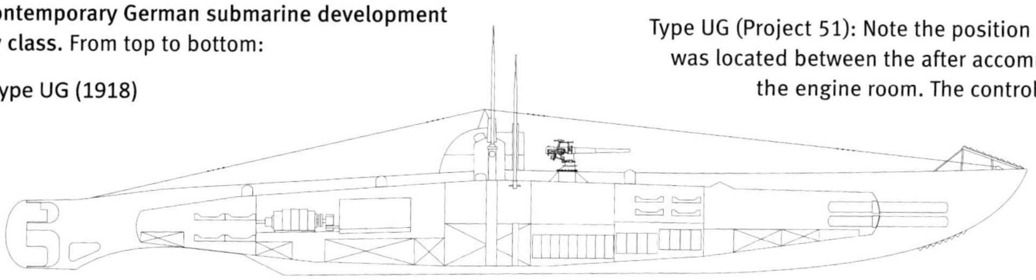

Type UG (Project 51): Note the position of the conning tower, which was located between the after accommodation/battery room and the engine room. The control room was close to the LCG.

E-1 (1930)

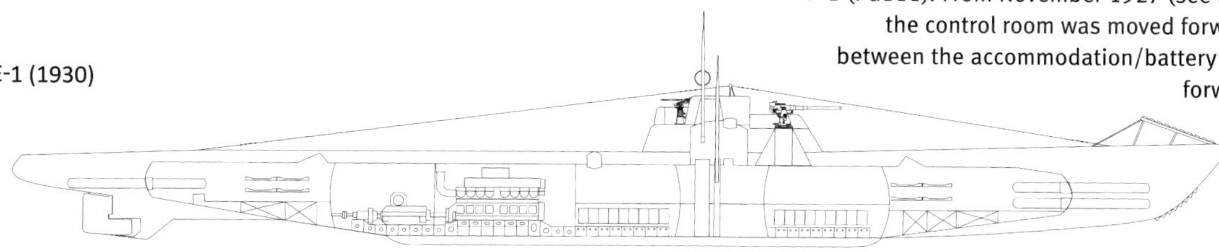

F-1 (Pu111): From November 1927 (see earlier drawing) the control room was moved forward and placed between the accommodation/battery compartments, forward of the LCG.

Type I (1933)

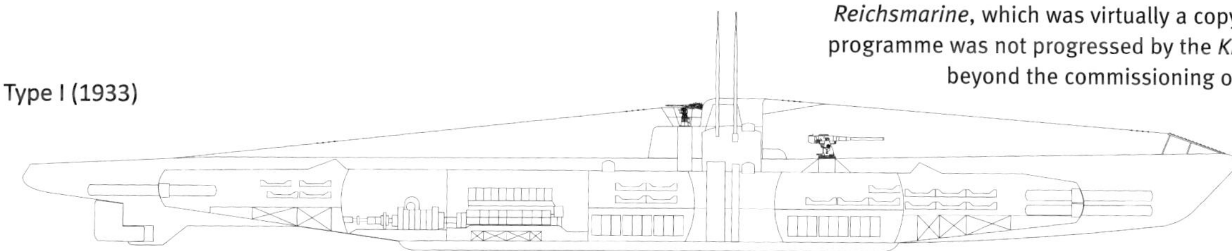

Type I: The mobilisation contingency type of the *Reichsmarine*, which was virtually a copy of *E-1*. The programme was not progressed by the *Kriegsmarine* beyond the commissioning of two boats.

E-2 (1933)

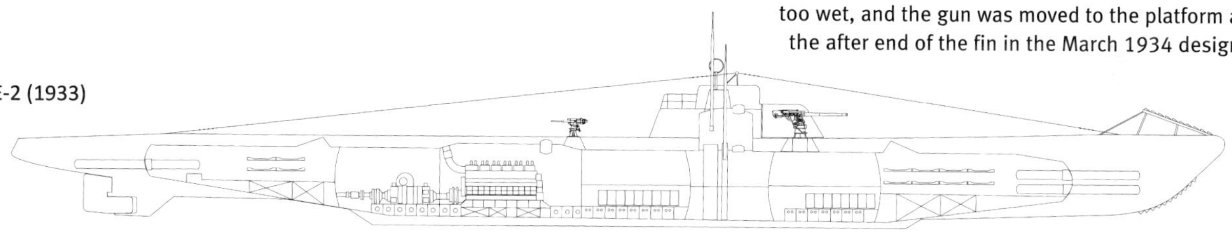

E-2 (Pr244): An enlargement of *E-1* and Type I. The adoption of heavier diesels placed the LCG farther aft in relation to the control room. Note that the AA gun is mounted on the main deck abaft the fin in this November 1933 design. This location was considered by the Soviets as too wet, and the gun was moved to the platform at the after end of the fin in the March 1934 design.

Type IX (1935)

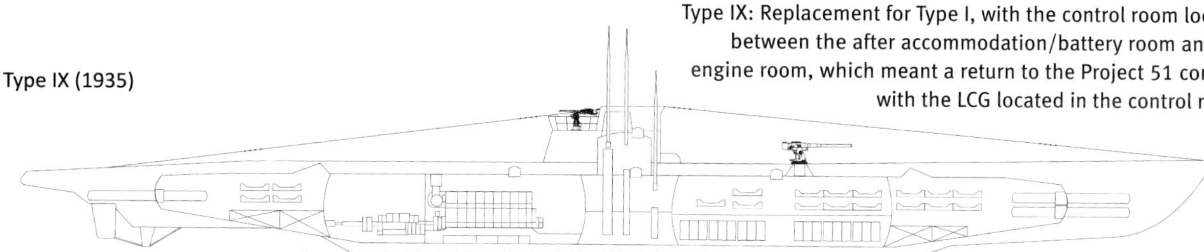

Type IX: Replacement for Type I, with the control room located between the after accommodation/battery room and the engine room, which meant a return to the Project 51 concept with the LCG located in the control room.

Type VII (1935)

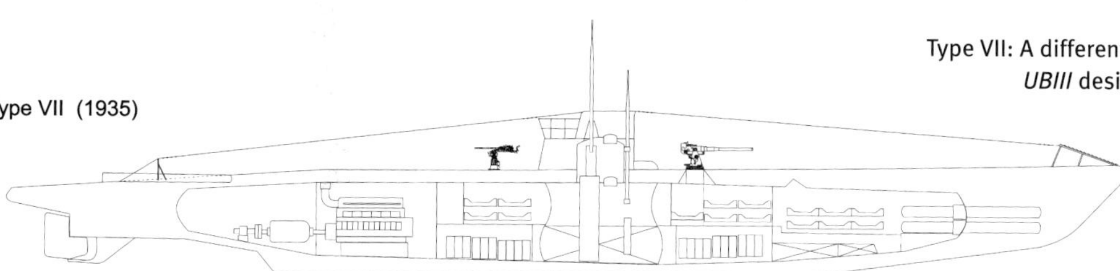

Type VII: A different layout, derived from the *UBIII* design of the First World War.

(Drawn by Jerzy Lewandowski)

tion of the air intake pipework protruding from the fin was changed. The production of other fittings also took longer than expected; this led to the order of supplementary Series X-bis boats (*Shch* class) in 1938 in an attempt to bridge the gap in delivery schedules.

S-3 was finally completed on 8 July 1938 and commissioned a few days later. The modified workshop documentation received the designation Series IX-bis and became the template for series production. The first batch of three boats was ordered at the end of 1935 from Works No 189 (Ordzhonikidze) at Leningrad to test the feasibility of production of the design and iron out any flaws. Large-scale series production was to take place at Works No 112 at Gorki (Krasnoe Sormovo), No 196 at Leningrad (Sudomekh) and No 198 at Nikolaev (Andre Marti). The lead shipyard (Works No 189) and Works No 194 at Leningrad (Andre Marti) were to build sections and supervise assembly at Works No 202 at Vladivostok (Dal'zavod). In order to streamline the production process, changes to the working drawings, even when based on operational experience, were forbidden.

The sliding fairing of the bridge was retained and no other changes were allowed except for small wave deflectors added on the initiative of the COs – the author suspects this may have been a trade-off in exchange for turning a blind eye to certain manufacturing defects. Other changes were to be introduced during routine or emergency shipyard repairs. At first, on some boats, a mushroom cap shield with scuttles was installed atop the fairing, with the sliding fairing retained. Later, the sliding fairing was removed and the bridge was plated up, with the compass column at its forward end (see the drawings of *S-9* and *S-56*).

In total 40 boats of Series IX-bis were laid down during the period 1936–40, of which 24 were completed to the original design by 1943. Further units were completed between 1944 and 1948 to the improved Series IX-bis2 design.

Series XVI (Proekt 97)

On 29 January 1940 the TsNII-45[31] (*Tsentral'nyj Nauchno-Issledovtel'nyj Institut* = Central Scientific-Research Institute) drew up the requirements for a new type of submarine based on the 'S' class coded 'SU' ('U' for *Uluchshennaia* = Improved). The list of changes included no fewer than fourteen improvements, the most important of which were:

- all-welded construction of the pressure hull
- number of reloads increased from six to eight
- wakeless firing of the torpedo tubes
- change to the 4D type of new-generation diesels with mechanical supercharging
- elastic pads for the diesels and key mechanisms, and flexible connections of internal systems
- ice strengthening of the bow (8mm plating at the waterline)
- strengthening of internal fuel tanks.[32]

TsKB-18 was opposed to the work involved in a new submarine project, and proposed instead a modification of Series IX-bis. The modified design was assigned the code Proekt 97 – it was also known as 'SM' (*Svarnaia Massovaia* = Welded Mass-produced). The design, except for the changes already mentioned, was practically a repetition of the earlier project with an enclosed bridge, which increased drag and reduced submerged speed by half a knot, and surface range was reduced to 8,200 miles,[33] which was nevertheless a significant improve-

S-1 in 1940 following her reconstruction: the fairing for the gun was removed and stiffening for the periscopes added. The salvage vessel *Kommuna* (ex-*Volkhov*) is visible in the background. (Boris Lemachko collection)

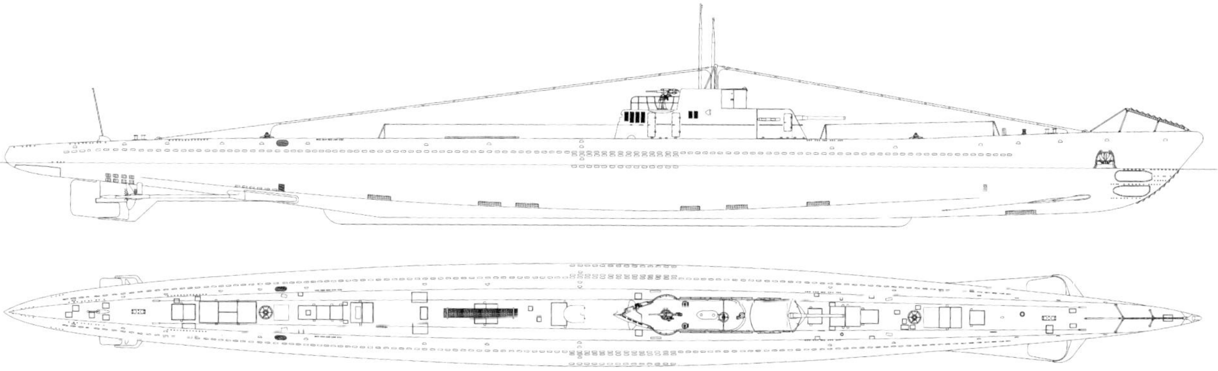

General Arrangement of the Series IX-bis boat, which featured an enclosed bridge. (Author's collection)

ment as compared with the performance of other Soviet boats, including the submarine cruisers of the 'K' class. The obsolescent 45mm 21-K gun, which was practically useless against modern aircraft, was replaced by a 0.5in DShK machine gun.

The new design was assigned the official naval designation Series XVI and was approved in November 1940. The intention was that it would replace Series IX-bis with immediate effect: twelve hulls had been laid down with a further eleven allocated to the Gorki, Leningrad and Nikolaev shipyards from December 1940 up to the German invasion in June 1941. In July 1941 construction of the hulls already laid down was suspended and the remaining boats were cancelled. The hulls at Nikolaev were sabotaged on the stocks by the retreating Soviets in August 1941. Further progress on the workshop drawings was hindered by the evacuation of TsKB-18 from Leningrad to Gorki, and detailed design work was not completed until 1942. Construction of the boats at Leningrad was not resumed due to blockade shortages,

Two close-ups of the bridge of *S-1* in 1940. In the first image, note the sliding bridge fairing with its lock mechanism at the side and the wave deflector above. In the second the Red Banner flag is prominent; it was awarded to the ship following her 'outstanding combat performance' achieved during the Winter War with Finland. (Author's collection)

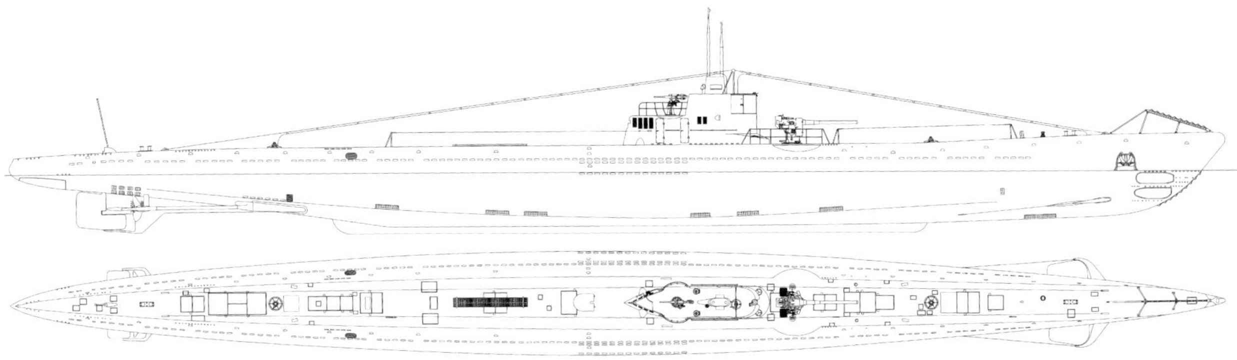

S-9, a Series IX-bis boat, as completed in 1940 without the main gun fairing, with the diesel exhaust moved forward and more prominent air intake pipes protruding from the fin. The early Gorki-built boats could be distinguished by four forward groups of the upper row of free-flood holes. (Drawn by Jerzy Lewandowski)

while the Gorki yard was too busy with other wartime orders to continue work on the prototypes. During 1942–43 Proekt 97 was improved and became Proekt 601, but was discontinued as it had been overtaken by technological progress. The incomplete hulls were dismantled after March 1947.

Series IX-bis2

Thirteen hulls of the Series IX-bis were completed during 1943–48 to the Series IX-bis2 configuration, which was aligned with Project 97 (but without the welded hull construction or the increase in the number of reloads). War experience and equipment delivered by the Allies was also incorporated. The most important changes comprised:

- a 4-torpedo salvo capability using the bow tubes and an electrical torpedo loading system
- elastic pads for auxiliary mechanisms, batteries and lighting

S-9 as completed, photographed in 1942 at Kronshtadt. (Boris Lemachko collection)

Another view of *S-9* at Kronshtadt. A *Proekt 7* destroyer can be seen in the background. (Vitalij Kostrichenko collection)

A Series IX-bis boat (probably *S-15*) under construction at Gorki, 1939. She was launched on 4 April 1940, transferred to Astrakhan in October 1941, and completed in December 1942 by Works No 638. On 24 April 1943 she departed Astrakhan for the Arctic via inland waterways, and was commissioned on 25 May 1943 into the Northern Fleet, becoming operational in November 1943. She was given the name *Kolkhoznitsa*, probably in 1944. (Author's collection)

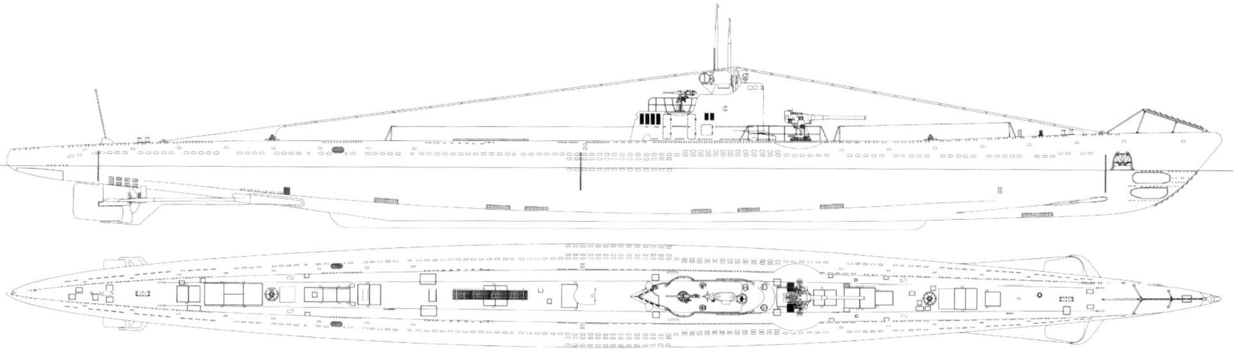

S-56, a Series IX-bis boat, in 1944, with an enclosed bridge. The Works 194 Leningrad boats could be distinguished by three rows of free-flood holes, the middle row in five groups, the lower one in two groups (one below the fin, the second at the bow). (Drawn by Jerzy Lewandowski)

- an enclosed bridge
- fresh water stowage increased to 7 tonnes
- five bunks added
- removal of the Sirius acoustic monitoring station and the Kingston valve of MBT no 1
- new systems added:
  - *Sprut* stabilisation for stationary boat
  - British Type 291 radar
  - British Type 129 sonar (Soviet codename *Drakon-129*).

Conclusions

In October 1939 the German naval high command concluded that the purchase of ten Soviet submarines could be of military significance when waging the war against British shipping. The prerequisite that the boats would be built to German plans made the 'S' class an obvious choice. Raeder also wanted to explore the possibility of building U-boats in the Soviet shipyards in the Baltic and Arctic, far from British air raids. Hitler

The hulls of *S-36* and *S-37* blown up on the slipway at Works 198, Nikolaev. The photograph was taken after the capture of the town by the Germans in August 1941. (Narodowe Archiwum Cyfrowe 3/2/0/-/890)

Table 6: **Production batches of the S class**[1]

YN	Series	Numbers allocated	LD	Comp	Number: LD	Number: Comp	Remarks
Builder: 189, Leningrad							
266 – 268	IX	N-1 – N-3[2]	1934	1938	3	3	
277 – 279	IX-bis	N-4 – N-6[2]	1936	1939	3	3	
284 – 286	IX-bis	S-51 – S-54	1937	1943	3	3	Assembled at 202, Vladivostok
Builder: 194, Leningrad							
403 – 405	IX-bis	S-54 – S-56	1936	1941	3	3	Assembled at 202, Vladivostok
Builder: 112, Gorki							
236 – 237	IX-bis	N-7 – N-8[2]	1936	1940	2	2	
241 – 242	IX-bis	N-9 – N-10[2]	1937	1940	2	2	
243 – 244	IX-bis	S-101 – S-102	1937	1941	1	1	
245 – 246	IX-bis	S-11 – S-12	1937	1941	2	2	
263 – 264	IX-bis	S-13 – S-14	1938	1942	2	2	
265 – 266	IX-bis	S-103 – S-104	1939	1942	2	2	Completed at 638, Astrakhan
271 – 274	IX-bis	S-15 – S-18	1937	1945	4	4[3]	3 completed at 638, Astrakhan, 1 (S-17) at 402, Molotovsk
295 – 299	IX-bis	S-22 – S-26	1940	1948	5	5[3]	
300	XVI	S-27	1940	–	1	0	
312 – 314	XVI	S-28 – S-30	1941	–	3	0	
315 – 319	XVI	S-39 – S-43	1941[4]	–	0	0	
321	XVI	S-44	1941[4]	–	0	0	
Builder: 198, Nikolaev							
347 – 350	IX-bis	S-31 – S-34	1937	1941	4	4	
360 – 363	IX-bis	S-35 – S-38	1940	1948	4	1[3]	S-35 completed
371 – 376	XVI	S-58 – S-63	1941	–	3	0	S-58 – S-60 laid down, S-61 – S-63 scheduled
Builder: 196, Leningrad							
132 – 134	IX-bis	S-19 – S-21	1939	1946	3	3[3]	2 completed at 638, Astrakhan, 1 (S-21) at 445, Baku
154 – 156	XVI	S-47 – S-49	1941	–	3	0	
157 – 158	XVI	S-50, S-57	1941	–	0	0	S-47 – S-49 laid down

Totals

	LD	Comp	Number: LD	Number: Comp
Series IX	1934	1938	3	3
Series IX-bis	1936	1942	40	24
Series XIV	1940	–	12	0
Series IX-bis2	1937	1948	0	13
Grand total			55	40

Notes:

1 The Ten-Year naval shipbuilding programme, presented on 6 August 1939 by *Flagman Flota 2-go Ranga* Nikolai Kuznetsov, envisaged construction of more than 140 medium submarines by the end of 1947.

2 Pennant changed 20 October 1937 to S.

3 Completed to the Series IX-bis2 project.

4 Scheduled.

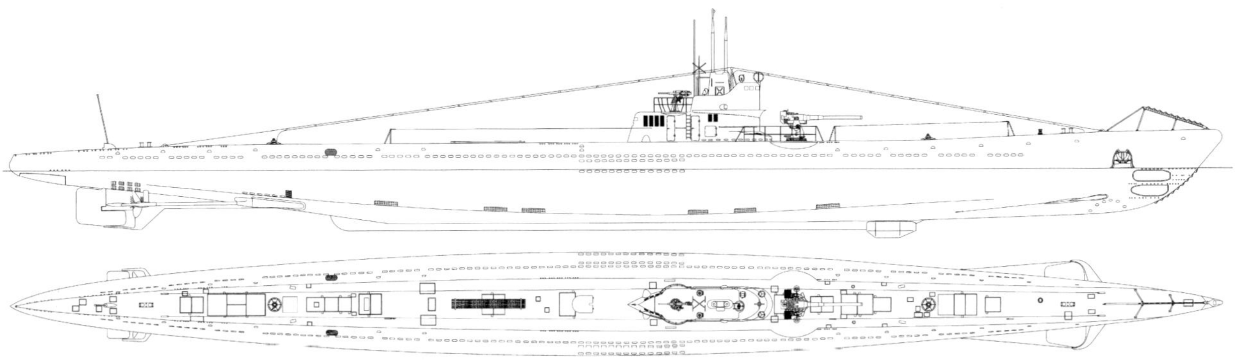

S-25, a Series IX-bis2 boat, as completed in 1948. Like early Gorki-built boats she had four groups of free-flood holes forward, but incorporated a number of changes typical of this series: Type 129 ASDIC dome, Type 291 radar antenna, a large wave deflector on the enclosed bridge, and the absence of the forward Kingston valve being the most prominent. (Drawing by Jerzy Lewandowski)

Table 7: **S Class Characteristics**

	S-1	**S-9**	**S-20**
Series	IX	IX-bis	IX-bis2
Date	1941	1943	1945
General characteristics:			
Displacement	840 / 1,075.2 tonnes	845 / 1,078 tonnes	840 / 1,075 tonnes
Length, oa	77.70m	77.70m	77.70m
Beam, max	6.40m	6.40m	6.40m
Draught, mean	4.00m	4.02m	4.35m[1]
Reserve of buoyancy	28%	20%	28%
Metacentric height	0.62 / 0.21m	0.54 / 0.185m	0.594 / 0.186m
Power plant:			
Number of shafts	2	2	2
Diesel engines	2 x MAN M6V 46/48	2 x Kolomenskij 1D	2 x Kolomenskij 1D
Unit power	2,000bhp	2,000bhp	2,000bhp
Electric motors	2 x Elektrosila PG72/35	2 x Elektrosila PG8	2 x Elektrosila
Unit power	550	550	550
Battery	2 x 62-cell AFA 38-MAK-360	2 x 62-cell Works No 236 LS	2 x 62-cell Works No 236 S
Speed	19.5 / 9.1 knots	19.4 / 8.51 knots	18.31 / 8.02 knots
Operational capabilities:			
Operational / crush depth	80 / 100m	80 / 100m	80 / 100m
Range surfaced	3,380nm at 19.5kts 9,860nm at 10ts	n/a 8,340nm at 9.5kts	3,995nm at 18.3kts n/a
Range submerged	9.1nm at 9.1kts 148nm at 3kts	10.2nm at 8.51kts 140nm at 3kts	9.06nm at 8kts n/a
Endurance	30 days	30 days	30 days
Continuous time submerged	72 hours	72 hours	72 hours
Diving time	68 secs	60 secs	60 secs
Oil fuel normal / max	39.4 / 100 tons	40.9 / 125.8 tons	30 / – tons
Fresh water	3.8 tons	3.84 tons	8.2 tons
Armament:			
21in TTs	4 bow, 2 stern	4 bow, 2 stern	4 bow, 2 stern
Reloads	6	6	6
Main gun	1 x 100mm/51 B-24 [200 rounds]	1 x 100mm/51 B-24 [200 rounds]	1 x 100mm/51 B-24 [200 rounds]
AA gun	1 x 45mm/46 21-K [500 rounds]	1 x 45mm/46 21-K [500 rounds]	1 x 45mm/46 21-K [500 rounds]
MG	1 x 0.3in DP 1 x 0.3in Maksim [500 rounds]	1 x 0.3in DP 1 x 0.3in Maksim [500 rounds]	1 x 0.3in DP 1 x 0.3in Maksim [500 rounds]
Communications:			
LW transmitter	Okun'	Okun'	Okun'
SW transmitter	Briz-MK	Briz-MK	Shchuka
LW receiver	Groza	Groza	Dozor
SW receiver	Purga	Purga	Purga
Sensors:			
Radar	–	–	Type 291[2]
Asdic[1]	–	–	Drakon 129[3]
Hydrophone set	Mars-12	Mars-12	Mars-12
Direction finder	Gradus	–	Burun
Sonic station	Sirius	Sirius	–
Complement	46	45	51
Officers/NCOs/men	8/16/21	8/16/21	8/?/?

Notes:

1 With ASDIC dome
2 British model
3 British ASDIC Type 129

Loading a torpedo on board *S-31* on 20 July 1944. The Soviets complained of the workload involved and the complexity of the torpedo embarkation procedure. (Author's collection)

rejected these requests, as he did not want Stalin to know the serious weakness of the *Kriegsmarine*. He was approached again with this request by General Wilhelm Keitel, Chief of the *Oberkommando der Wehrmaht*, but this request was likewise rejected on the grounds of the poor quality of the Soviet-built submarines. The latter judgement was based on the observations of the German naval attaché in Moscow, *Fregattenkapitän* Norbert von Baumbach, and his view was borne out by the numerous accidents that plagued the Soviet submarine force.[34]

In June 1941 the Germans seized *S-1* almost intact, the boat having been scuttled ineptly by her crew at Liepāja (Latvia). She was raised and towed to Kiel, where she was used as a test boat by the *Schiff- und Maschinenbau-Versuchsanslalt des Kriegsmarinearsenals Kiel* during a series of trials held 20–27 June and 26 July–7 August 1943 before being finally sunk by depth charges. The trials did not reveal any surprises: the pressure hull reacted elastically to shocks, as did the compressed air cylinders mounted on it. Construction was considered weaker than that of the German Type VIIC and Type IXC boats; however, this is unsurprising, as they were built taking war experience into account. The advantages of mounting key equipment on flexible pads was confirmed during tests, as even electrical fuses remained intact when housed in fuse boxes mounted on steel springs.

The wartime performance of the 'S' class submarines was significantly better than that of their Soviet-designed counterparts. Although numerically they made up only 24 per cent of the Soviet medium submarine force, they accounted for 39 per cent of sinkings. They did particularly well in the Arctic, where their comparatively large size and range counted, with 60 per cent of sinkings. Achievements were less impressive in the Black Sea, where they were few in number. The most successful boat of the Soviet submarine force was *S-13* which, under *Kapitan 3-go Ranga* Aleksandr Marinesko, sank over a

S-31 of the Black Sea Fleet in 1947. Note the distinctive fairing of the bridge, which retained the original sliding system topped by a mushroom-like cap with scuttles. (Vitalij Kostrichenko collection)

S-24 of the Series-IX-bis2, with the modified fin and the British Type-291 radar antenna. The photo was taken in 1949 during transfer from the Arctic to the Far East via the Northern Sea Route. (Author's collection)

Close-up of the bridge of *S-25* of Series IX-bis2. Note the prominent wave deflector around the forward edge of the fin. (Boris Lemachko collection)

period of 11 days the German liners *Wilhelm Gustloff* and *General Steuben* of a combined 40,000grt, with the loss of more than 9,000 lives. That became possible only in 1944/45, when a favourable strategic situation was achieved in the Baltic by an advancing Red Army, air supremacy, and German shipping focused primarily on evacuation. During the last seven months of the war, Soviet submarines of the Baltic Fleet sank as many ships as during the preceding 36 months.

A by-product of the secret arms programme of the *Reichsmarine*, the 'S' class caused difficulties for Russian historians once the origins of the design had been declassified. The Russians attempted to ignore the essentially illegal nature of the activities[35] and the participation of the *Reichsmarine* in the construction of *E-1*. Secondly, they exaggerated the contribution of Soviet designers to the *E-2* type project. And thirdly, they stressed the importance of the 'S' class for the development of German submarine technology(!). They even seemed to take pride in the contribution these boats made to the improvement of the Type VII and Type IX U-boats.

In reality, after the ex-British *L-55* (see *Warship 2020*), the 'S' class was the second breakthrough in Soviet submarine design and construction, broadening the knowledge of designers and the experience of shipwrights, and influencing product development in Soviet industry. The experience gained on each of them shaped the programmes that followed: in the case of *L-55* during the period 1928–1934 and for the 'S' class the period 1935–1945. The third impetus to development took place in 1944/45 with the influx of Western technology: weaponry, sensors, engines delivered by the Allies under Lend-Lease, documentation, devices, ships and construction blocks looted in Germany, and German naval architects abducted to the Soviet Union. All this provided the foundations for a new *U-bootwaffe*, this time a Red one, which in less than decade was to menace the West.

Acknowledgments:

The Author wished to thank: R-Adm Andrzej Ogrodnik, CO of the 3rd Flotilla of the Polish Navy, and Lt-Cdr (Ret) Edward Olewiński, former CO of the submarines *Orzeł* and *Bielik*, for their support with an understanding of submarine steering; Prof Serhat Güvenç of the Kadir Has University at Istanbul for his information on *Gür* and the submarine service of the Turkish Navy; Alberto Mantilla Pérez (www.buques.org) for the photographs from the Echevarrieta y Larrinaga shipyard; Burak Reis for sharing illustration and information on the German-built Turkish submarines; Marek Twardowski for his support on the British types of Asdic; and last, but not least, Päivi Vestola of the Sotamuseo, Helsinki, for researching the photographs; and Jean Moulin for revising the photo caption on page 150 and supplying details of the fates of the submarines.

The IvS and shipyard drawings are copies of the originals published in the book *'Ski' v boiu* by Miroslav Morozov and Konstatin Kulagin. They were adapted for publication by Jerzy Lewandowski. Drawings of the pressure hulls of Pu111, Project 51, Pu111 and Pr224 are based on drawings in Eberhard Rössler, *Die deutschen U-Boote und U-Bootentwüfre zwischen den Weltkriegen 1922–1939*. Drawings of the Type I, VII and IX U-boats have been based on drawings published in Erich Gröner, *Die deutschen Kriegsschiffe 1815–1945*.

Endnotes:

1 Karl Bartenbach (1881–1949) was one of the first COs of *U-1*, the first submarine of the German Navy. In 1915 he was appointed commander of the *U-Flotille Flandern*. He was retired in 1920 but, as the leading German submarine expert at that time, he was involved in supporting and promoting sales of the IvS-designed submarines.

2 Dr Hans Techel (1870–1944) was from 1907 the Director of Construction at the GW, where he supervised the design and construction of more than 100 submarine types. He was an advocate of the introduction of the diesel engine. From July 1922 to August 1938, he was the technical director of IvS with more than 400 designs for submarines drawn up under his leadership. In 1940, on his 70th birthday, Techel was decorated by Hitler with the *Goethe-Medaille*.

3 Urlich Blum (1892–1937) was the executive officer of *U-1* when commanded by Bartenbach. In 1914 he became Military Liaison Officer in the *Uboot Inspektorate*, and after the war was instrumental in devising the concept of the IvS. He retired from the Navy in 1920.

4 Son of a powerful German shipping magnate and a prominent opponent of the arms restrictions imposed on Germany

S-102 with an 'L'-class submarine minelayer in the background. Note the 'mushroom' cap atop the fin with sliding fairings divided by the compass column. (Boris Lemachko collection)

S-51 in 1947. Note the removal of the net cutter and the peacetime paint scheme, comprising a white band at the waterline and white framing of the bridge windows. (Vitalij Kostrichenko collection)

by the Treaty of Versailles, he worked from 1920 to 1927 as head of the maritime transport department 'to combat and mitigate the Entente conditions with all the means and abilities at his disposal.'

5 Head of the *Abwehr* from 1935.

6 Friedrich Schotte did not belong to the submarine design office of the Imperial German Navy and probably knew little about this top-secret project. From 1929 he was significantly involved in the construction of *E-1* in Spain. He succeeded in overcoming considerable difficulties in the Andalusian shipbuilding yard, so 'deserved' special mention in the Nuremberg War Tribunal proceedings.

7 The figure of 600 tons had been proposed during negotiations at the Geneva Naval Disarmament Conference (February–August 1927) as the limit for 'small submarines' which did not have to be counted in fleet force strengths.

8 In 1929 the shipyard merged with Wilton to become Wilton-Fijenoord, which would build the IvS-designed Turkish submarines *Birinci Inönü* and *Ikindci Inönü*.

9 Papenberg had held the same post on board the first German submarine *U-1* under Bartenbach 22 years previously.

10 The Library of Congress system has been used for the transliteration of Russian names. Only the names of well-known historical personages and geographical places are given using their normal English renderings.

11 The Naval Scientific-Technical Committee (the succesor to the Naval Technical Committee of the Imperial Navy) was formed by the *Revvoensovet* on 30 September 1923 with Pyotr Leskov (Rear Admiral of the Imperial Navy) in charge. The responsibility of this warship research group encompassed ships and weaponry designs, planning of defence facilities and naval bases, and the tracking of foreign naval progress. The most prominent members at that time were Aleksej Krylov, Iurij Shokal'skij, Aleksej Berg, and Iulian Shimanskij.

12 The *Profintern* (Red International of Labour Units) was established in 1921 on the initiative of Grigory Zinovev with the intention of spreading communist activities within Western trade unions; the aim was to support the interests of the Soviet Union. The *Profintern* would be dissolved in 1937.

13 The bureau was manned by the staff of OKTB No 2, which was a combination of the OGPU *sharashka* (employing prisoners) with the technical bureau TB-4 of the Baltic Works (employing free men).

14 The office for the design of shipyards and ship repair enterprises, which in 1930 replaced *Sudotrest*, an organisation grouping all machinery and steel construction activity formed in 1921.

15 Delegations to foreign countries were lucrative. Participants received hotel allowances, a budget for food and pocket money, which they scrupulously saved by using shared rooms instead of singles, and by eating food brought from home or taking advantage of the hospitality of the hosts. With the money saved delegates purchased goods that were difficult to obtain in the USSR, which they sold on their return at vastly inflated prices. This system operated until the end of the 1980s, and the author was himself a beneficiary during the late 1980s, when the savings from one day's posting were equivalent on the black market to a junior engineer's monthly salary.

16 Deutsche Schiff- und Maschinenbau Aktiengesellschaft was a cooperative of eight German shipyards in the period 1926 to 1945. The leading company was the AG Weser shipyard in Bremen.

17 Zel'ting graduated from the Naval Academy in 1925, and in 1928 was appointed executive officer of the submarine *Revoliutsioner*. From 1929 he served on the staff of the Naval Forces of the Black Sea, was commander of the 2nd Submarine Flotilla from 1931 and commander of the 2nd Submarine Brigade of the Pacific Fleet from 1934; he was promoted Captain 1st Rank in 1935. Arrested in 1937, he was sentenced to imprisonment in the *Gulag*, and was released after the war.

18 Some of these were previously in Japan in a similar role, taking part in the design and construction of the submarines of the *I 1* and *I 24* classes.

19 From 30 June 1936, in an effort to keep its existence secret, it was designated Works No 189.

20 IvS branch office 'E' was set up on the premises of AG Weser for the preparation of construction of the two Type IA submarines *U-25* and *U-26*.

21 At that time, it was quite a risky business for a foreigner to work in the Soviet Union. In October 1933 two German engineers were accused of leading a sabotage group in the Marti Yard at Nikolaev, where three submarines were damaged by fire on 22 October 1933, with one of them, *M27*, becoming a constructive total loss and having to be laid down a second time.

22 The class-name *Stalinets* is not confirmed, despite its unofficial use. The boats were also nicknamed *Eski* (transliterated pronunciation of the consonant 'S').

23 They were renumbered with the 'S' pennant in 1949 only.

24 This key information was hidden in a long list of technical issues, and it is possible that the censor, bored with the incomprehensible text, did not reach this line.

25 'Istoriia Otechestvennovo Sudostroeniia v piati tomakh', *Sudostroenie*, vol IV, 140 (St Petersburg, 1996).

26 It was not acceptable by Soviet standards to allow them out of the country to travel to The Hague and to participate in secret negotiations.

27 During the diving operation, when Doenitz was on board, a faulty connection in the electrically-operated forward hydroplanes caused the boat to dive immediately as far as the gun position.

28 This was confirmed by the experience of the Turkish Navy, which operated several submarines designed by IvS (including *E-1*). Prewar standards of training (with plenty of time and thoroughness) apparently compensated for this inconvenience.

29 Kolomenskij had had a licence agreement with the German MAN company since 1924.

30 In service they had a tendency to overheat, causing the pistons to seize.

31 Founded in 1931 as the Scientific-Research Institute of Naval Shipbuilding (NIVK) based on the Experimental Basin and its associated laboratories. In 1938 the institute was transferred to the People's Commissariat of the Defence Industry and renamed TsNII-45.

32 The internal fuel tanks tended to leak after shocks (*eg* depth-charging) were sustained, but this was apparently the result of poor Soviet welding technology. Interestingly, the Germans considered welded joints to be more oil/watertight but weaker in terms of strength than riveted ones; the Soviets held the opposite view.

33 MTBs nos 1 and 2 were not to be of the double ballast/fuel tank type.

34 During the first three quarters of 1940, 54 accidents and breakdowns happened in the submarine service of the Baltic Fleet; 36 of these were the result of poor training.

35 Confirmed in the indictment at the Nuremberg trials.

THE GERMAN FLAK SHIPS

PART II: THE FORMER DUTCH AND DANISH HULLS

In this sequel to their article published in *Warship* 2023, **Aidan Dodson** and **Dirk Nottelmann** outline the history of the final four of eight planned state-of-the-art anti-aircraft batteries converted from old cruisers and coast defence ships during the Second World War.

The Dutch hulls

Following the invasion of Denmark and Norway in April 1940, the Netherlands were attacked by Germany on 10 May, and surrendered on the 15th. Much of the home-based element of the Dutch Navy had already been evacuated to the UK, meaning that only a few vessels were seized by the occupying Germans, and many of these had been sabotaged to a greater or lesser degree. These included a pair of coast defence ships and a cruiser, all three being converted to Flak ships.

Ariadne (ex-*Vlierede*, ex-*Hertog Hendrik*) and *Undine* (ex-*Ijmuiden*, ex-*Jacob van Heemskerck*)[1]

A series of nine coast defence ships (formally 'armoured ships') had been initiated by the three ships of the *Evertsen* class, launched in 1894, with progressive developments culminating in *De Zeven Provincien* of 1909.[2] Most had been disposed of between 1914 and the early 1930s, leaving just the home-based *Hertog Hendrik* (1902) and *Jacob van Heemskerck* (1906), and *De Zeven Provincien* based in the East Indies.[3] All were primarily employed in training roles.

Hertog Hendrik had, like most of the series of ships, originally been armed with single 24cm guns fore and aft, with two 15cm and two 75mm on each beam, and three 45.7cm torpedo tubes. Machinery consisted of six Yarrow-De Schelde watertube boilers, which supplied steam to two triple-expansion engines each of 3,200ih, providing a maximum speed of about 16.5 knots. Until the 1920s, she had been deployed principally to the East Indies but, with the commissioning of the new cruisers *Java* and *Sumatra* during 1925/26, she was converted into a training ship in 1926, with the torpedo tubes and after 24cm gun removed, and deckhouses erected aft. In 1928, a large latticework crane was added behind the mainmast to handle the two aircraft that were now being carried; this was removed during the 1930s. As well as undertaking training voyages, *Hertog Hendrik* spent time as station ship in the Caribbean, standing in for her half-sister *Jacob van Heemskerck*. She replaced the latter in the cadet training role in 1933 (see below), but escorted Dutch merchantmen through the Straits of Gibraltar during the Spanish Civil War from March 1937 to September 1938.

Hertog Hendrik then reverted to training duties, returning from a final cruise to the West Indies in the summer of 1939 and paying off. However, with war in Europe looming, she soon returned to service as a floating battery to protect the Stortemelk, the waterway between Vlieland and Terschelling. She was moored at Vlieland under the name *Vlierede* from 3 August to 4 November 1939, when a shore battery of three 15cm guns took over the role. On 11 November *Hertog Hendrik* paid off once more at Den Helder, to become an

Hertog Hendrik on patrol duty in 1937 during the Spanish Civil War. (Netherlands Institute of Military History)

Jacob van Heemskerck during the early 1930s, with the new destroyer *Banckert* in the background. Both she and *Hertog Hendrik* had been fitted with aircraft cranes aft during the 1920s, but *Hendrik*'s had been removed in the mid-1930s. (Netherlands Institute of Military History)

accommodation ship for the Dutch Royal Naval College, whose shore quarters had been taken over as a hospital.

Jacob van Heemskerck was a slightly enlarged version of *Hertog Hendrik*, with two more 15cm and 7.5cm guns but one fewer torpedo tube. Unlike her half-sister and most other major Dutch warships, she was never deployed to the East Indies, spending her career in the western hemisphere. Her principal role was as station ship in the Caribbean, based at Paramaribo in Suriname, but also deployed on occasion into the Atlantic, the Baltic and the Mediterranean. Like *Hertog Hendrik*, she was fitted with a crane abaft the mainmast to handle a pair of aircraft during the late 1920s. Extensively refitted during 1919–23, she was subsequently frequently used as a seagoing training ship for cadets and seamen. This included an annual cadet training cruise into Scandinavian waters, the last ending in August 1933; the ship paid off in October and was replaced by *Hertog Hendrik*. In the spring of 1939, now partialy dismantled, she was renamed *IJmuiden* (to free her original name for a new cruiser, launched on 16 September 1939) and brought forward to serve as a floating battery at the eponymous port, whence she was towed on 19 April 1939.

IJmuiden was scuttled there on 14 May 1940 by opening her seacocks, settling by the bow in shallow water. Refloated by the Germans on 16 July, she was towed to Amsterdam on the 21st for initial repairs. *Hertog Hendrik* was similarly scuttled at her berth at Den Helder, settling in shallow water. The ship was further damaged on 21/22 June by a near-miss during a raid by nine Lockheed Hudson aircraft of 206 Squadron, Royal Air Force. Progressive flooding left her on the bottom with a heavy list to starboard. She was refloated in October 1940.

It was as early as 14 May that a decision was made to convert the two coast defence ships – and the cruiser *Gelderland* (see below) into Flak ships 'along the general

The floating battery *IJmuiden* (ex-*Jacob van Heemskerck*) is towed out of Den Helder on 19 April 1939, en route to the port for which she had been renamed. Laid up since 1933, she has been stripped of most tophamper, including the mainmast that had formerly supported her aircraft crane. (Royal Netherlands Navy Museum)

Vlierede (ex-*Hertog Hendrik*) at Den Helder in late June 1940, after being successively scuttled and bombed, and *IJmuiden* on 1 May 1940, after being scuttled at her eponymous port. (Royal Netherlands Navy Museum/Netherlands Institute of Military History)

lines of *Arcona*, *Medusa*, *Nymphe* and *Thetis*'. However, it was explicitly stated that the conversion work on the ex-Dutch vessels was to be on an 'as-and-when' basis, keeping yard workforces occupied when no other jobs were available. This was in part owing to the lack of manpower to crew the vessels; however, the ships' guns were removed immediately, the 24cm weapons being installed for coast defence in Batterie Rozenburg (XV) on

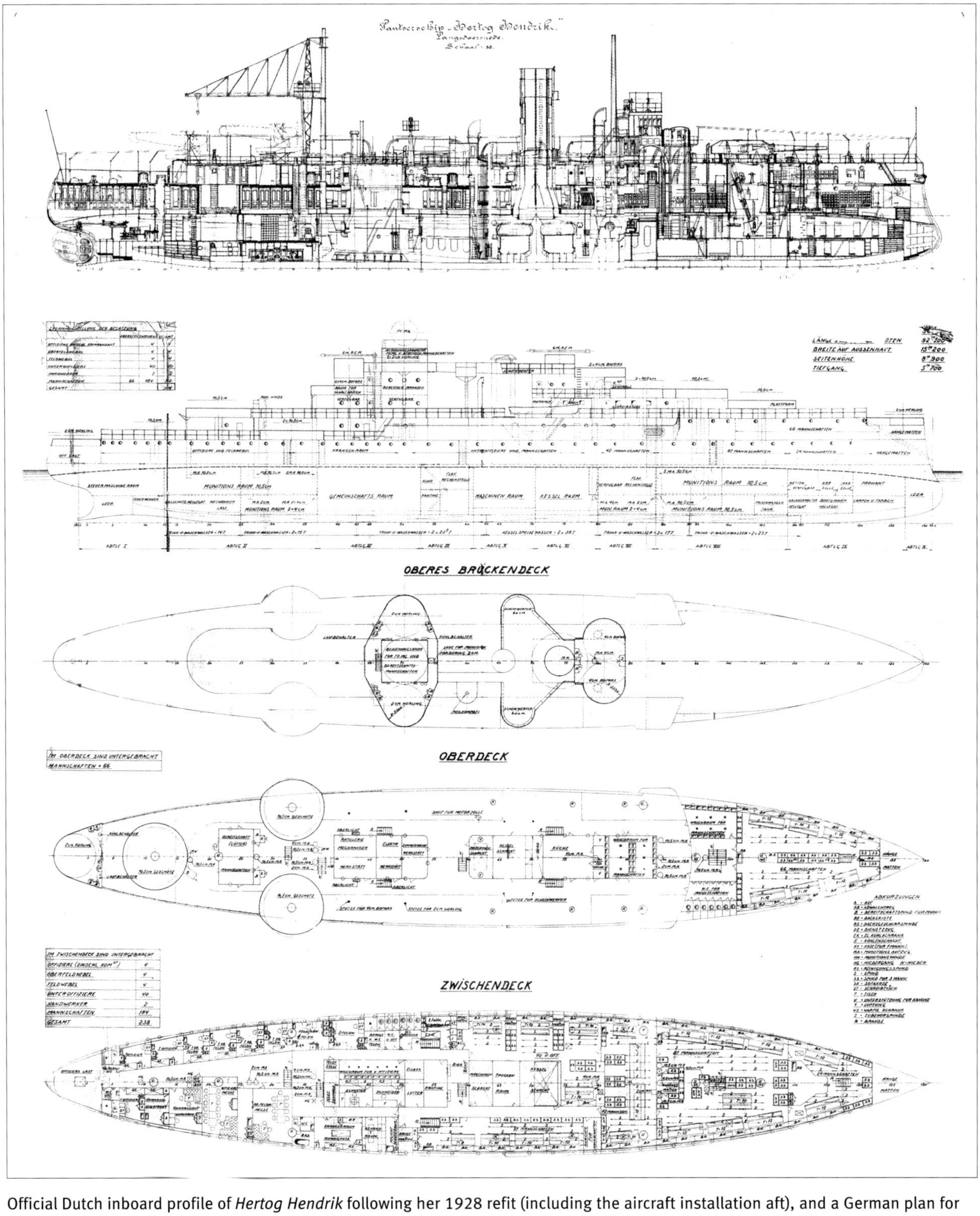

Official Dutch inboard profile of *Hertog Hendrik* following her 1928 refit (including the aircraft installation aft), and a German plan for her AA conversion approved on 20 December 1942. It shows the twin fire control positions and the installation of a Würzburg antenna, together with the short funnel originally envisaged to give the latter uninterrupted sky arcs. (AD collection/DN collection)

the Insel Beer peninsula, on the southern side of the Maas estuary at the Hook of Holland.

An initial proposal for the conversions was completed by Kiel Dockyard on 30 October 1941, followed by two variants in December and April 1942 respectively. The latter proposal was accepted on 19 May 1942, with the explicit remark 'No preparations for self-propulsion!' Meanwhile, *Ariadne* – as the former *Hertog Hendrik* was to be renamed, following the 'tradition' begun by the Norwegian conversions (employing the name of a sister of the German conversion *Medusa*, sunk at the battle of Heligoland Bight in August 1914) – had been allocated to the Mercantile shipyard, Antwerp, on 6 June 1941, where preparatory works would commence in August.

The former *Jacob van Heemskerck* was towed to Hamburg in March 1941, and was converted under the name of *Undine* (after a sister of *Arcona* sunk by torpedoes in 1915) at the local Howaldt yard from 5 September 1942 to 28 August 1943. Unlike her half-sister and the ex-*Gelderland*, a degree of priority was applied in her case.

Both ships were rebuilt to the same basic design, the detail work being executed by Howaldtswerke, with contributions from the Kiel Dockyard and the van der Giessen yard at Krimpen-aan-den-IJssel. Their armament was agreed in September 1942 to consist of the usual mix of heavy and light AA guns but, compared with the earlier conversions, the larger deck area of the ex-Dutch coast defence ships allowed the mounting of two additional heavy Flak guns, giving a total of eight single 10.5cm/45.

Single 10.5cm gun mountings were fitted on the forecastle and quarterdeck, three being placed on a superimposed platform forward and one in a corresponding position aft. The final pair of 10.5cm guns was installed just abaft the break in the forecastle. Pairs of single 4cm guns were fitted on platforms behind and above the superimposed 10.5cm guns, while 2cm quadruple (*Vierlinge*) mountings were placed at the bow, stern and amidships. The latter guns were placed on platforms on the sides of a tower erected amidships to support the antenna of a Würzburg (FuMG 39 TD) radar. The former 24cm magazines were adapted to hold 10.5cm ammunition, the conning-tower trunk being used for hoisting all rounds from the forward magazines, which included a space for 4cm and 2cm ammunition in what had originally housed auxiliary machinery directly forward of the boiler rooms. The armament was controlled by two *Kleinkommandogeräte* (*Kleinkog*) directors, incorporating 6-metre optical rangefinders, and a pair of G 60(cm) searchlights, installed on an after extension of the forward 4cm gun platform, on either beam just forward of the funnel.

The designers were under instructions to minimise modifications to the hull structure and bulkheads: only the superstructures were open for substantive modification. Regarding the ships' machinery, it was reported that a considerable amount of 'cannibalisation' had taken place following their prewar decommissioning (especially in *Undine*, which had been out of seagoing service since 1933), especially of components of non-ferrous metals. Since the latter were already in short supply in Germany, mobility was not to be reinstated. Rather, most of the main machinery was to be removed, in part to gain some recreation space for the crew, as had been done in *Medusa* in the spring of 1942 (see Part I).

Thus, the engine room was cleared for this purpose, the after boiler room being adapted to house such machinery as was now needed, leaving just the forward boiler room to supply steam for this and other auxiliary machinery. The forward steam-powered capstan is explicitly mentioned, and to increase its flexibility a new hawsepipe was fitted through the stem, presumably to aid secure mooring for the ships' intended new role. Electrical power was to be provided by steam- and two 19kVA diesel-generators. A new, smaller, funnel was fitted, initially terminating below the top of the Würzburg tower, to provide the radar antenna with uninterrupted all-round coverage. However, early experience in *Undine* required the addition of a narrow stovepipe to prevent the upper platforms of the superstructure being smoked out.

Initial plans had envisaged only a single director/rangefinder, but early in the conversion process, in November 1942, it was recognised that this would be inadequate for operating the armament as two independent batteries. Accordingly, a second director was added abaft Würzburg tower. The location of this second position differed between *Ariadne* and *Undine*, and served as the principal means of distinguishing the near-sisters. In *Ariadne*, it was located about station 50, while in *Undine* it was around station 57 – some 5 metres closer to the funnel. Together with other detail differences in the design of the two conversions, this made the latter's superstructure appear more congested. Although nowhere explicitly stated, it seems that this was intended to clear the arcs of the after superimposed 4cm guns,[4] those of *Ariadne* being almost abreast the director, whereas in *Undine* they were fully aft of it.

Broadside photographs of *Ariadne* (top) and *Undine*, showing the differences between the two vessels. Apart from the slightly different layout of their superstructures, *Ariadne* was the only one of the pair to carry an external degaussing coil. She was also never camouflaged. (DN collection/AD collection)

As further distinguishing features, *Ariadne* had a crowsnest, a slightly wider stovepipe, an external degaussing coil and the bow 2cm mounting slightly farther forward. Both ships were intended to be camouflaged, but only *Undine* ended her career with a patterned paint scheme; *Ariadne* seems to have been painted grey overall for her entire German career.

Undine was commissioned on 29 September 1943 at Hamburg, and immediately afterwards was towed to Cuxhaven for ammunitioning. On 5 October her guns were test-fired for the first time. At the same time her future was discussed within the *Oberkommando der Wehrmacht* (OKW – High Command of the Armed Forces). Originally, the intent was that she be anchored within the Putziger Wiek, supporting the defence of the seaward approaches to Gotenhafen (Gdynia). However, with reference to the Peenemünde missile test centre, it was noted that '[an] augmentation of heavy (AA) batteries will not produce the desired strengthening of AA cover, because the location of the place to be protected on a peninsula provides for comparatively few effective defence positions, which will not, especially during attacks in daylight, hinder the dropping of bombs before fire can be opened. Therefore, it is requested that enquiries be made as to whether the Naval High Command (OKM) might provide a floating AA battery for improving aerial protection'.[5] On 9 October, the OKM agreed and *Undine* was moored in a position off the island of Usedom, southeast of Peenemunde, while officially attached to the nearby naval base at Swinemünde. She would remain there, with some breaks, until November 1944, when the protection of the production plant for synthetic petrol at Pölitz (Police), close to Stettin, became even more important than Peenemünde. During one of these breaks, in or close to the port of Stettin, *Undine* would receive some damage during an air raid, albeit remaining operational.

For the protection of the Pölitz plant, she was moored in the confined delta of the Oder river until February 1945, when Navy Commander-in-Chief Karl Dönitz decided to employ her slightly farther north at Swinemünde, which had become the focal point for the mission to rescue civilians from the Baltic. However, this would still not be her final location, as she was transferred one last time to be moored at Travemünde, north of Lübeck, on 17 April 1945. This seems to be the location where *Undine* was surrendered in May. Nevertheless, beyond this sketchy information, her exact movements during these troubled days still require further research.

Ariadne, giving an idea of the excellent arcs enjoyed by her gun battery. (AD collection)

In view of the deteriorating air defence capability in Germany in 1943, it was decided on 11 November of that year to accelerate the conversion of *Ariadne* and ex-*Gelderland*. The former was planned for use in the Baltic, while the latter was to be sent to Norway. Exactly one month later, the OKM approved the preliminary list of crucial equipment for both ships, which were to be repeats of *Undine* (see above). However, despite all attempts at acceleration, it was not until 31 March 1944 that *Ariadne* was completed at Antwerp, to be formally commissioned on 4 April 1944.

Ariadne remained at Antwerp until towed to Brunsbüttel in May 1944; she was anchored off the locks of the Kiel Canal from 20 May to 30 June 1944, when she was relieved by *Arcona*. *Ariadne* was then transferred to Putziger Wiek on 6/7 August (as had been intended for *Undine* the year before), replacing *Niobe* (ex-*Gelderland*: see below) there. *Ariadne* would remain in this theatre of operations until she was withdrawn westward in March, ultimately joining her sister *Undine* at Travemünde on 17 April. Both were towed to Wilhelmshaven by the end of May 1945.

On 12 August 1945, the two ships were moored

Undine very early in her career – indicated by the lack of a 'stovepipe' extension to her funnel – under tow, with a tug alongside. She would retain this camouflage scheme until after the end of her German service. Compared with her sister, her superstructure appears much more compact, with the platform aft supporting two single 4cm mountings clearly visible. (DN collection)

Ariadne and *Undine* moored side by side at Wilhelmshaven after the end of the war. (AD collection)

Hertog Hendrik (top) and *Neptunus* during their final decade of service as accommodation ships at Den Helder. In their final configuration they had a large deckhouse added abaft the funnel. (Netherlands Institute of Military History)

together in the Nordhafen opposite the IV. Entrance, where they were later joined by *Arcona* and *Medusa* (see Part I), while awaiting tow back to the Netherlands. It is unclear exactly when *Ariadne* was returned home, but *Undine* arrived back at Den Helder in the tow of the tugs *Tyne* and *Amsterdam* at the end of February 1946. Curiously, contemporary press reports refer to her as *Hertog Hendrik*: clearly there was confusion as to which of the old coast defence ships was which when first reacquired.

However, identities were soon confirmed and *Ariadne* reverted to her own original name of *Hertog Hendrik*. She was taken in hand by the Wilton-Fijenoord shipyard in Schiedam, to be rebuilt as an accommodation ship. Restored to Dutch service on 21 October 1947, *Hertog Hendrik* was initially berthed at Amsterdam and formally classified as a guardship with the pennant number HW 4 (changed to A 888 on 15 October 1950). She was moved to Den Helder in April 1951. On 27 September 1968 she was removed from service, and sold for scrap on 28 August 1972.

Undine, however, now became *Neptunus*, since the

cruiser that had received her original name had also survived the war. She likewise became an accommodation ship, being rebuilt as such at Amsterdam Dockyard. Generators and ventilation equipment were fitted in her former boiler rooms, with facilities for 800 men of the Dutch Royal Navy College being installed. She recommissioned on 23 February 1948 and was moored in front of the college until 1974 (bearing the successive pennant numbers HW 3 and A 881), when new shore facilities were completed. *Neptunus* was taken out of service on 11 September 1974, stricken on 4 October, and sold for scrap on 11 February 1975 to E Kaptein, of Nieuwendam, Amsterdam.

Niobe (ex-*Gelderland*)

Also found by the Germans at Den Helder, but left unharmed by both Dutch sabotage and the British air raid, was the protected cruiser *Gelderland*. She had been the last to complete of the six 4,000-ton *Holland* class, launched between 1896 and 1899. They had been built as a general-purpose design for service both at home and abroad, and were broadly inspired by the British *Apollo* class (launched 1890–91). *Gelderland* commissioned on 15 July 1900, originally armed with two 15cm, six 12cm, six 7.5cm and eight 3.7cm guns and two 35cm (from 1907 45cm) torpedo tubes. Initially assigned to the East Indies, while outbound she was diverted to South Africa to carry Boer President Paul Kruger into exile in October 1900. Reboilered in 1906–07, the funnels were fitted with prominent caps up to their original height. Service in the Atlantic followed during 1907–09, and then employment as a cadet and boys training ship from 1911.

Gelderland was the last of her class in seagoing service, her sisters all having been withdrawn between 1911 and 1924.[6] She had, however, been refitted as a gunnery training ship in 1920, when the funnel caps were removed. The rebuilt ship was armed initially with four 15cm and eight 7.5cm guns, but from 1922 she mounted ten 12cm, two 7.5cm and four 3.7cm guns, two 7.5cm mortars and a pair of machine guns. She served on into the 1930s, including a landing exercise in September 1933 in company with *Jacob van Heemskerck*, paying off in August 1939 when relieved as gunnery training ship by the brand-new sloop *Van Kinsbergen*. *Gelderland* was then laid up at Den Helder.

In German hands, the ship initially remained at Den Helder as an accommodation ship, but in August 1941 she was taken in hand by the Van der Giessen de Noord shipyard at Krimpen-aan-den-IJssel, near Rotterdam for low-priority conversion (see above) to a Flak ship, to the same basic scheme as *Ariadne* and *Undine*. Unlike the latter, however, her machinery was still intact and operable, so it was decided that the ship would be self-propelled, like the ex-Norwegian *Nymphe* and *Thetis*. However, while her original engines remained in place, her original twelve Yarrow boilers were reduced to six (three in each boiler room). Accordingly, it proved possible to trunk the uptakes into a single funnel. Her armament was identical to that of the other ex-Dutch ships. However, her live ammunition outfit of 950 shells included 150 non-HA rounds, given the greater likelihood of being involved in a surface action as a seagoing unit.

Gelderland as a gunnery training ship. (Netherlands Institute of Military History)

Two detail images of *Niobe* in the port of Rotterdam, shortly before delivery. She would never receive any splinter protection for her guns, a significant omission which is difficult to explain. Her original steering position can just be made out below the forward 4cm platform. (DN collection)

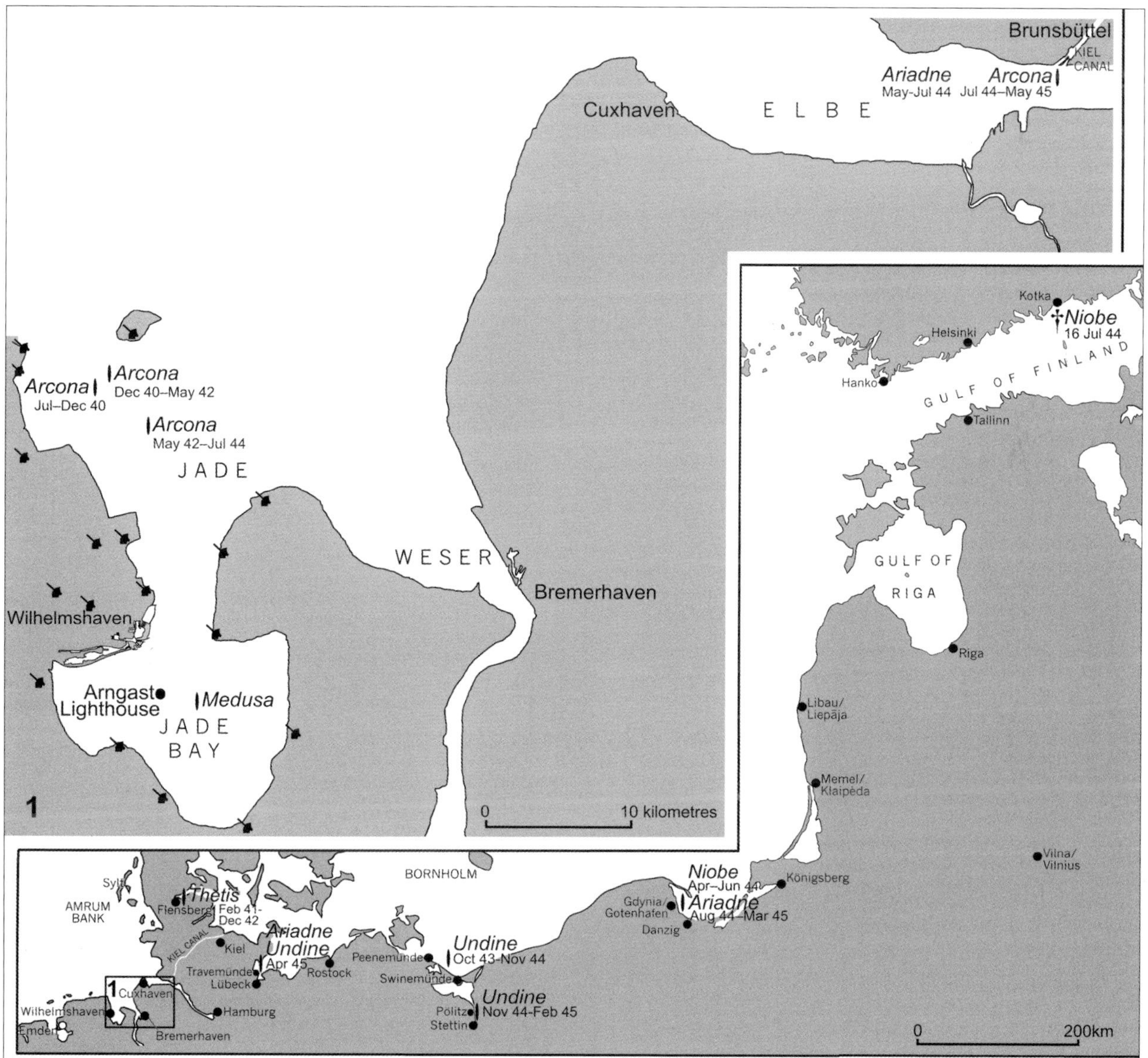

The principal duty stations of the ex-Dutch AA battery conversions. (AD graphic)

Conversion work was completed at the end of February 1944, and the ship commissioned as *Niobe* on 1 March 1944, being assigned to *Marine-Flak-Abteilung* 282. Her German name had been allocated provisionally some time earlier, but there had been a complication when the original *Gazelle*-class cruiser of that name had returned to German service in September 1943. However, the latter would be lost in the Adriatic three months later, meaning that the name was once again available.

Niobe commissioned with the very short funnel and light mast of *Undine*, but concerns regarding stability are suggested by the fact that, while still at Rotterdam, the originally mounted quick-launch davits for two motor launches on both sides of the funnel were replaced by lighter, hand-operated davits serving pulling cutters. This may have been with a view to the major modifications approved on 25 March 1944 and carried out at the Schichau yard at Elbing.

The principal changes comprised increasing the height

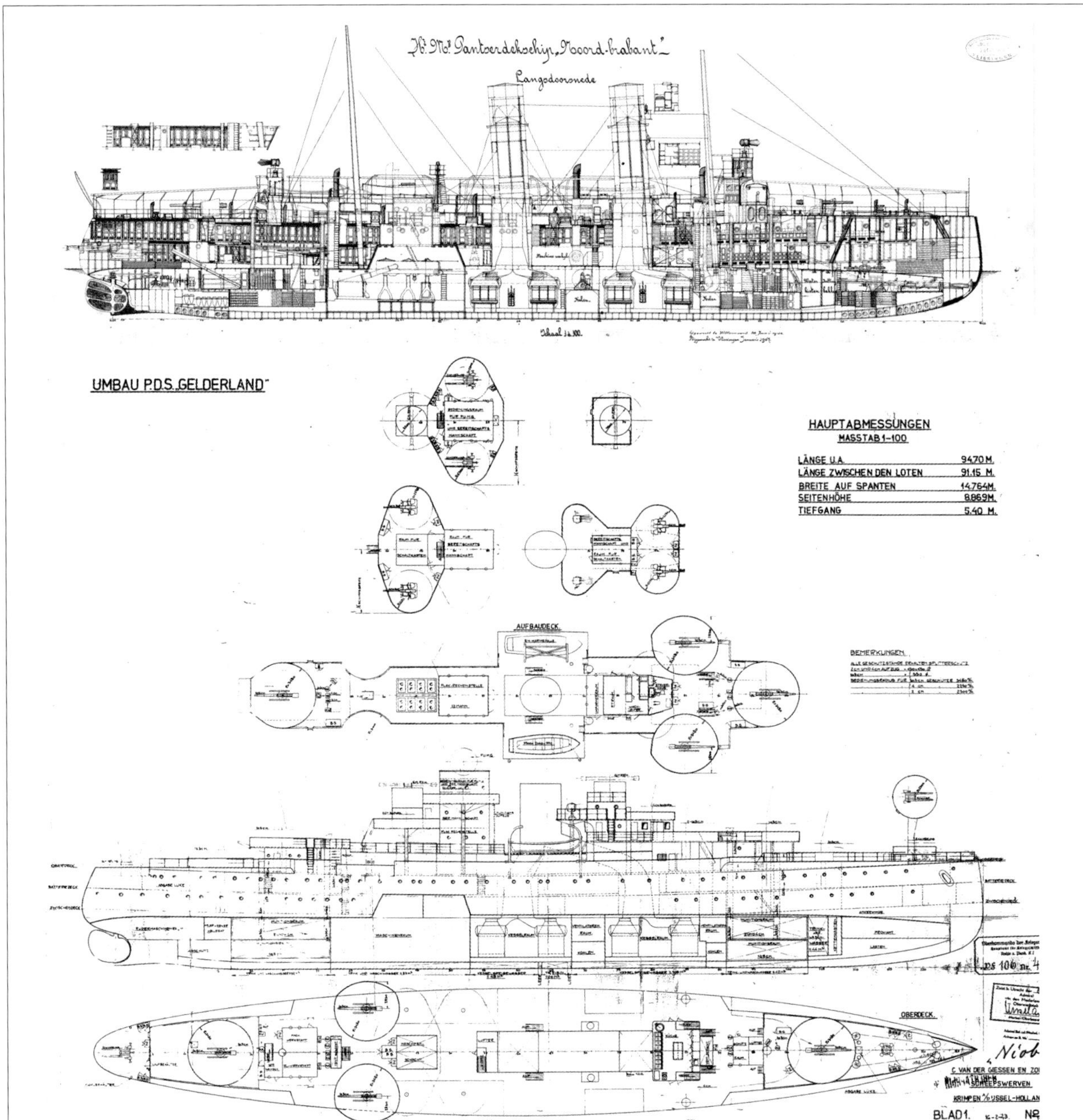

Official Dutch inboard profile of *Gelderland*'s sister *Noord Brabant*, and an April 1943 plan for *Gelderland*'s conversion. At the time no German name had been allocated, even though the old German cruiser *Niobe* was then still part of the Italian Navy (having been sold to Yugoslavia in 1925 and captured from them in 1941), thus her eventual name was still available. The latter was once again German from August 1943, but had been lost well before the ex-*Gelderland* finally commissioned. (AD collection/DN collection)

of the funnel, raising the level of the searchlights, improving the splinter protection of the guns and heightening and strengthening of the mast, which was converted into a tripod. In addition, the steering position was moved from the superstructure deck to a position a deck-and-a-half higher, on the front of the tower supporting the 6-metre director/rangefinder. This gave a far better view than its original location, where sight lines were hampered by the surrounding three 10.5cm guns. Although originally intended for Norwegian deployment, *Niobe* remained in the Baltic for her short German operational career.

The ship's planned functions had included the escort of coastal convoys, but in practice she was first employed in a static role in Putziger Wiek, arriving there on 12 April. However, Soviet advances on the Finnish front meant that, in June 1944, she was transferred to Kotka on the Gulf of Finland, a base for light forces guarding the minefields across the gulf. The Finnish air force having withdrawn local air cover to meet an even more urgent requirement, *Niobe* was to be used to bolster the shore-based anti-aircraft batteries.

She arrived on 10 July and her presence was spotted by the Soviets on the 12th. However, they misidentified her as the Finnish coast defence ship *Väinämöinen*, long a priority target: an immediate strike was therefore ordered. Thirty Petlyakov Pe-2 dive bombers, covered by 24 Yak-9 fighters, dropped some seventy 500kg and 1,000kg bombs, but no hits were scored.

A new airstrike (Operation 'Hurricane') was launched on the 16th, this time with no fewer than 132 aircraft – the largest single Soviet air attack of the war. The first group comprised 23 Ilyushin Il-2 ground attack aircraft armed with bombs, which all missed. Next came 28 Pe-2s, each carrying two 250kg bombs, accompanied by 16 fighters; again, no hits were scored. The third group was a mix of four Douglas A-20C, each with a FB 1,000kg bomb and flying at low level, preceded by several high-flying Pe-2s to deceive the defenders. While the dive-bombers missed again, the deception worked and at around 17.00 the cruiser was hit by at least two 1,000kg bombs from the A-20Cs, which hit below the waterline to starboard. The crew was unable to contain the damage and *Niobe* sank with a heavy list to starboard, coming to rest with part of the superstructure and funnel above water. While 260 men were rescued, some 70 were dead; in exchange, according to Soviet sources, all four A-20 were shot down.[7]

The ship's guns and much of her ammunition were soon salvaged, and seven of the 10.5cm weapons taken into Finnish service (the eighth gun was damaged beyond repair). These remained in service until the 1990s. The wreck remained in place until 1953, salvage being made difficult by the presence of residual ammunition. Finally,

Niobe in the Rotterdam area after commissioning. Note the heavy motor launch on its davits abreast the funnel; it would be replaced (like the one on the port side) as a first step towards her major refit. (AD collection)

Niobe after her refit at the Schichau yard at Elbing, during which her funnel was heightened, a light tripod mast fitted, and her steering position moved. (AD collection)

A view of the ship's stern, taken on the same occasion. (AD collection)

The sunken *Niobe* at Kotka, with all major fittings already removed. (AD collection)

however, she was refloated by the Hamburg firm of Taucher Beckedorf and scrapped.

The Danish Project

In 1943 it had been planned to undertake a reconstruction of another coast defence ship, the Danish *Peder Skram*, as a Flak ship. She had been the final unit of the three-strong *Herluf Trolle* class, all designed and built by Copenhagen Dockyard between 1897 and 1909. *Peder Skram* was laid down on 2 April 1905 and launched on 2 May 1908. Her armament consisted of a 24cm/43 gun fore and aft with four 15cm/50 on the upper deck amidships, backed up by ten 7.5cm/55 and two 3.7cm weapons. Machinery comprised six watertube boilers plus one cylindrical auxiliary boiler, supplying steam for VTE engines of 5,400ihp, giving 16 knots.[8]

Peder Skram joined the fleet for the first time in 1910 for a cruise to Norway, and afterwards was employed with the fleet up to the First World War. The war itself was spent lying within the Sound, between Denmark and Sweden, protecting Danish neutrality. In 1916 the 7.5cm guns atop the turrets were exchanged for anti-aircraft weapons of the same calibre. Following the end of the war, the ship was laid up at Copenhagen for two years, before being reactivated in October 1920 to serve as command vessel for that winter's training squadron, remaining active until late 1922, when she was laid up in the wake of major cuts in naval spending. She was not recommissioned again until 1929, when she served, alongside the coast defence ship *Niels Juel*, as a training ship for reserve cadets. Paid off in August 1929, she was again in commission during August/September 1934, following a refit during which her turret-top 7.5cm guns were replaced by 8mm machine guns and the searchlights abreast the bridge swapped for twin 2cm AA mountings. In May 1935 *Peder Skram* escorted the Danish royal yacht to Stockholm; she was again laid up on her return. During the 1930s she was equipped with a heavy crane on the port quarterdeck for potential submarine salvage purposes.

The ageing vessel was recommissioned in May 1939 to form once again the core of a training squadron with *Niels Juel*, but paid off on 7 July. *Peder Skram* was brought back into service on 20 September 1939, with her twin 2cm mountings replaced by single 4cm Bofors. She formed part of a squadron carrying out neutrality patrols at the northern entrance to the Great Belt, including the convoying of German merchant vessels through Denmark's territorial waters in the Kattegat and protecting them from attacks by British submarines. On 9 April 1940 the country was occupied by the Germans but remained nominally self-governing, with the armed forces remaining in existence. On that day, *Peder Skram* was moored at Frederikshavn in northeast Jutland, sailing from there to the small port of Horsens in mid-Jutland, where she was laid up from 13 April until 11 June 1941. The ship then transferred to Copenhagen, to be used as a seagoing training ship for a variety of tasks, before being laid up again at Copenhagen in 1943, decommissioned but serving as an HQ ship.

Owing to rising tensions between the Danish government and the German occupiers, all Danish warships began to prepare for scuttling, which finally occurred on 29 August 1943. This followed civil unrest that had culminated in Denmark being declared 'enemy territory' the previous day. As her sinking was carried out solely by opening her seacocks – as a decommissioned vessel she had not been issued with scuttling charges – salvage of *Peder Skram* was straightforward: refloated in November 1943, she was stripped of her four 15cm guns,

Peder Skram: her evident low freeboard explains why the German Navy saw her only as a stationary AA battery, leaving aside any issues around restoring her water-damaged machinery. (Geirr Haar collection)

which became a battery on the island of Fanoe, protecting the entrance to the important port of Esbjerg, but her two main guns remained in place. No information has survived regarding the fate of her 4cm Bofors, but as the preferred close-defence AA weapon of the German Navy it seems likely that those guns remained on board when the hulk was towed to Kiel during the winter of 1943/44.

In parallel with the salvage operations, consideration was given to potential uses for the hull. A December 1943 sketch for a Flak ship conversion was accompanied by a note from the Naval Leadership (Skl) that noted that 'with regard to the ship's low freeboard, it can be employed in calm waters only. The armament may follow that of the Dutch conversions, but, owing to the shorter hull, two 10.5cm guns are to be omitted. The Würzburg installation has been positioned forward of the funnel to avoid it being affected by exhaust gases. [Constructor departments] K II and K III will provide information regarding progress of salvage and possible restoration of mobility'. The anticipated time for conversion was given as 8–12 months, with the work probably given to the Mercantile yard at Antwerp, which was already converting *Ariadne*.

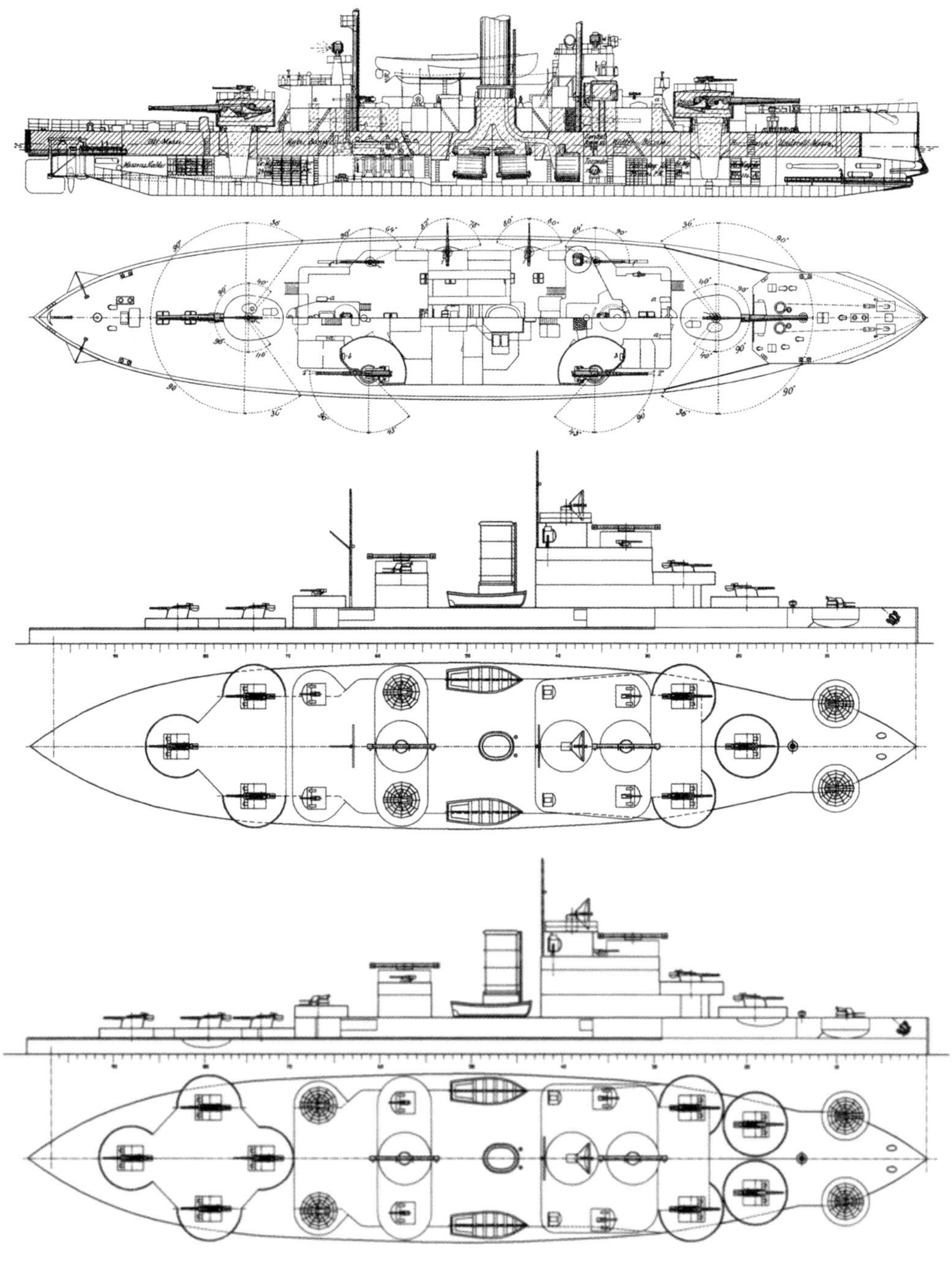

Official Danish GA plans of *Peder Skram* as built, and the two German proposals for her conversion into *Gefion*. The second was clearly too ambitious and would have resulted in far too cramped a vessel. (AD collection/ DN drawings)

Peder Skram scuttled in Copenhagen Dockyard. The tulip-shaped platforms each supported a 4cm Bofors AA gun. (AD collection)

In spite of all the elaborate plans, no substantive work was carried out by the Germans on *Peder Skram*, as can be seen from this view taken on her arrival at Odense for scrapping in 1950, with her main guns and fittings intact. (AD collection)

The constructors responded in March 1944 that, in view of the scarcity of manpower and materials, *Peder Skram*'s engines should be dismantled and the ship permanently moored. Skl subsequently approved the proposed design, omitting, however, a planned after mast. On 6 June 1944, a meeting was held at the OKM, chaired by naval Chief Counsellor Wilhelm Hadeler,[9] regarding possible improvements to the current design. It was argued strongly that the 10.5cm battery should be increased to eight guns and the positions of some of the smaller guns changed. This was rejected by the central staff constructors in July on account of the already cramped placement of the guns. It was also noted that no free capacity was available in any German shipyard to commence the conversion. Nevertheless, the worsening general situation of Germany is reflected by the fact that as late as October 1944 plans were ongoing for the dismantling of *Peder Skram*'s machinery, which would include the removal of the main engines (including tail-shafts and propellers) and five of the seven boilers. The remaining pair would provide steam for heating, two generators, some pumps and the anchor windlass.

A November 1944 calculation of manpower requirements highlights another alternative armament: six single 10.5cm, two single 4cm, six single 2cm and one quadruple 2cm. However, although the ship had been moored at Kiel since early 1944, no work had yet been carried out, not even tidying up her hull paintwork, which still showed the water/oil marks from her 1943 scuttling. Rather, she was moored off Friedrichsort, on the north side of the entrance to inner Kiel fjord, with her remaining original guns still on board, apparently serving as a training ship and AA battery. Surviving files suggest that the ship would have been renamed *Gefion* (another late-1890s cruiser name): there is no external corroboration for the statement in Erich Gröner's *Deutschen Kriegsschiffe* that she was to become *Adler*.

In April 1945, *Peder Skram* was damaged aft during a bombing raid, and was put aground with her stern flooded. In late August, the Em Z Svitzer salvage company was contracted to retrieve the ship, the salvage vessel *Garm* succeeding in refloating her and preparing her for the tow home within three days of arrival. *Peder Skram* arrived back in Copenhagen on 3 September 1945 and was laid up at the dockyard. She was sold for scrap on 1 April 1949 and broken up from 1950 by HJ Hansen at Odense.

Archival Sources:

Deutsches historisches Institut Moskau, *Akte 143. Anweisungen und Schriftverkehr des Oberkommandos der Kriegsmarine (OKM) und der anderen Marinebehörden über den Umbau der Beuteschiffe in die schwimmenden Flak-Batterien in der Zeit*

Table 1: **German Flak Ships: Building Data**

Name	Built	Launched	Converted	Recomm
Ariadne (ex-*Vlierede*, ex-*Hertog Hendrik*)	Amsterdam DYd	7 Jun 1902	Mercantile (Antwerp)	4 Apr 1944
Undine (ex-*IJmuiden*, ex-*Jacob van Heemskerck*)	Amsterdam DYd	22 Sep 1906	Howaldtswerke (Hamburg)	29 Sep 1943
Niobe (ex-*Gelderland*)	Fijenoord (Rotterdam)	28 Sep 1998	Van der Giessen de Noord (Krimpen-aan-den-IJssel)	1 Mar 1944
[*Gefion*] (ex-*Peder Skram*)	Copenhagen DYd	2 May 1908	Intended: Mercantile (Antwerp)	?

Table 2: **German Flak Ships: Characteristics**

Ariadne* & *Undine

Displacement:	4,371 tonnes
Dimensions:	96.6m (oa) x 15.2m x 5.8m (*Ariadne*); 98.0m x 15.2m x 5.7m (*Undine*).
Machinery:	Inoperative
Armament:	Eight 10.5cm/45, four 4cm, sixteen 2cm (4 x IV) AA
Fire control:	Two directors (2 x 6-metre rangefinders)
Radar:	One FuMG 39 TD (Würzburg)

Niobe

Displacement:	3,512 tonnes
Dimensions:	93.3m (wl) 94.7m (oa) x 14.8m x 5.4m metres
Machinery:	Twelve Yarrow boilers; 2-shaft VTE; 9,867ihp = 16 knots
Armament:	Eight 10.5cm/45, four 4cm, sixteen 2cm (4 x IV) AA
Fire control:	Two directors (2 x 6-metre rangefinders)
Radar:	One FuMG 39 TD (Würzburg)

[*Gefion*]

Displacement:	3,735 tonnes
Dimensions:	84.0m (pp) 87.4m (oa) x 15.7m x 5.0m
Machinery:	Inoperative
Armament:	Six/eight 10.5cm/45, four 4cm, sixteen 2cm (4 x IV) AA
Fire control:	Two directors (2 x 6-metre rangefinders)
Radar:	One FuMG 39 TD (Würzburg)

vom Juli 1941 bis September 1944, Bestand 500 Findbuch 12453 – Oberkommando der Kriegsmarine (OKM)

Endnotes:

1. The authors are grateful to Leon Homburg of the Royal Netherlands Navy Museum, Den Helder, for the provision of information on and photographs of the ships in Dutch service.
2. J Mulder and WF Ruygrok, *De Nederlandsche Zeemacht:Pantserschepen, Pantserdeckschepen, Monitors*, Lanasta (Emmen, 2004).
3. Renamed *Soerabaja* in 1936 and sunk by the Japanese in February 1942.
4. The 4cm Bofors gun, installed initially on *Nymphe* and *Thetis*, would become the Navy's principal choice of close-defence AA gun in preference to the much-disliked 3.7cm.
5. M Kanetzki, *Operation Crossbow: Bomben auf Peenemu?nde*; Ch Links (Berlin, 2014), 106–7.
6. All were scrapped except for *Noordbrabant* (1899) which became an accommodation ship at Vlissingen; she was burned by her crew on 17 May 1940 to prevent her use by the Germans.
7. R Höfling, 'Die Versenkung des deutschen Fla-Kreuzers *Niobe*', *Jet & Prop: Flugzeuge von gestern und heute im Original und im Modell* 5/02 (2002), 29–31.
8. R Steen Steensen, *Vore Panserskibe 1863–1943*, Strubes Forlag (Lyngby, 1968); T Wismann, 'The Coastal Battleship *Peder Skram* (1908)', in B Taylor (ed), *The World of the Battleship*, Seaforth Publishing (Barnsley, 2018), 113–28.
9. Hadeler (1897–1987) had been one of the leading constructors responsible for the unfinished aircraft carrier *Graf Zeppelin*.

Ariadne

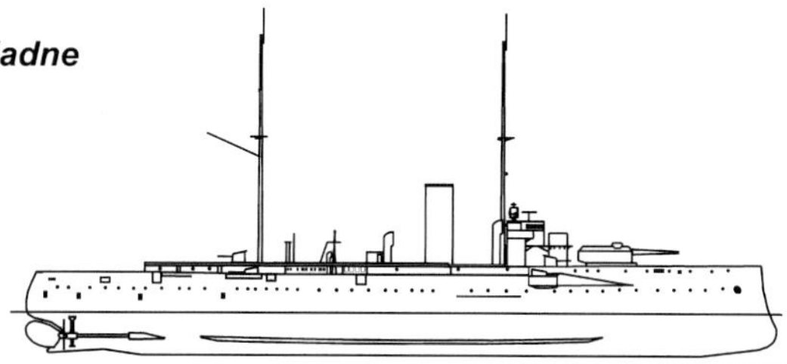

Hertog Hendrik 1937

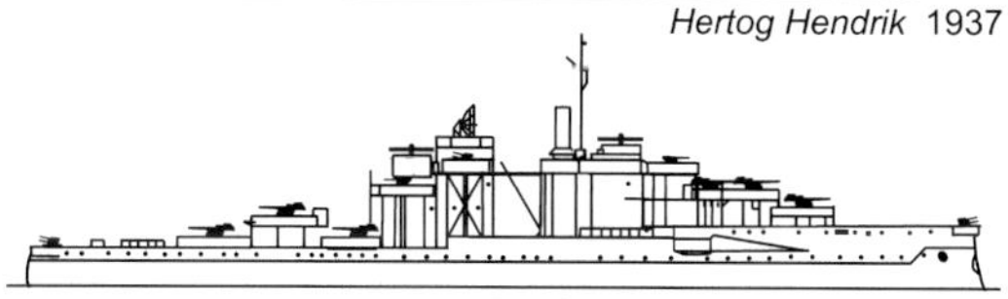

Ariadne (ex-*Hertog Hendrik*) 1944

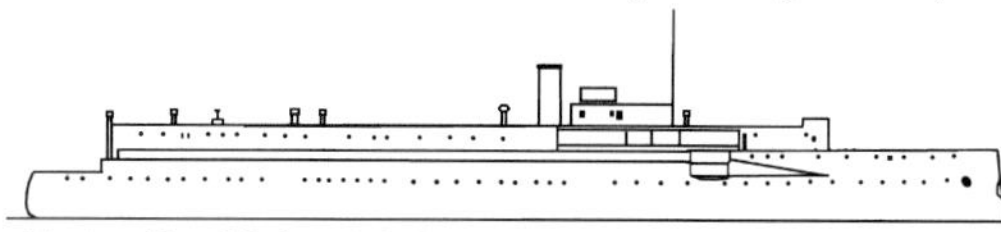

Hertog Hendrik (ex-*Ariadne*, ex-*Vlierede*, ex-*Hertog Hendrik*) 1947

Undine

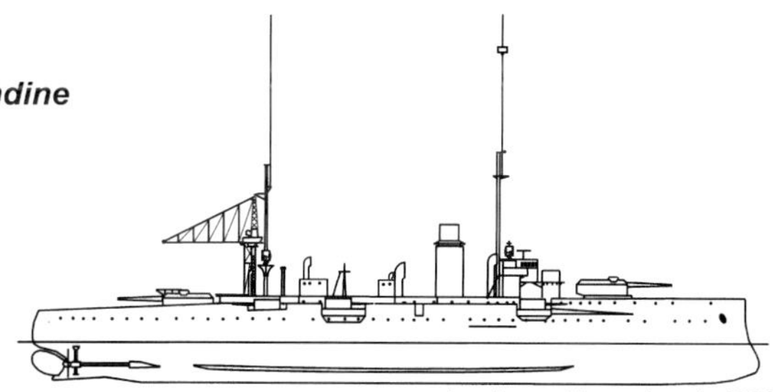

Jacob van Heemskerck 1933

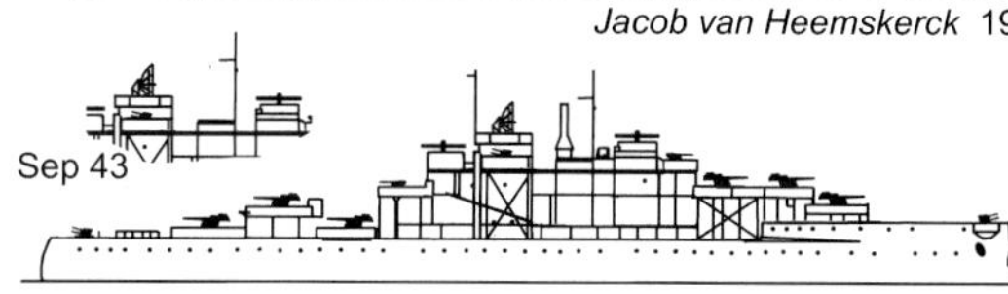

Undine (ex-*IJmuiden*, ex-*Jacob van Heemskerck*) 1944

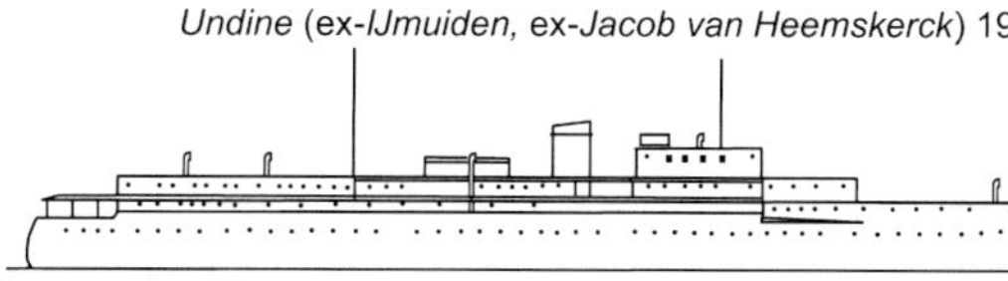

Neptunus (ex-*Undine*, ex-*IJmuiden*, ex-*Jacob van Heemskerck*) 1947

Niobe

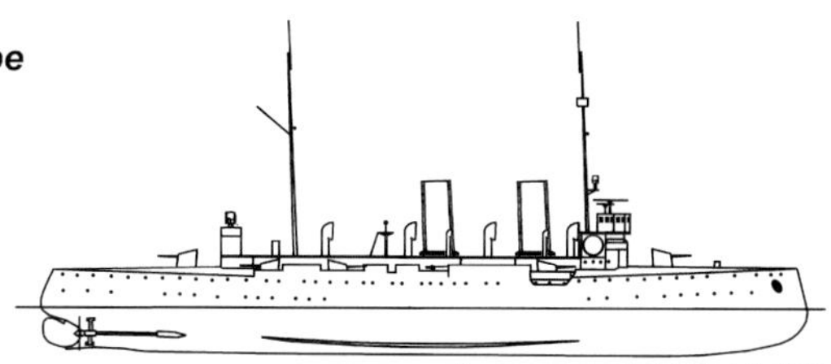

Gelderland 1939

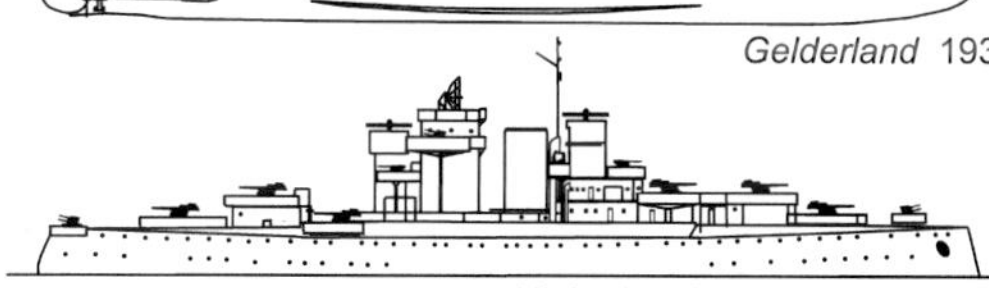

Niobe (ex-*Gelderland*) Mar 1944

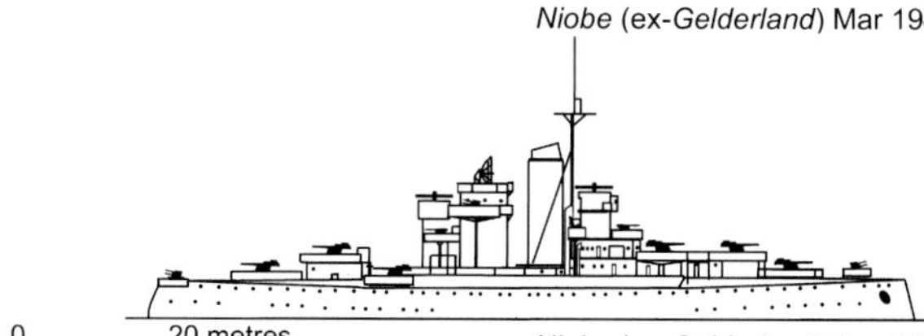

Niobe (ex-*Gelderland*) Apr 1944

WARSHIP NOTES

This section comprises a number of short articles and notes, generally highlighting little known aspects of warship history.

THE PROTECTION SYSTEM OF THE FRENCH BATTLESHIPS OF THE SECOND WORLD WAR

Following a recent correspondence, **John Jordan** is able to provide new information on the protection system of the fast battleship *Dunkerque*.

The author's book *French Battleships 1922–1956*, with Robert Dumas, was first published by Seaforth in 2009. A second, corrected edition was published the following year, followed by a slightly amended softback edition in 2020. However, it needs to be acknowledged that, regardless of how many editions a book goes through, the information provided in any secondary source must always be regarded as 'provisional'. Documentation held in the various archives can be lost or mislaid, and plans may be unsuited to digitisation due to their large size or poor condition. Robert completed his research on these ships in the late 1980s, leading to a series of French-language monographs published by Marines Editions in the early 1990s, and the English-language book was based on, and informed by these. By the time *French Battleships* was being prepared, plans of these ships had been posted for free download from the French *Service Historique de la Défense* website. However, only the plans for *Dunkerque* and *Richelieu* were present, and a number of key plans were missing.

In late May 2024 the author received an email from Mate Mityok via Tony di Giulian of navweaps.com. Mate requested information on the thickness of the lower edge of the (tapered) armoured belt in *Dunkerque.* I was unable to supply this, as the necessary section plans showing the detailed structure of the hull (normally headed '*Coupe au maître*') had not been made available when Robert and I compiled the Seaforth book. However, Mate persisted with his quest: a promising reference was located, and the SHD managed to locate some early plans dating from 1934. It took two weeks for the creases in the originals to be straightened out(!), but Mate finally received high-resolution scans and kindly sent them to me for analysis and comment. The present note is based on the resulting new information, which not only adds to our knowledge about *Dunkerque* but contains one surprise.

Armoured Belt

The battleships *Dunkerque*, *Strasbourg* and *Richelieu* had a modern 'all-or-nothing' protection scheme, with an armoured *caisson* over the ships' vitals that comprised an inclined belt of uniform thickness, a heavy upper armoured deck seated on the upper edge of the belt, and a light 'splinter' deck below comprising a central flat with inclined sides that joined the lower edge of the belt (see drawings). The plates for the belt were in a single strake 5.65–6 metres high and were tapered beneath the waterline. The belt for *Dunkerque* had a maximum thickness of 225mm (9in) and was angled at 11°30' to the vertical;

French WWII Battleships: Protection Scheme

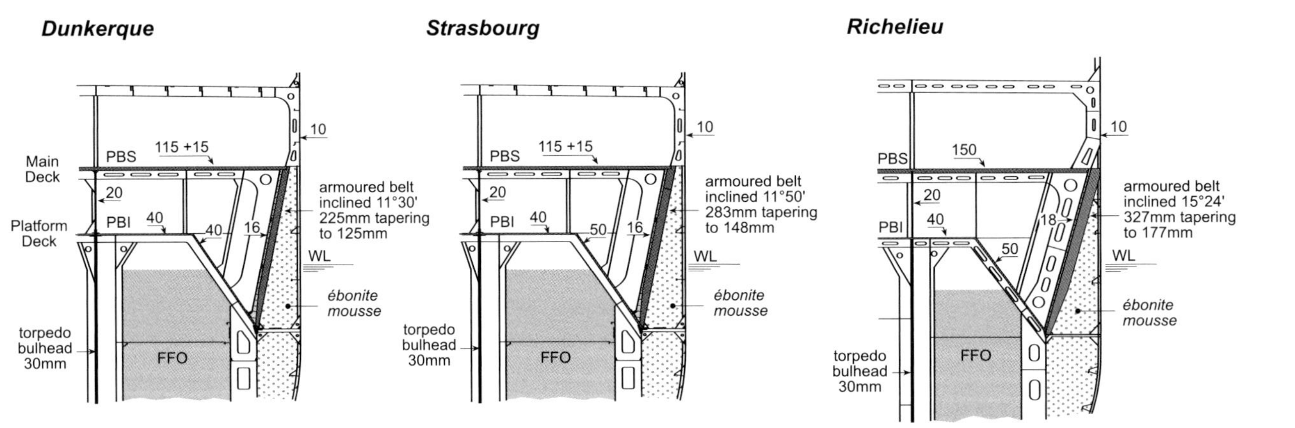

Dunkerque
Height of belt above waterline: 3.42m
Depth of belt below waterline: 2.24m

Note: The armour plates of the belt on *Dukerque* had a rebate in the upper edge into which the main armoured deck was seated. This was not the case in *Richelieu*, in which welding was used extensively.

0 5 METRES

© John Jordan 2024

Richelieu
Height of belt above waterline: 3.38m
Depth of belt below waterline: 2.58m

for the later *Strasbourg* the thickness of the plates was increased to 283mm (11in) and the angle slightly increased to 11°50'; and in the larger and more powerful *Richelieu* the thickness of the plates increased to 327mm (13in) and the angle of incline to 15°24'.

Outboard of the belt was a compartment filled with *ébonite mousse*, a water-excluding rubber compound. This served to preserve the watertight integrity of the ship above and below the waterline if the light outer hull, which was of 10mm steel plating, was penetrated by shell. It was therefore comparable in function to the sealed tubes adopted by the RN and IJN outboard of their own inclined belts during the 1920s. It appears to have been effective, but its presence complicated repair, and when impregnated with oil, as occurred with *Dunkerque* following the damage sustained at Mers el-Kebir in July 1940, could be set ablaze by oxyacetylene cutters and welding torches.

Drawings located earlier showing how the foot of the armour belt in *Strasbourg* and *Richelieu* was secured had established that the thickness of the lower edge of the belt in these ships as completed was 148mm and 177mm respectively. The newly-discovered plan of *Dunkerque* shows that the corresponding thickness was 125mm. A comparison with the maximum thickness of the belts gives the following: *Dunkerque* 55.5%, *Strasbourg* 52.6% and *Richelieu* 54.5%. This suggests a further slight tapering of the belt in *Strasbourg* to fit the original structure of the design, which was essentially unchanged. The point at which the belt was tapered was an estimated 0.870 metres beneath the waterline in *Dunkerque*, 0.855 metres in *Strasbourg*, and 0.925 metres in *Richelieu*, which had a slightly deeper belt. The armour plates for the former ships had a continuous flat (hardened) outer face, whereas in *Richelieu* the continuous face was on the inboard side and the tapering outboard. In all three ships the armoured belt was mounted on a 60mm teak backing (*matelas*), and

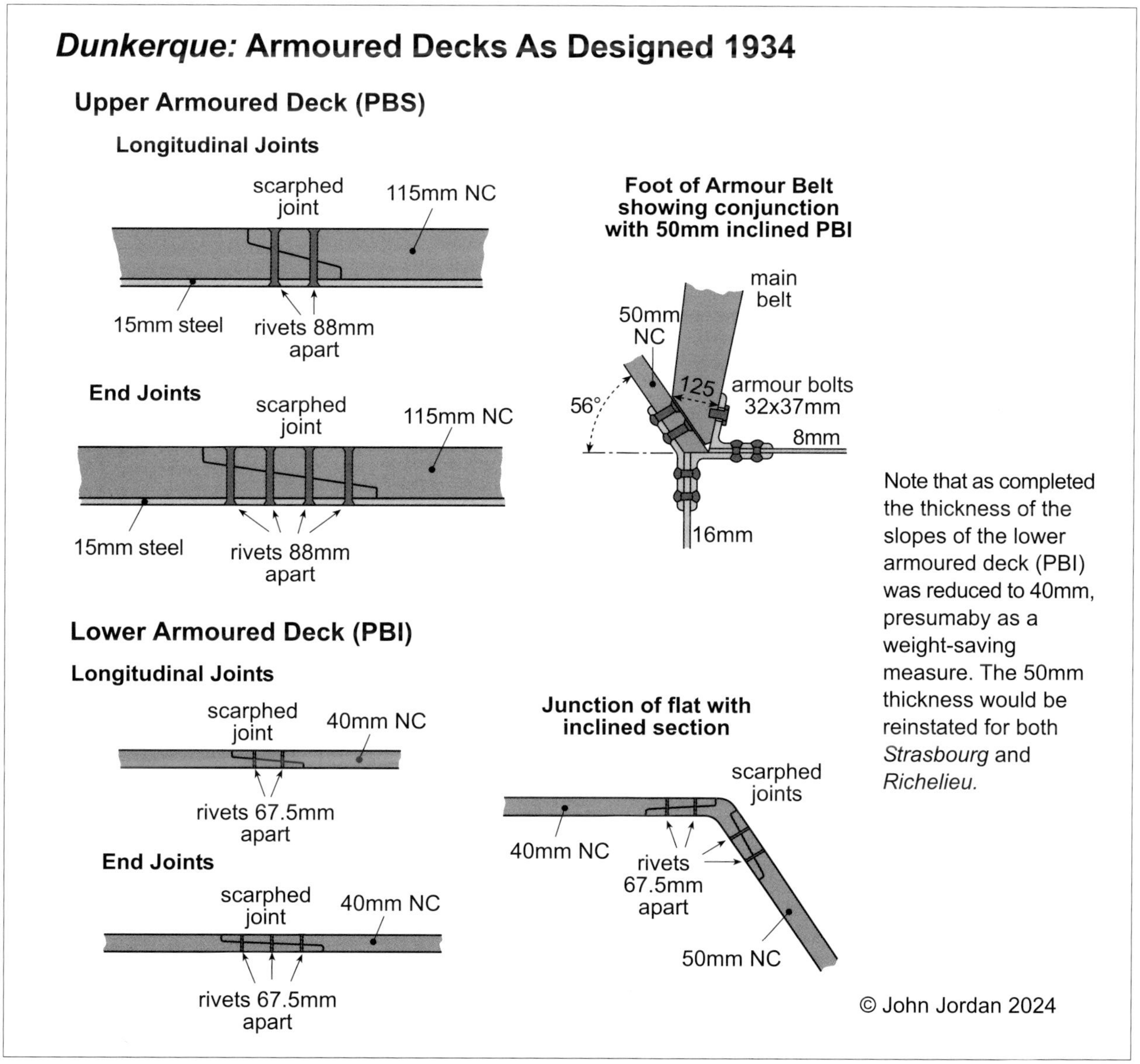

the individual plates were secured to an inclined inner bulkhead (16mm in *Dunkerque* and *Strasbourg*, 18mm in *Richelieu*) by rows of 60mm (upper) and 45mm (lower) armour bolts.

Armoured Decks

The upper armoured decks (PBS) of *Dunkerque* and *Strasbourg* were assembled from plates of 115mm (4.5in) non-cemented (NC) armour secured to a backing of 15mm (1.2in) construction steel, as in the earlier generation of French battleships. The plates were laid lengthways and had scarphed, riveted joints at the sides and ends (see drawings).

In the first edition of *French Battleships* it was incorrectly stated in the text that a similar arrangement was employed in *Richelieu* and her sister *Jean Bart*. This error was corrected in the second edition, in which it was made clear that the thicker 150mm (6in) armour plates were assembled without a backing but with butt straps reinforcing the scarphed joints (see *French Battleships* p 116 for detailed drawings).

The book also also stated – correctly – that the lower armoured deck (PBI) of *Dunkerque* had a uniform thickness of 40mm (1.6in) on the flat and the slopes, whereas for the two later ships the thickness of the inclined sides was increased to 50mm (2in). The newly-discovered plans of Dunkerque show that the thickness of the inclined plates was originally to have been 50mm – an identical arrangement to that of her sister *Strasbourg* and *Richelieu*. The plan shows that the 40mm and 50mm plates were laid towards the junction of the flat and the inclined sections, and joined by a connecting piece with scarphed, riveted joints at either end (see drawing).

Underwater Protection

Underwater protection for all three ships was on the air/liquid/air 'sandwich' principle introduced by the US Navy at the end of the First World War. Arrangements were virtually identical: an outer compartment 1.5 metres wide (1.95m for *Richelieu*) filled with *ébonite mousse*, a watertight compartment 0.9 metres wide, a fuel tank 3.9 metres wide (3.4 metres in *Richelieu*), a watertight compartment 0.7m wide, and a torpedo bulkhead of 30mm (1.2in) special steel backed by a holding bulkhead abreast the machinery compartments with a maximum thickness of 9/10mm reducing to 6mm. There was a slight reduction in the depth of the system in *Richelieu* (6.7m from the outer skin to the torpedo bulkhead compared with 7.3m for the earlier ships). This was still an impressive figure; the slight reduction appears to have been related to the need to accommodate three boilers side by side in the two boiler rooms versus only two (in three boiler rooms) in *Dunkerque* and *Strasbourg*; in most contemporary foreign battleships the depth of the underwater protection system was approximately 5 metres.

Acknowledgement: The author wishes to thank Mate Mityok and the French Tech Tree Sources' Discord server for their assistance in obtaining the plans of *Dunkerque*.

HMS *AGAMEMNON* AS A RADIO-CONTROLLED TARGET SHIP

This Warship Note by **Cdr Martin Marks RN (Ret), OBE, BSc (Eng)**, was originally published in the Newsletter for *Steam Pinnace 199*, which has been painstakingly refurbished under the aegis of the National Museum of the Royal Navy (NMRN).

Experiments involving radio control for small boats ('drones') began in the Royal Navy in 1917, when a secret Distantly Controlled Boat (DCB) section was set up at Calshot, near Southampton. Coastal Motor Boats (CMB) and Motor Launches (ML),[1] together with the coastal submarine *C4*,[2] were converted for radio control from 'mother' aircraft using a specially-developed transmitter. The DCB section also developed a transmitter for remote-controlled mines to be employed in the raid on the mole at Zeebrugge in April 1918.

In September 1918, the Commander-in-Chief of the Grand Fleet, Admiral David Beatty, had called for a large target to be provided that would allow realistic gunnery practice for the battleships of the Grand Fleet, which had seen little action since the Battle of Jutland in 1916. Tests against armour plate in 1919 demonstrated that firing 15in guns at any pre-dreadnought battleship would sink her quickly, but the use of such a ship for target practice against guns of 6in calibre or smaller seemed practical. At first, the battleship *Hibernia* was suggested for target duties, but ultimately HMS *Agamemnon* became available and was selected instead. *Agamemnon* had been part of the British squadron that went to Constantinople in November 1918 following the Armistice. She returned to the UK in March 1919, where she paid off at Chatham Dockyard and went into reserve on 20 March.

Using the experience acquired during the late-war period with radio-controlled drones, *Agamemnon* was modified at Chatham Dockyard for use as a target ship between 6 December 1920 and 8 April 1921. The ship was rewired for radio control and stripped; the 12in turrets remained aboard, but all of her guns and their associated fittings were removed, as were her torpedo tubes and handling gear, flying deck, sea cabins, main derrick and boats, lower conning tower, masts and yards, and most of her crew amenities and other equipment. Unnecessary hatches, coamings, scuttles and lifts were removed and plated over, and ballasting arrangements modified. It was not intended to sink her, so she was assigned a crew of 153 to maintain and operate her when she was not under fire.

Agamemnon's first target service took place before her modifications were completed. On 19 March 1921 she was exposed to a cloud of poisonous gas to determine its effect on a battleship. It was found that gas could pene-

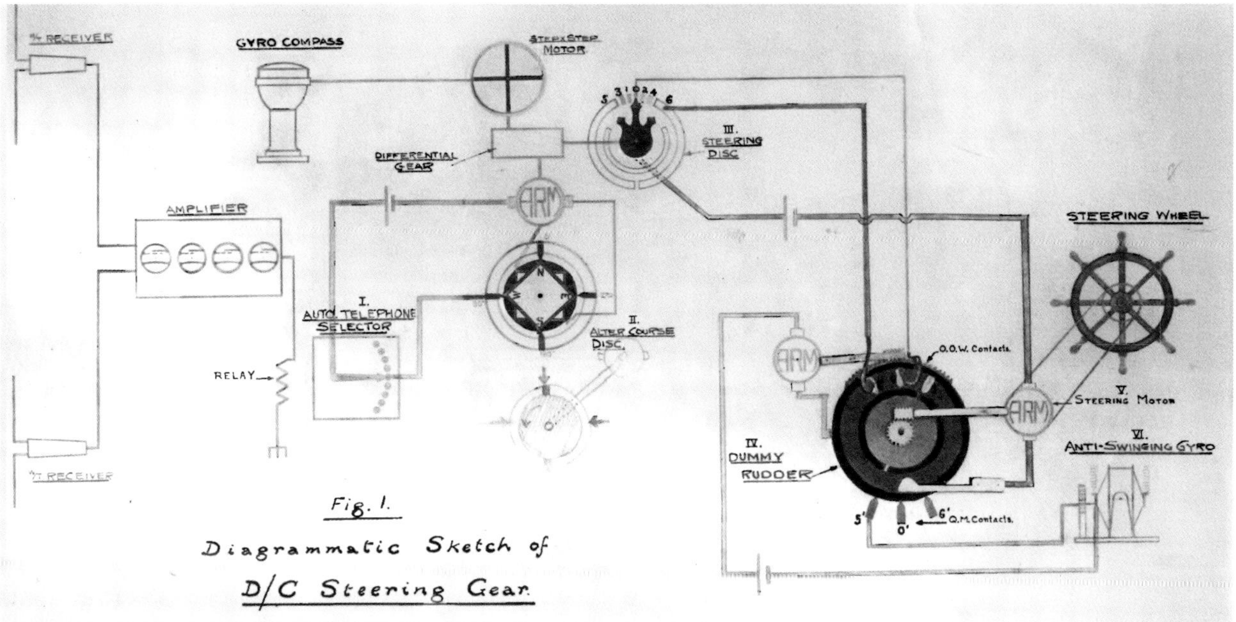

Schematic showing *Agamemnon*'s DC steering gear system. (All images accompanying this article are published by kind permission of the Trustees of the National Museum of the Royal Navy.)

Above: The steering motor and wheel.

Right: The main steam valve control.

Far right: Amplifier No 5.

trate the ship via her various openings, but the ship had not been sealed against gas before the trial and no accurate results applicable to a commissioned battleship could be obtained.

HM Signals School and Portsmouth and Chatham Dockyards provided the gear for the steering system; Chatham produced and fitted all the gear for controlling the engines and oil fuel valves. *Agamemnon* sailed from Chatham for Portsmouth on 25 July 1921, and carried out trials for the steering gear on passage without the wireless control element. Aircraft bombing and machine gun trials were planned for late August, but the three weeks of practical testing requested by the Signals School were 'dispensed with'. On 23 and 24 August the trials went well except for the failure of a few 'pea valves'. A few more failed on the 25th, and on the 26th all the valves failed. The bombing trials were successful. Every available pea valve was then collected and the mounting panels modified. These were trialled successfully on 31 August. As no more pea valves were available, the Signals School commenced a trial to produce its own valves. On 7/8 September HMS *Snapdragon*, a 'Flower'-class sloop of 1,250 tons, was fitted out for control, and produced excellent results except that more pea valves failed and the trials had to be stopped to ensure sufficient valves were available for machine gun trials on the 9th. This was the first time that *Agamemnon* was completely unmanned and the trial was successful. On 21 September *Agamemnon* was subjected to machine-gun fire by strafing aircraft. These trials showed that such strafing could harass a battleship but could not impair her fighting or steaming capabilities, and helped to determine protection for bridge personnel.

The Signals School produced a specification for new

Agamemnon as a target ship, photographed in Grand Harbour, Valletta, during the early 1920s. (RNM1968/58h)

valves (D1X) and these were put out to tender. After a trial on 18 October it was decided that *Agamemnon* was ready to proceed north for her role as target ship for the Atlantic Fleet. There was still concern that the eight sets of receivers and aerials would not stand the shock of incoming shells. As a result they were hit with a sledge-hammer and a 200lb projectile repeatedly dropped on the deck directly above the control rooms. It was considered that the controlling ship might have to circle *Agamemnon* to ensure contact with the remaining serviceable aerials. An 'indicating device' was fitted to *Agamemnon*'s conning tower to give a visual signal when a control order was received. The trials report indicates that she was quite difficult to steer, with 5 to 8 degrees of overshoot after course changes which took five minutes to steady. Various trials of the five engine and oil fuel pump controls were satisfactory, but it was noted that they had been designed for gradual increase and decrease to avoid fluctuations in water levels and steam pressure, as the auxiliary machinery was not under control. It was

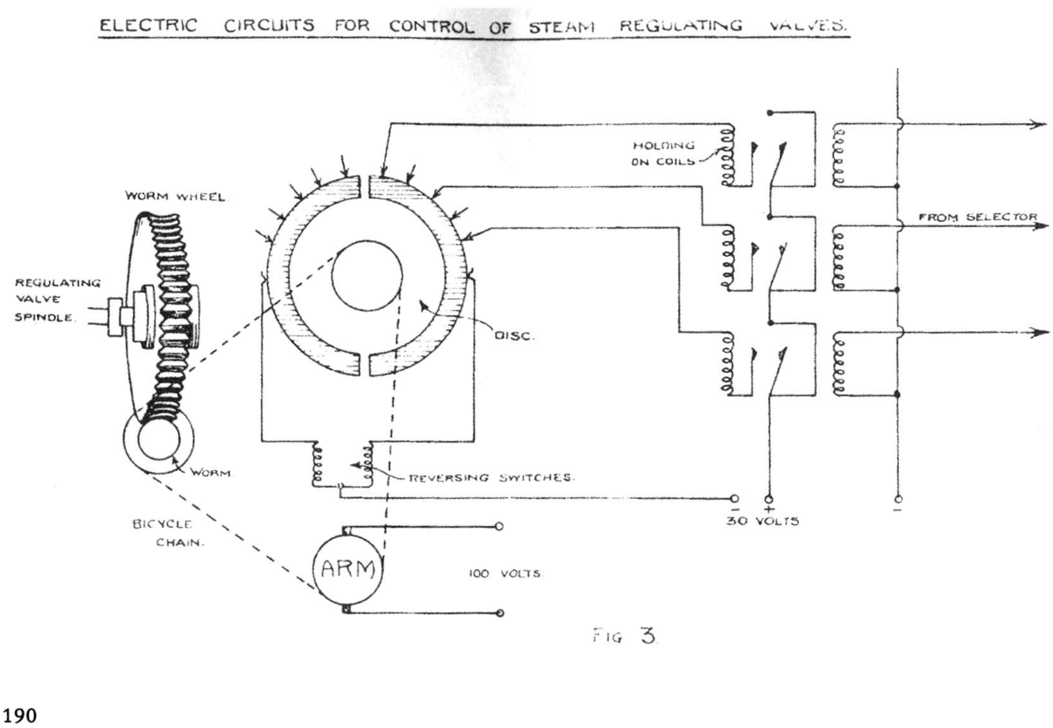

Schematic showing the electrical circuits for steam.

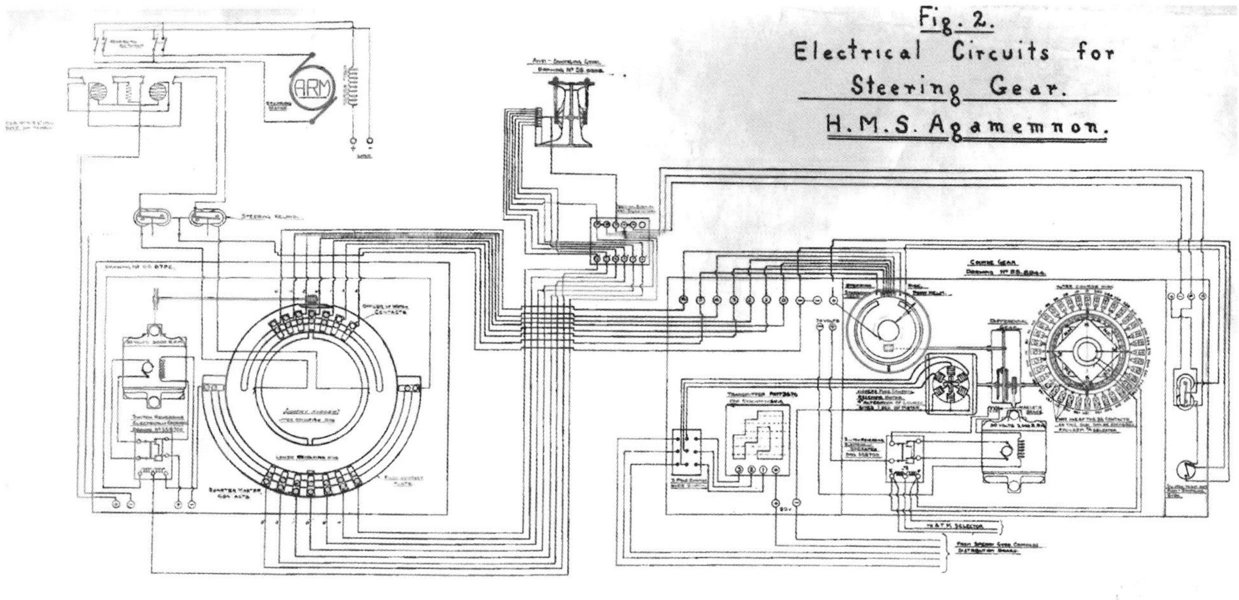

Schematic showing the electrical circuits for steering.

also noted: 'When steam in all boilers any delay between abandoning ship and commencing to work up the speed of main engines should if possible be avoided'!

Experienced seamen and engineers are no doubt by now envisaging 17,000 tons of battleship at 15 knots steaming up the east coast completely out of control following a signal failure. The trials team did too! *Agamemnon* was fitted with a clockwork mechanism that ran for fifteen minutes after the last control signal was received. A further incoming signal reset the fifteen-minute cycle again. Failure to reset caused the steam to the main engines to be shut off. (Experienced engineers will also realise the potential effect of this!) This system was later modified to reduce the boiler pressure to a minimum to prevent all the safety valves being floated by the earlier system. After three minutes a Very flare was fired and after a further three minutes the main engine regulating valves were closed. In the event of a main electrical system failure, a battery bank was used to initiate the shut-off.

The project used a control system wavelength of 10 metres, still in its infancy in 1922 but found to be less susceptible to interference. The transmitting aerial current in *Snapdragon* was reported as 1.5 amperes, giving a range of about two miles.

During her service as target ship, *Agamemnon* was also was used to test the vulnerability of battleships to 6in, 5.5in and 4.7in rounds fired at her by ships such as the battlecruisers *Renown* and *Repulse* while she manoeuvred under steering motor and wheel radio control. These tests showed that ships as well protected as *Agamemnon*, such as the later dreadnoughts, would suffer damage to their upper works if struck by such shells, but would not have their steaming or fighting capability seriously impaired even by numerous smaller-calibre hits.

Agamemnon was relieved as target ship in 1926 by HMS *Centurion* when this *King George V*-class dreadnought battleship was converted to Remote Radio Control. By then the last British pre-dreadnought battleship in existence, *Agamemnon* was sold to J Cashmore of Newport, South Wales, on 24 January 1927 for scrap, and departed Portsmouth Dockyard on 1 March to be broken up at Newport.

Sources:

Steve Mills, *The Dawn of the Drone: from the Back-Room Boys of World War I*, Casemate Publishers (Oxford, 2021).

HR Everett, *Unmanned Systems of World Wars I and II*, MIT Press (Cambridge Ma, 2015).

National Museum of the Royal Navy: Special Collections.

Acknowledgements:

The author wishes to thank Heather Johnson, Librarian (Special Collections) National Museum of the Royal Navy, who kindly provided information and references on early radio control of warships. Sincere thanks are also due to Lt-Cdr Clive Kidd, Honorary Curator of the HMS *Collingwood* Heritage Collection, who pointed the author to a folder on *Agamemnon* in the Collingwood Heritage Collection that is available as a 19-page PDF.

1 The only First World War radio-controlled boat to survive is *CMB 9/DCB 1*. Surviving radio control equipment for these drones is in the archives of the Imperial War Museum. The Germans successfully used remote-controlled boats to attack monitors off the Belgian coast during the Great War. Their control was by wire paid out from the shore and guided by aerial observation.

2 This may be the reason why *C4* was the only boat of her class not to be scrapped shortly after the Great War; she survived until 1922, when she was broken up (see Mills, *op cit*).

THE *FAIDHERBE* AND THE MARCHAND EXPEDITION

In his second Warship Note on the French colonisation of Africa, **Ian Sturton** writes about the Expedition of Captain Jean-Baptiste Marchand during 1896–98.

The Sudan, part of Egypt from 1820, was conquered between 1881 and 1885 by fanatical Dervish troops under Muhammed Ahmad, the Mahdi (died June 1885), and his lieutenant and successor the Khalifah Abdullah. The Khalifah set up a theocratic Islamic government that ruled ruthlessly and remorselessly, relentlessly massacring opponents. Egyptian Government troops (Egypt was under Ottoman suzerainty – nominal ultimate control by the Turkish Sultan in Constantinople – but a *de facto* British protectorate from 1882 after the bombardment of Alexandria and the defeat of Arabi Pasha at the Battle of Tel-el-Kebir) were slowly advancing south down the Nile, reconquering the country incrementally while spending a minimum of money.

The other European Powers with African interests, having regard to the post-1884 'Scramble for Africa' (Berlin Conference 1884–85), considered the Sudan was not under effective control so was 'up for grabs'. France decided on expeditions from the southwest (Captain Jean-Baptiste Marchand in the French Congo) and from the east (from Gore in Abyssinia, setting out in early December 1897); the latter expedition gave up and turned back on 22 June 1898 after reaching the upper White Nile, as the route was difficult and there was no news of Marchand, who was still struggling across the Congo/Nile watershed. The Abyssinians under the Emperor Menelik II sent an expedition from Gore operating in parallel with the French, but rather half-heartedly and lacking enthusiasm, as Menelik II would not fully commit himself until he could see the likely victor. Italy, crushingly defeated at Adowa (March 1896), was out of the running. Imperial Russia was interested and sent a mission, but would not support the French expeditions.

A Belgian approach from the Congo Free State to the south led by Baron Dhanis started to disintegrate when mutinous troops killed some officers at Dirfi on 15 February 1897, and after another disaster (18 March) collapsed completely near Lake Albert. A British attempt from the southeast (Uganda) under Captain JAL Macdonald likewise failed when troops mutinied; their commander diverted them towards Lake Rudolf and Abyssinia.

Only Marchand and the British from Egypt succeeded in penetrating the Sudan effectively.

The *Faidherbe* and the Marchand Expedition

Marchand reached Libreville (Gabon) on 20 June 1896 with a force of 13 French officers (including himself), NCOs and support staff with 120 *tirailleurs* (Senegalese soldiers armed with rifles). *Faidherbe*, a small sectional (*démontable*) screw steam launch belonging to the French Congo colonial authorities and named after the outstanding French colonial explorer and administrator General Louis Faidherbe (1818–1889), was commandeered by Marchand and joined his expedition in May 1897 at Kouango on the Ubangi river, a major tributary of the Congo. From there she was steamed upriver to the impassable rapids at Ouango, where she was dismantled and dragged on tree-trunk rollers past the rapids, a task

Faidherbe under way in the Sudan; note the French *Tricolore*. (Musée national de la Marine)

All hands hauling *Faidherbe* out of the Ubangi or an Ubangi tributary bow-first; a French colonial flag flies from the ensign staff. An impassive local watches from a convenient rock. (Private collection)

Faidherbe pulled ashore on an Ubangi sandbank. Personal weapons are stacked in the foreground; modern Lebel rifles provided formidable firepower, although completely outclassed by the Maxim gun. (Private collection)

Faidherbe again seen on the Ubangi in a gelatin print, sheltered by trees and with native craft in the foreground. (Private collection)

which took eight days. The work was supervised by Engineer Souyri, specially recruited by Marchand for the purpose. *Faidherbe* continued up the M'bomu river to where the M'boko joined, the latter being chosen for the further route as bringing the expedition closer to the Congo/Nile watershed. On 11 Sepember she reached the point where the Méré, yet another tributary, joined the M'boko. This was as far as she could go under steam; she had covered 1,500km from Kouango and steamed where no steamship had ever been before. Marchand was 3,300km from Brazzaville, his starting point.

Faidherbe had then to be moved overland 160km to Khojali or Kodgalé, the nearest point on the Soué (or Sué), a tributary of the White Nile, where she could be floated in what was left of the rainy season. Once pulled ashore, dismantling began on 12 September and the first pieces were soon on their way. Preliminary work at Khojali had already begun (20 August) and the bare hull was refloated on 20 September. The hull framing, in five sections each lifted by twelve porters, reached Khojali by 14 October. The cast-iron boiler, in two parts each weighing around 1,000kg, was moved overland by carefully lowering each part into a light steel whale-boat, fixing it securely for a total weight of around 3,000kg, and then dragging each ensemble by brute force on tree-trunk rollers along a path hacked through the bush. The boiler parts reached Khojali on 2 and 5 December, too late for reassembly in the rainy season. Engineer Souyri, hitherto so successful but overworked and depressed, became ill and returned to Ubangi. Immediate further progress was halted by the *sudd*, the vast reeking morass of rotting intertwined vegetation that filled the Bahr-el-Ghazal, a saucer-like depression extending over several White Nile tributaries; the *sudd* was insect- and snake-infested and nearly impenetrable. Marchand's immediate goal was Fashoda (French 'Fachoda'), a town on the White Nile about 1,000km south of the capital Khartoum where east/west and north/south African trade routes crossed.

As the water remained too low for the *Faidherbe*, Marchand, impatient to reach his destination, managed to by-pass the *sudd* and on 4 June start downriver in small boats, propelled by makeshift sails and long punting poles. He reached Fashoda on 10 July, ceremonially raising the *Tricolore* on the 12th. His force totalled eight white men, the 120 *tirailleurs* and 50 Yakoma boatmen brought from the mouth of the Congo, armed with muskets ineffective because of damp powder.

The Dervishes in Khartoum, learning of the arrival of unknown white men at Fashoda, sent a strong force comprising the gunboats *Safieh* and *Tewfikieh* (surviving

Faidherbe seen on the Ubangi in a gelatin print. (Private collection)

Moving half of a *Faidherbe* boiler on land. The formidable technical issues involved are discussed in the text. (ehne.fr)

A substantial contingent of the Europeans in Marchand's mission (white sun-helmets) are among those at work on *Faidherbe*; her bow (machinery space) is away from and the stern (the frames) nearer to the camera. (fachoda.blogspot.com)

Gordon side-wheel steamers that had been given some light protective plating) towing seven lighters with *c*1,000 men, to investigate and if necessary eject the intruder. On arrival this force was driven off by the newcomers on 25 August, accurate French rifle fire damaging the gunboat boilers before the troops could be landed. A second attack by land on 29 August that would probably have overwhelmed Marchand's tiny force was called off when the *Faidherbe* was seen to arrive. The Dervishes, with a healthy respect for armed rivercraft, withdrew a few kilometers, set up a base, licked their wounds and awaited new instructions.

Faidherbe was kept busy. From 1–14 September she was despatched on a mission up the Sobat river, a fast-flowing tributary of the White Nile, to check for traces of the French Gore expedition (and any Abyssinian outposts); nothing was found, the river degenerated into swamps and the mission was aborted. Two days later she was sent south back up the White Nile, retracing Marchand's route to order replacements and supplies from the Ubangi colonial authorities, and thus missed the Kitchener-Marchand confrontation.

Kitchener, Marchand and Fashoda

Sir Herbert Kitchener, the Sirdar (or Commander-in-Chief) of the British-controlled Egyptian Army, destroyed the Dervish army at the Battle of Omdurman on 2 September 1898, ending the Khalifah's regime. He was about to depart for Cairo and a hero's welcome in Britain when, on 7 September, he received information that a party of white men, nationality as yet uncertain, was in Fashoda, 600 miles upriver. The information came from the captain of the Dervish sidewheel gunboat *Tewfikieh*, who was bringing news of the intruders back to higher authority in Khartoum and found himself reporting not to the Khalifah but to the Sirdar. The latter lost no time, started south on the 10th in the screw gunboat *Sultan* with the sternwheel gunboats *Fateh*, *Nasr* and *Abu Klea* and the Egyptian post-boat *Dal*. With him in the gunboats and in towed lighters were a company of Cameron Highlanders (100 men), two Sudanese battalions (2,000 men), an Egyptian artillery battery and four Maxim guns – an overwhelming force.

On his way south (15 September) he shelled the Dervish base, seizing fuel and supplies and badly damaging the sidewheel gunboat *Safieh* (later salvaged), reaching Fashoda four days later. His gunboats (excepting the one carrying the Highlanders) flew the Egyptian flag and he transferred to the post-boat before arrival. He had immediate meaningful discussions with Marchand, who conceded the physical weakness of his own position – at the time he had just seven Europeans, the 120 *tirailleurs* and the boatmen – and his lack of official French support. Both decided that the future of Fashoda must be settled by higher authorities in London and Paris. They went on to toast the Queen and President Faure in Kitchener's lukewarm whisky and soda; Marchand is supposed to have remarked later that one of the greatest sacrifices he ever made for France was drinking *cet affreux alcool enfumé* (that frightful smoking alcoholic liquid)! The Sirdar hoisted the Egyptian flag ashore at 1pm, leaving Marchand and the *Tricolore* in place, before departing for Khartoum at speed the same afternoon. Before leaving, Kitchener ordered the establishment of posts to the south on the Sobat and at Meshra-er-Rek on the Bahr-el-Ghazal, in effect surrounding Fashoda. Major Jackson was left with one Sudanese battalion, four guns and a gunboat to keep direct watch over the French.

Jackson's task was unenviable, as the climate was very hot, very wet and very humid; mosquitos, other insects and further unpleasant creatures abounded, the whole place stank and the *Faidherbe* could run rings round heavily-laden British gunboats like a gadfly. Ignoring Jackson's orders to the contrary, the French fished when and where they pleased. Worst of all, perhaps, were his sessions with the prickly Marchand.

The Sirdar reached Cairo on 6 October and Marseille on the 26th, passing through France without recognition before arriving at Dover on the 27th to a rapturous hero's welcome. Marchand, enraged by the lack of support from France, decided, strictly against his orders, to take his final report to Paris in person, leaving Fashoda on 24 October and reaching Cairo on 3 November, the day the French *chargé d'affaires* there was informed that France was going to back down and withdraw from Fashoda. His journey now pointless, Marchand went no farther, and turned to negotiating the date and route of his withdrawal with the British. He rejected the rapid, straightforward route down the Nile – it would mean much too much cooperation with his new and greatly-disliked masters – and decided to pursue his original plan of continuing to Djibouti on the Red Sea through Abyssinia.

To the relief of the sorely-tried Jackson, Marchand finally left Fashoda during 11–13 December in the *Faidherbe* and any craft that could be pressed into service, taking eight Frenchmen, 100 *tirailleurs* and the Yakoma boatmen (two Frenchmen and 50 *tirailleurs* had left earlier to find a land route to Abyssinia). The little flotilla passed safely from the White Nile to the Sobat before being halted some 200km upstream on 5 January by a series of waterfalls; *Faidherbe* hit a rock and started sinking. A large hollow was dug in the river bank and all rivercraft abandoned on 7 January, the French continuing their journey on land. Five weeks' hospitality at the court of Menelik II in Addis Ababa preceded a gentler descent with every comfort to French Djibouti. The British later

Data (British from Brassey and Laird Clowes):

Faidherbe

The only data found for the launch are: draught 60cm, speed 14 knots. Comparison with contemporary sectional steel steam launches suggests figures of *c*14m length x *c*2.2m beam, and cranage weight of bare hull *c*4.5 tonnes.

Sultan

140t; draught 60cm; twin-screw, 1–12pdr, 1 howitzer (about 10cm), 4 Maxims (Melik Society website: 134t; 145ft x 24ft 6in x 2ft draft; twin-screw compound machinery, 900ihp = 12kts).

Fateh, Nasr

128t; draught 85cm full load; stern-wheelers, 1–12pdr, 2–6pdr, 1 howitzer, 3/4 Maxims

Abu Klea

Stern-wheeler, 1–22pdr, 2qf

found the remains of *Faidherbe* and removed the hull, useful as scrap.

Marchand finally reached Djibouti on 16 May 1899, almost three years after starting from Libreville, and left in the cruiser *D'Assas* (which had departed from Toulon on 3 April expressly to bring him home), arriving at Toulon on 30 May and (by rail) Paris on 2 June, without ceremony at any stage (in contrast to his assistant Captain Baratier, who had been given a great welcome when he brought a preliminary Fashoda report to Paris while the crisis was at its height). He found a new era in Franco-British relations was approaching. France had for centuries been weaker than Britain at sea and since 1871 weaker than Germany on land; the Dreyfus case had made her internal divisions very apparent. Britain also had discovered the limitations of 'Splendid Isolation' in 1896, when simultaneous crises with the United States

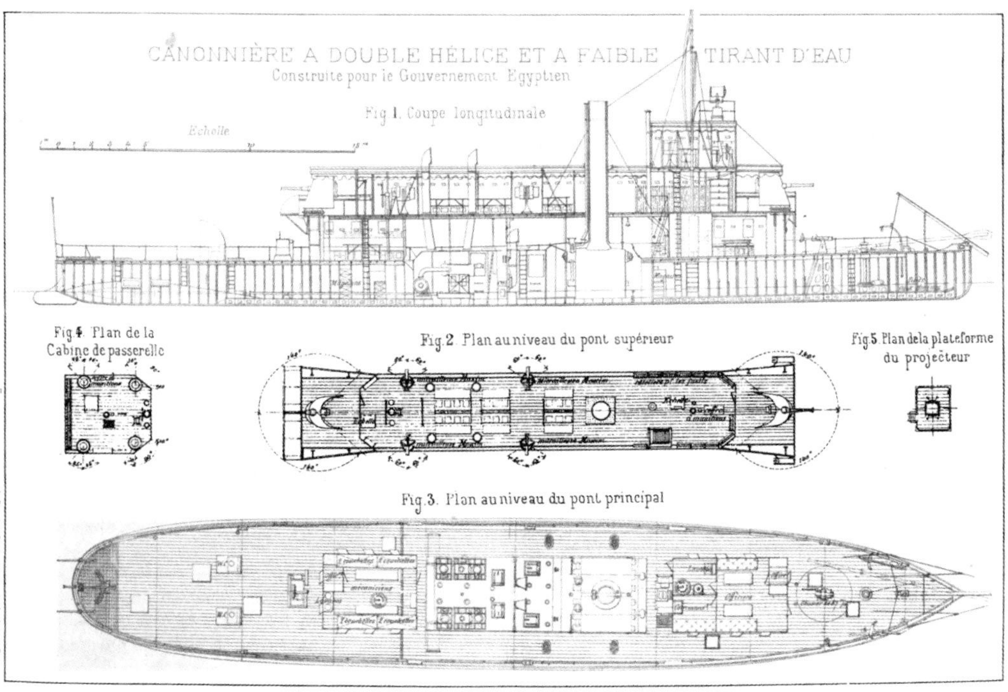

A contemporary drawing of HMS *Sultan*. (Le Génie Civil).

and Germany had coincided with general European enmity over the Jameson raid. Tirpitz's Navy Laws made future German intentions at sea very plain. It was becoming clear that the Kaiser's Germany was likely to be the future adversary of both France and Britain.

Foreshadowing the Anglo-French *Entente* (April 1904), the Fashoda incident would be quietly 'brushed under the carpet' and Marchand's epic exploits, no longer regarded as important or relevant, forgotten. He himself was sidelined on a military mission to China and then retired on half pay. He was recalled to the colours in 1914 and finished the war as a succesful Lieutenant-General, dying in 1934.

Afterwords

An Anglo-Egyptian Condominium over the Sudan was proclaimed 19 Jan 1899. It was formally terminated on 1 January 1956 when Sudan became an independent republic. South Sudan split from Sudan in 2011.

The Sirdar continued to advance in the Egyptian and then in the British service. In 1899 he was the first Governor-General of the Anglo-Egyptian Sudan. As Field-Marshal the Earl Kitchener of Khartoum, Secretary of State for War, he drowned when HMS *Hampshire* was mined off the Orkneys in June 1916.

The boundary between the British and French spheres of interest was fixed at the Congo-Nile watershed by the Anglo-French Agreement of 21 March 1899. In 2023 the watershed is the boundary between the South Sudan on the Nile side and the Central African Republic on the Congo side (and, further north, between the Sudan and the Republic of Chad), an area of extreme instability and seemingly perpetual racial strife.

The fugitive Khalifah and his principal advisers were hunted down and killed by Sudanese troops on 24 Nov 1899 at Om Dubreikat in Kordofan, some 300km south of Khartoum.

Fashoda was renamed Kodok at the start of the Anglo-French *Entente* (1904) and so remains. Today it is in the Republic of South Sudan.

Sources:

D Bates, *The Fashoda Incident of 1898*, Oxford University Press (Oxford, 1984).

DL Lewis, *The Race to Fashoda*, Bloomsbury (London 1988).

P Magnus, *Kitchener, Portrait of an Imperialist*, John Murray (London 1958): Chapter 8, Ruler of the Sudan 1898–1900.

TA Brassey (ed), *The Naval Annual 1899*, Griffin (Portsmouth 1899): Chapter VIII, The Operations of the Gunboats during the Soudan Campaign.

WS Chalmers, *Life of Lord Beatty*, Hodder and Stoughton (London 1951): Chapter II, 1896–1898 The Nile. Beatty commanded the gunboat *Fateh* at Omdurman and at Fashoda.

J Randier, *La Royale, L'Éperon et la Cuirasse*, Éditions Marcel-Didier Vrac (Le Touvet 1997): p 49 photo.

S Roskill, *Earl Beatty, the Last Naval Hero*, Collins (London 1980): Chapter 1, Early Years.

W Laird Clowes, *The Royal Navy, Vol VII 1857–1900*, Chatham Publishing, London 1997 (reprint): 364–370 (Gordon and Khartoum) and 449–451 (Kitchener and Fashoda).

WRITING ON NAVAL HISTORY: SOME NOTES FOR BEGINNERS

In response to a suggestion by *Warship*'s editorial team, regular contributor **Conrad Waters** has submitted some helpful advice for prospective authors.

The future of naval research is heavily reliant on a new generation of writers putting their work forward for the benefit of a wider readership. This, however, can present a daunting prospect for the uninitiated, as the process of securing publication in print has always been surrounded by a certain air of mystique. These notes have been put together at the suggestion of Stephen Dent, *Warship*'s Assistant Editor, in order to shed light on various aspects of the writing process. It is hoped that they might encourage new writers to throw their hats into the ring.

Background

Like many readers, I have had a life-long interest in naval history. I began writing for magazines when I was sixteen, and have subsequently written more than 150 articles and chapters on, mainly, 20th and 21st century warships and navies. I have edited the naval annual *Seaforth World Naval Review* since its initial 2010 edition, but only had my first standalone book published in 2018. My particular interest is in Royal Navy cruiser design of the Second World War period. To date, I have written histories about two of these classes of cruiser.

Shaping a vision

All potential authors will have certain specific areas of interest and ideas about the subjects on which they feel competent to write. The more fortunate will be sufficiently capable of writing their first book without further ado. Others will need time to hone their thoughts and skills as they shape their book's vision. As an author I fall firmly into the latter category.

My own route to becoming a published author started several decades ago, writing short articles for the magazine *Ships Monthly*. This allowed me to develop research and writing techniques while also making a start on developing that all-important name recognition. A subsequent opportunity to contribute a couple of short notes for *Warship* further refined my skills and provided me with the confidence to produce more complex articles. The guidance I received from the annual's editorial staff during this time was particularly important. It is hard to overstate how valuable this experience was in assisting my future writing career.

This process of self-development is ongoing. Before embarking on my two cruiser class histories, I was fortunate enough to be approached to write one of Seaforth Publishing's series of books focused on the ship plans held by the National Maritime Museum (part of Royal Museums Greenwich). This book, focused on the Second World War cruiser *Birmingham*, compelled me to develop a detailed knowledge of virtually all elements of contemporary warship design so that I could describe what was shown in the plans accurately. Any similar opportunities presented to authors in their own fields should be seized with both hands.

Finding a publisher

Unless intending to follow the self-published route – one about which I have little direct experience but understand to have many practical challenges – a prospective author will need to find a publisher if they are to make their work available to a wider audience. Given that naval history is a fairly niche subject, there are only a limited number of possible options to explore. Of these, *Warship's* publisher Osprey, Pen and Sword and, in particular, their specialist imprint Seaforth Publishing, are the main British publishers currently active in the field. Their main counterpart in North America is the US Naval Institute Press. Other British specialist historical publishers also commissioning the occasional naval title include Amberley, Helion and Whittles.

The point at which it is best to approach a potential publisher to gauge their interest in a particular project is debatable. My personal perspective is that it is probably best to do this at an early stage, as this allows the practicalities of publication to be addressed before too much effort on research or writing has been expended. Publishers are inevitably driven primarily by commercial considerations and will have a clear view as to whether a particular book is 'marketable'. These discussions can sometimes lead to events taking an unexpected direction. My first meeting with Seaforth was to explore the possibility of writing a book on the British Type 45 destroyers; the subsequent conversations ultimately resulted in my editing the *Seaforth World Naval Review* (*WNR*) series.

When you approach a publisher it is vital to have a clear idea of your book's subject matter, its proposed length and structure, and its target readership. If a publisher is interested, further discussions will follow. These can form the start of a long-term relationship so it is important that they are conducted in a friendly, practical and flexible manner.

Research

Like many authors, I find the research aspect one of the more interesting parts of the writing process. I tend to start by reading as many published sources in my field of interest as possible, carefully noting the sources previous authors have used for their own research. I then follow up in the various archives any of these that are relevant to my own work; living within easy travelling distance of The National Archives at Kew and the Royal Museums Greenwich archive of material at the Brass Foundry in Woolwich is of considerable help in this regard. In the course of my own research I found it fascinating to read the then Chancellor Winston Churchill's original handwritten annotations to discussions on cruiser policy in the 1920s, and the detailed briefings on strategy that preceded the various Naval Treaty negotiations. However, time at the archives is precious and it is important to have a firm plan as to which documents are relevant and need to be ordered and/or photographed. It goes without saying that all archives have their own rules as to how documents can be accessed and copied, and that it is therefore important to understand these requirements before embarking on a visit.

The staff members of the various archives often have an extensive and detailed knowledge of the material in their collection. It is worth tapping in to this knowledge, particularly as it can sometimes be tricky to track down information using formal catalogues. At the same time, it is important to be aware that these people have many demands on their time and to respect these. They often tell me that the prospective visitor needs to be as specific as possible in setting out the purpose of a visit or the information required, given the vast extent of the material in their collections.

The books I have written on Second World War cruiser design have been broad in their scope, and it is impos-

Some of the author's books, referred to in this article. From left to right: *Cruiser Birmingham* (2018), the first edition of *World Naval Review* (2010), *British Town Class Cruisers* (2019) and *British Fiji Class Cruisers* (2024).

sible to be an expert on all aspects – ranging from amphibious aircraft through to wireless telegraphy – that fall within their ambit. I have therefore relied heavily on the goodwill of experts in these various specialised fields to make good my own lack of knowledge. It has been heart-warming to find out how readily many have made considerable efforts to share their expertise or otherwise help with my research. For me, the ability to correspond with and even meet childhood 'heroes' such as the late Alan Raven has been a particular highlight.

One word of caution. The joys of research are very seductive, and it can be difficult to know when to draw a line. There are many potentially excellent books that have not been published because their authors have never been quite able to conclude their research.[1]

Writing the book

Given the hurdles that have already been overcome, the actual writing of the book can often seem (at least to me) to be one of the easier parts of the process. Each author will have their own approach that best suits their own method of working. I tend to use the synopsis previously provided to the publisher as a framework for the overall production process, tracking progress against the required timeline and word count. I find this helps to keep things on track and prevent undue distraction from the inevitable temptation to pursue attractive but irrelevant tangents.

Where it is possible, it can also be helpful to seek help from trusted friends working in the same area to review text as chapters are finished. This process can provide valuable feedback at an early stage, affirming that things are on track or identifying potential weaknesses (such as lack of clarity) that can assist later parts of the book. When writing my two cruiser class histories, I was fortunate enough to have the entire text reviewed by a well-known naval author focused on a different but related area, with a number of specific chapters also being read by relevant subject-matter experts. As well as avoiding a number of errors, this process provided valuable ideas on areas for further explanation or focus.

Illustrating the book

Illustrating a book – particularly one that relies heavily on visual content – can be a major challenge. I have built-up an extensive library of photographic material over many decades to support my research, and actively seek out images for sale on eBay and elsewhere to support future projects. Fortuitously, I also have the legal background necessary to understand some of the more obvious pitfalls associated with the complex and sometimes opaque subject of copyright law. Many writers, however, will need to licence the use of images from archival collections. Regrettably, some of the major British national museums demand fees that would make a medieval moneylender blush; an approach which the writer considers a significant (and arguably scandalous) obstacle to the work of writers and researchers in the naval field. Writers focused on other countries, particularly the United States, may encounter a more benign attitude. Some American websites – that of the US Navy's Naval History and Heritage Command being a good example – hold significant volumes of freely downloadable material covering many countries. The author has also found many British regional and local archives can be much more flexible in making material available to writers on cost-effective terms than their national counterparts. There are undoubtedly other sources available, dependent on the subject matter being covered. Authors are probably best served by seeking guidance from their publisher or from others researching in the naval historical field as how best to meet their own needs.

Two notes of caution may be helpful. One is to be attentive to the requirements of copyright and not knowingly breach creators' rights. This can be easier said than done given the previously mentioned complexity of the law; however, a good publisher should be a source of guidance. The other is to supply material that is of sufficient quality for print reproduction. Photos posted on the internet can look fine on a screen but, when downloaded, can prove to be of too low a resolution for publication. This has relevance to the broader subject of layout, discussed further below.

Design and layout

When finishing my first book – the initial edition of what turned out to be the *WNR* series – I naively thought that nearly all my work was done. This proved to be a major misapprehension. While the design and layout of a publication is rarely within an author's direct control, its importance in determining the presentation of the fruits of his or her research to the intended audience will be readily apparent. This is particularly the case for a technically complex book that might be heavily reliant on the use of photographs or other visual material. I cannot stress how important effective design is to realising an author's vision.

Publishers – wrongly, in my opinion – tend to discourage authors' involvement in the design process. However, it is often possible to provide – via the publisher – a note guiding a designer on particularly important aspects of the layout, such as photographs that are significant and should be given particular prominence. Over time, it might even be possible to develop a working relationship with a particular designer, which presents an invaluable opportunity. In any event, it is far easier to influence the layout at an early stage than when proofs are received. Significant change then will be difficult and expensive to implement, producing resistance from the publisher. This is something I only learned the hard way through experience.

It goes without saying that proofs need to be carefully checked, as the publisher cannot reasonably be relied on to spot factual errors. I personally find the task a time-consuming and tedious process. Despite considerable effort, it is surprising how many glaring mistakes can escape the net. Realistically, all one can do is to chalk this up to experience and try to do better in the next book.

Sales and marketing

In line with commercial pressures, publishers are tending to pass more and more responsibility for marketing a particular title down to the book's author. At the least, this will involve providing substantial information on the book and potential marketing leads to an online platform. The enthusiasm with which an author engages with this process will inevitably vary from individual to individual. However, it is probably important for most new authors to engage as fully as possible with the process. This will drive sales and, hence, name recognition, thereby paying dividends in facilitating the publication of the author's subsequent books.

Concluding thoughts

Writing a book on naval history for publication can seem a daunting process. There are many factors to be considered and many hurdles to clear on the way to success. It will certainly never be a financially lucrative career. Despite these challenges, prospective writers should be assured that it can be a fascinating and rewarding experience, with many sources of help and encouragement available to guide them on their way.

Note: Conrad Waters' book British Fiji Class Cruisers and Their Derivatives *was published by Seaforth in late May 2024, RRP £50.00.*

[1] The Assistant Editor, glancing across his study at a huge pile of un-utilised papers in the corner, can readily agree.

A's & A's

A MYSTERIOUS GERMAN BATTLECRUISER PROJECT (*WARSHIP* 2024)

Contributor **Toby Ewin** writes in response to Kenneth Fraser's query about British knowledge of German battlecruiser projects during the First World War:

Kenneth Fraser's note in *Warship 2024* (p 198) observed that one of the plans for some projected German battlecruisers, in a British wartime Naval Intelligence report, was remarkably similar to one of those (D48 of 1913) considered by the Germans in the design process for the never-completed *Mackensen* class. He is surely right to say that, given the limited surviving Naval Intelligence records, we can probably never *know* how the Royal Navy became aware of this German design.

We do, however, have a good candidate for the source of such intelligence. In November 1914, a former German naval engineer officer named Karl Krüger[1] walked into the British legation in The Hague and offered his services as a source of intelligence. From then until the late 1930s he worked as an agent for the Foreign Section of the Secret Service Bureau, an organisation known during the First World War as MI1(c) and later as the Secret Intelligence Service (SIS or MI6). Unlike MI5, many of whose older files have been sent to the National Archives (where they form the KV series), SIS has generally not released papers, but in Krüger's case we have a file for the principal customer for reports on naval matters: Naval Intelligence. ADM 223/637 contains some 60 reports from Krüger, dating from April 1915 to early 1919, plus a handful of reports from other agents.[2] We also have two books written with access to MI1(c)'s papers of the era,[3] and a book by a former officer in MI1(c)'s Rotterdam station that was first published in 1938.[4]

Krüger was 'embittered by an earlier court martial and greedy for money';[5] he visited the Netherlands roughly monthly, and was met at a series of 'safe houses'. In the intelligence reports, he was identified variously as 'TR/16', 'R16', 'H16', and 'VII'. One Richard Tinsley was the head of the Rotterdam station, from which the prefix 'TR' derived; the later 'H16' may refer to Holland or The Hague.[6] All Krüger's surviving agent reports focus on the identity and condition of ships in port, those building or under repair, the state of naval and other dockyards, and occasionally battle damage – with some particularly valuable reports coming in a few weeks after the battle of Jutland. A handful of drawings and maps are present in ADM 223/637, including drawings of the battleship *Bayern* and cruiser-minelayer *Bremse*,[7] but several surviving reports mention additional drawings that are no longer present. It is clear that the bulk of Krüger's information, in the surviving reports, came from his direct observation of ports and shipyards, conversations with naval and dockyard personnel (facilitated by his own naval engineering background), and occasionally information overheard. None the surviving reports explores long-term plans for future ship designs, but given the gaps in both dates and (where they are provided) between report numbers, it is entirely possible that TR16 *did* generate such materials but that they have simply not survived.

Henry Landau, who worked in MI1(c)'s Rotterdam station and knew of Krüger's work – though apparently not his real name, background or nationality – wrote of him:

> I knew him as 'The Dane'. What his name was, or where he came from, I do not know, although I met him several times. Slight of build, fair, with blue eyes, he looked the reserved, well-bred Scandinavian of cultured and professional interests. ... When I came to know him better, however, I realised why he was so successful. He was a marine engineer of exceptional quality; he was a man without nerves, always cool and collected; nothing escaped his aus-

> terely competent eye; and he was possessed of an astounding memory for the minutest detail of marine construction.
>
> ... In my opinion, he was undoubtedly by far the most valuable agent the Allies ever had working in Germany... He covered every shipbuilding yard and every Zeppelin shed in Germany... he rendered such outstanding services to our military sections as well as his own, and his reports were so brilliant ...
>
> As a representative of a Danish shipbuilding yard, which was supplying the Germans from time to time with tugboats and marine equipment, he was allowed to travel freely to Kiel, Wilhelmshaven, Hamburg, Bremen, Emden, Lübeck, Flensburg and other shipbuilding centres ... His popularity with German clients and their trust in his apparently candid nature was unbounded ... With his extraordinary memory, he was able to sit down, when in Holland, and write out page after page of reports, giving an exact description of the ships which were under construction or repair, and supplying us with the invaluable naval information on which the Admiralty relied absolutely ...

Krüger also provided details of the ultra-long range 21cm (later 24cm) guns used to fire on Paris.[8]

The official historian of SIS, the late Professor Keith Jeffery, concurred with Landau's judgement: Krüger's reporting was 'on balance MI1(c)'s most significant contribution in the western theatre of operations during 1914–18'.[9]

Endnotes:

1 The 1910 edition of the *Rangliste der Kaiserlich-Deutschen Marine* records a *Marine-Stabsingenieur* (an engineer-officer rank equivalent to *Kapitänleutnant*) surnamed Krüger who had joined the Navy in 1885 and was serving in the Inspectorate of Coastal Artillery & Mine Warfare. Entries for this Krüger do not mention his forename, but a naval engineer of this seniority does not appear in the 1914 *Rangliste*.

2 The title of the file is 'Agents' reports: naval', and apart from Krüger's information there is a handful of reports from other agents: two about the battle of Jutland, and several about Russia after the 1917 Revolutions.

3 Alan Judd, *The Quest for 'C': Mansfield Cumming and the Founding of the Secret Service* (Harper Collins, 1999; see 339–343); Keith Jeffery, *MI6: The History of the Secret Intelligence Service 1909–1949* (Bloomsbury, 2010; see 83–85, 97 and 300). Krüger's case is also described in Andrew Boyd, *British Naval Intelligence through the Twentieth Century* (Seaforth, 2020), 146–147, 177–179, 212–213; Michael Smith, *Six, a History of Britain's Secret Intelligence Service: part 1, Murder and Mayhem, 1909–1939* (Biteback, 2010), 113–118; and his execution or suicide following arrest by the Nazi authorities is summarised in Christopher Andrew, *The Defence of the Realm: The Authorized History of MI5* (Allen Lane, 2009), 245–247.

4 Henry Landau, *The Spy Net* (Biteback, 2015), 180–186; first published by Jarrolds in 1938.

5 Gill Bennett, 'The Spy and the State', 17 December 2020 (https://engelsbergideas.com/essays/the-spy-and-the-state/, viewed 19 May 2024). The court martial was apparently the result of Krüger's striking another officer, a relative of the Kaiser's. The friction over status between the Imperial German Navy's executive and engineer officers (for which, see Holger Herwig, *The German Naval Officer Corps: a Social and Political History 1890–1918* [Clarendon Press, 1973], Chapters V and VIIA) may perhaps also have played a part, both in the incident leading to the court martial, and in Krüger's reaction to the latter.

6 For more information about Tinsley's activities and associates, see Frans Kluiters, 'R B Tinsley, a biographical note' of January 2004 (https://nisa-intelligence.nl/PDF-bestanden/Tinsley.pdf; viewed 20 May 2024).

7 A drawing of *Bayern* is attached to a report of 14 March 1916; a more stylised one of *Bremse* is attached to a report of 20 July 1916 (ADM 223/627 items 13 and 24). The drawing of *Bayern* is reproduced in Prof Jeffery's book.

8 See report CX 038561 of 27 June 1918.

9 Jeffery, *MI6*, 97.

NAVAL BOOKS WRITTEN BY AIDAN DODSON

Contributor **Aidan Dodson** has asked us to inform readers of *Warship* that he has set up a website from which they can download corrections and additions to his naval books. It can be accessed at https://navalmatters.blogs.bristol.ac.uk/.

THE YEAR 2027 MARKS *WARSHIP*'S 50th ANNIVERSARY!

2027 will see the fiftieth anniversary of the first publication of *Warship* (initially as a quarterly and then, from 1989, as an annual). This is a notable occurrence, especially for a specialist journal, and is one that the publishing and editorial teams intend to mark with a one-off commemorative 'special' (as well as the regular annual). A number of ideas for this are being considered, including that of having an expanded 'A's & A's' covering all fifty years' worth of the publication. Just as with the regular feature, we therefore invite readers to write in with submissions for this. The past half century has seen much new information become available, through the release of numerous previously-classified documents, the opening up of new archival sources, the discovery and exploration of wrecks, not to mention the arrival of the internet with all the benefits (and pitfalls...) that it has brought.

While small corrections to things like dates and names will be accepted, we will particularly welcome submissions of a more substantial nature, providing additional information relating to previously published articles or updates to stories covered in *Warship* since 1977. Readers are invited to get in contact via the usual address. The editorial team are, as ever, happy to offer help and advice if necessary.

NAVAL BOOKS OF THE YEAR

Brian Lavery

Two Navies Divided: The British and United States Navies in the Second World War

Seaforth Publishing, Barnsley 2023; hardback, 623 pages, c130 B&W illustrations; price £35.00.
ISBN 978-1-3990-4721-0

Brian Lavery's 'Two Navies Divided' aims to provide a comparative description and analysis of the Royal Navy and US Navy as they fought alongside each other during the Second World War. The title is derived from playwright George Bernard Shaw's reported statement that 'England and America are two countries separated by a common language', and the adaptation is appropriate given the many differences revealed in the two navies' management and materiel, despite significant commonality in their heritage and culture.

The book's challenging objective is matched by the breadth of its coverage, which embraces politics, culture, personnel, logistics, equipment and intelligence. Following an initial preface and explanatory foreword, the main body of the book – some 562 pages – is divided into 19 separate chapters examining these and other themes. They commence with an analysis of 'The Structure of Naval Power' as it then pertained to the two nations: essentially an overview of political, economic and command factors. The following five chapters largely examine societal and human elements. These include assessments of recruitment, training and living conditions pertaining both to officers and ratings, as well as analysis of specialist branches such as the medical services and marines. The author's attention then turns to logistics and equipment, including overviews of bases, construction, warship design and fleet composition. Again, specialist arms such as naval aviation, submarines, coastal navies and amphibious forces are accorded their own chapters. These are followed by two chapters providing inevitably brief assessments of enemy and other Allied navies before a succinct conclusion draws the analysis to a close. A comprehensive list of footnotes, an extensive bibliography, and both general and ship name indices are also provided.

Physically, the book adopts a 24cm x 17cm portrait format. A reasonable selection of black and white photographs is interspersed throughout the text. While often reproduced at relatively small size, these are effective in illustrating the written material. The same cannot be said of the book's rather uninspired jacket design, which does little to advertise the undoubted interest of the detailed analysis contained within.

Brian Lavery has enjoyed a long career as a distinguished naval historian. The depth of his knowledge is reflected in the near-encyclopaedic level of detail he provides, which also allows some interesting and informative conclusions to be drawn. This is, perhaps, most frequently demonstrated in the book's earlier chapters assessing cultural and human factors. To give just one example, his analysis of the two navies' comparative officer training systems and career paths – with earlier entry into the service for Royal Navy officer cadets and a subsequent tendency towards specialisation – leads to the interesting conclusion that American naval officers tended to have the broader outlook. One particular benefit was the greater political awareness they enjoyed when they rose to senior levels and needed to walk the corridors of power.

Unfortunately, there are some detractions from the book's overall excellence. These include a not-inconsiderable number of minor errors and mistakes that are probably inevitable in a work of this scope, but which could have been reduced by a more thorough editing process. The breadth of analysis is also sometimes counterbalanced by its relative shallowness, with some questionable assertions justified by limited – and often anecdotal – evidence. Readers might question the author's concluding assertions that 'The navies could learn from one another, but mostly the Royal Navy from the USN'. This ignores the detailed, hard-won operational lessons from nearly two and a half years of hostilities that were continuously fed through to the US Navy by its British counterpart before America entered the conflict, thereby considerably easing its transition to the realities of current naval warfare.

These criticisms aside, 'Two Navies Divided' presents an informative assessment of two navies as they faced a challenge that was, perhaps, unparalleled in their respective histories. Whilst hedging his bets as to which navy best adapted to wartime circumstances, the author notes positively that: 'It is a striking paradox that, in the case of both navies, the war began with professional sailors conducting themselves in a rather amateurish way ... but proceeding later in the war with largely amateur forces that fought with great professional skill.' It is a tribute to both navies – and the great democracies that sustained them – that this transformation was so successfully achieved.

Conrad Waters

GH Bennett

The War for England's Shores

US Naval Institute Press, Annapolis 2023; hardback, 273 pages, illustrated with maps and photographs; price £25.00/$39.95.
ISBN 978-1-55750-375-6

While there are numerous books about the *Schnellboote* of the *Kriegsmarine* during the Second World War, GH Bennett's latest offering is perhaps the first to properly focus on their most dangerous campaign: that around the

English Channel and North Sea. Between 1940 and 1945, *Schnellboote* (abbreviated to S-boats but known to the Allies as E-boats) effectively took on the role performed by coastal U-boats in the First World War, intercepting British coastal shipping to disrupt trade and supply. The subject has been covered before in Nick Hewitt's excellent *Coastal Convoys* (Pen & Sword, 2009), but while that author focused on the Allied ships, Bennett's emphasis is very much on the German S-boats.

Bennett sets the scene well, exploring German maritime strategy from the late 19th century and how it influenced the evolution of fast coastal craft, before homing in on the evolution of the S-boat. With the outbreak of the war he follows the fortunes of the various North Sea and Channel flotillas and their impact on British shipping, along with the British response and the development of the Motor Gunboat (MGB). This leads the reader into the crucial year of 1943, which Bennett views (correctly) as the turning point. After that, S-boat activity starts to decline during 1944, and although there is a brief resurgence in the winter, by the spring of 1945 they are a spent force. A conclusion looks at the effectiveness of the campaign, while the single appendix has a detailed list of every vessel known to have been sunk by S-boats.

The book makes extensive use of English and German sources, both official contemporary records and later publications. Memoirs are also used, although the book is very much a technical and operational history rather than a dramatic retelling of small-boat actions. However, this does not detract from an excellent reference book and analysis of an oft-overlooked campaign.

Stephen Fisher

Brian Lane Herder, illustrated by Paul Wright
British Pacific Fleet 1944–45

Osprey Publishing, Oxford 2023; softback, 80 pages, illustrated with 33 photographs, 3 original paintings and 6 colour maps and diagrams; price £15.99.
ISBN 978-1-4728-5677-7

This book forms part of Osprey's new 'Fleet' series and follows their familiar layout. The text is interspersed with photographs and specially prepared colour images, in this instance three excellent paintings by Paul Wright, a well-known maritime artist.

The author describes the development and operations of the British Pacific Fleet and the politics involved, up to the successful conclusion of the Second World War. The first of four main sections covers the intended mission and rationale for the deployment of a fleet to the Pacific. The subsequent sections explore the composition of the Fleet, its operations, and finally combat and analysis. Each of these is illustrated with relevant photographs, charts and diagrams, interspersed with Paul Wright's paintings.

Some of the politics involved, both local and global, was new to this reviewer, although he was aware of the US Navy's local leadership's growing admiration for the RN fleet and its supporting supply train. There is one glaring omission: the converted 'amenities ship' HMS *Menestheus* receives no mention; at the time at least one USN officer went on record as applauding the provision of a floating brewery, something the US Navy had never contemplated.

That apart, this is an interesting book, a product of much research on the part of the author; it is to be hoped that this material can be utilised in a more substantial work in the future.

W B Davies

Geirr Haarr & Tor Jørgen Melien
The Sinking of the *Blücher*: the Battle of Drøbak Narrows, April 1940

Greenhill Books, Barnsley 2023; hardback, 419 pages, more than 200 B&W photographs and maps; price £29.95.
ISBN 978-1-78438-875-1

Geirr Haarr has written a number of substantial and well-received titles for Seaforth (see reviews in *Warship* 2010 and 2015) covering the early part of the Second World War, and in particular the German invasion of Norway in 1940. Here, as well as switching to a different imprint (though still part of the Pen & Sword stable), he has combined with Tor Jørgen Melien, a retired officer of the Norwegian Navy, former editor of the *Norsk Militaert Tidsskrift* (Norwegian Military Journal) and now a researcher at the Norwegian Institute of Defence Studies, to recount in great detail one of the pivotal events of the campaign, looking not only at the destruction of the unfortunate *Blücher* but also at the wider context, in terms of both the background and also the very considerable consequences, both immediate and more long-term.

The first part of the book sets the scene, looking at Norway's efforts to maintain its increasingly tenuous neutrality as conflict developed all around, and the motivations of those who sought to influence the country for their own ends. None of the major participants emerge from this too well: the Norwegians themselves were at least arguably naïve, the British and their French allies altogether too complacent (at one crucial moment First Sea Lord Pound had gone fishing while his counterpart Churchill was 'well dined'...), and the Nazis over-confident and, as was to become commonplace, riven by internal disagreements. The plan to invade a large and geographically challenging country was knocked together in just a few weeks, and was much influenced by faulty intelligence and misjudgement.

The authors then focus on the task group that had the Norwegian capital, Oslo, as its target, and in particular the titular cruiser. *Blücher*, one gets to feel, was almost doomed from the start, neither the ship nor the crew being remotely battle-ready, and then being sent off on a high-risk mission that relied as much on luck as anything more substantial to succeed. As it was, the ship's sinking, and the consequent disruption of the plan to occupy

Oslo, allowed King Haakon and his government to escape and so begin five years of committed opposition to Nazi rule. The description of *Blücher*'s first and last action makes compelling reading, with the ship already in a dire situation even before the two fatal torpedo strikes, courtesy of the second heavy shell to hit her, which ignited a partially-fuelled Arado floatplane on the catapult and turned the midships section into a raging inferno.

The final part is probably the least successful, as the story divides into several strands, some of which are left hanging more than others. However the importance of *Blücher*'s sinking in terms of both the invasion itself and the subsequent course of the war overall is clearly explained.

The book is well printed on substantial, good-quality paper. The photographs are well chosen and equally well reproduced, and the maps excellent. However, as is so common these days, less-than-thorough editing has resulted in numerous minor typos and other mistakes. The dual authorship also makes itself felt in terms of occasional moments of disjointedness, with characters in the story sometimes appearing without having been introduced and unexpected jumps in the narrative from one subject to another. These minor criticisms aside, this is a well-written, wide-ranging and thoughtful piece of top-quality naval history, and is thoroughly recommended.

Stephen Dent

Ian Johnston

Clydebank Battlecruisers: Forgotten Photographs from John Brown's Shipyard.

Seaforth Publishing, Barnsley 2023 (reissue of 2011 hbk); softback, 192 pages. 218 B&W photographs and nine other illustrations; price £16.99.
ISBN 978-1-1399-0248-1 (pbk), 978-1-7834-1936-9 (e-book)

Five out of Britain's 13 battlecruisers (*Inflexible, Australia, Tiger, Repulse,* and *Hood)* were built at John Brown's of Clydebank, a town that did not exist until the yard was established, and this book by the author of several authoritative works on Clyde shipbuilding, now reissued in softback, provides an informative account of their construction. There is no attempt to describe their later careers (amply described elsewhere) or the pros and cons of the concept of the battlecruiser.

The author's extensive knowledge of both the technical and commercial aspects of shipbuilding is evident throughout. He remarks, for instance, that apparent sister-ships were not really identical, since lesser details of construction were left to the builder's discretion. He shows how Brown's negotiated with the Admiralty over difficulties that had come about during the building process – an unexpected problem arising when the Admiralty, having denoted the wing turrets of the *Invincible* class as 'B' and 'C', had to change them to 'P' and 'Q' because the sounds of the original letters were too similar and were causing confusion – and over modifications that it had asked for after the design was finalised (one with ominous implications was the addition of anti-aircraft guns to *Tiger).* The building of *Repulse* generated an unprecedented problem because Lord Fisher had insisted she be completed within 18 months of the order (compared with 30 for the *Tiger).* This required building to start three months before the plans were completed. Conversely, *Hood* took 45 months because her plans had to be substantially recast after Jutland to incorporate 4,000 tons of extra armour.

The most distinctive feature of the book is the wealth of photographs, which occupy more than half of it. Brown's commissioned photographs of each ship at various stages of construction. Apart from their intrinsic interest as records of a lost industry, they show many details of the ships which would be invaluable to model-makers: boats, davits and derricks, searchlights, sirens, ventilators, hatch covers and so on. Here also the author demonstrates his interpretative skills, as when he notices, through the window of a temporary cabin on the deck of *Inflexible,* a bowler hat, showing that it must have been used by a manager.

A final short chapter describes the correspondingly short career of the unnamed G3 battlecruiser ordered from Brown's in October 1921, suspended in November and cancelled under the terms of the Washington Treaty in February 1922. The appendices include: a summary of the main particulars of each ship; graphs of the numbers of men employed each week on *Repulse* and *Hood;* tables comparing the periods of construction of each ship; transcripts of some of the weekly reports to the management on the progress of the ships; and extracts from the official accounts of the trials of *Repulse* and *Hood.*

The last ship built by Brown's for the Royal Navy was the assault ship *Intrepid,* completed in 1967; their last merchant ship was completed in 1972; and there remains only the massive 'Titan' crane – now a tourist attraction – to remind us of Clydebank's vanished days of greatness.

Kenneth Fraser

Roger Knight

Convoys: The British Struggle Against Napoleonic Europe and America

Yale University Press, New Haven & London 2022; hardback, 387 pages, illustrated in colour and B&W; price £25.00.
ISBN 978-0-300-24697-1

As with many people, the reviewer's previous reading of the period has been centred around the great fleet actions, the frigate battles and numerous biographies of Nelson, as well as the novels of CS Forester, Patrick O'Brien, Julian Stockwin *et al.* Roger Knight, on the other hand, views the convoys and protection of merchant shipping as the keystone of the naval war, contending that, without an effective system for protecting trade, the British efforts in fighting Napoleonic France would have been unsuccessful. In chapters titled 'Convoying Before 1803', 'The Admiralty and Trade Protection, 1803–1815', 'Warships and

Merchantmen', the author presents the reader with descriptions of troop convoys, convoys in coastal waters, the Western Approaches, the North Sea, the Mediterranean, the East and West Indies, and the North Atlantic 1812–1815, together with the all-important Baltic trade. All these accounts make for fascinating reading, while the Epilogue offers some telling comparisons between the convoys of the Napoleonic Wars and those of the Second World War. In neither war were there sufficient escort ships or sufficient seamen; navigational aids were lacking; merchant captains were unused to sailing in formation, objected to the slow speeds prescribed and resented naval command; and in both wars the principal enemies were the weather and the sea. Despite this, in both wars the safe and timely arrival of convoys carrying essential war materials was crucial.

Roger Knight's book has already received many well-deserved plaudits. The Society for Nautical Research awarded him the 2023 Anderson Medal for the best maritime history book, with the judges praising it as '... a thoroughly researched study of a relatively neglected but hugely important subject ...', written in a style accessible to both specialists and a more general readership. In their award, the SNR also praised Yale University Press for the quality of the production of the book. A number of recent naval history books, all produced by Yale UP, have been excellent in this regard, and it is to be hoped they continue to produce good quality and affordable books. This particular title is highly recommended.

Andy Field

Terry Crowdy
Formidable: Arthur Flint's War against *Tirpitz* and the *Kamikazes*.

Pen & Sword Maritime, Barnsley 2023; hardback, 256 pages, illustrated with more than 150 photographs; price £25.00.
ISBN: 978-1-3990-8766-7

It was the reminiscences of his grandfather Arthur Flint late in life that prompted military historian Terry Crowdy to research the wartime history of the aircraft carrier *Formidable* and her crew. Flint had joined her in June 1944 as an officer's steward, serving during her second commission and her operations with the British Pacific Fleet in 1945. Although he left no diaries or photographs, his grandson has researched widely to tell the story, using official archives and reports, ship logs, recordings from others who served and a raft of published sources.

Formidable's early years are covered only briefly, with most of the book focused on air operations against *Tirpitz* in her Norwegian fjord and then against the Japanese. Chapter 1 describes the mishap at her launch on 17 August 1939 at Harland & Wolff's Belfast shipyard. Partly due to inadequately prepared launch ways (she was the first carrier that H&W had built), the timber supporting structures failed, allowing the ship to slide into the water prematurely and showering shipyard workers and spectators with debris, killing one. Completed in November 1940, *Formidable*'s first deployment was to the Mediterranean; bombed by German Stukas she was sent to the US for repair. Back in service early in 1942, she was first allocated to the Eastern Fleet and then took part in the North African and Italian landings, later covering Arctic convoys.

Having completed his training (partly in the USA), Flint joined the ship after a six-month refit, her CO now being Captain Philip Ruck-Keene, a disciplinarian anxious to rid his ship of troublemakers and get his aircrews worked up – they went to the USA for training on new American aircraft. *Formidable* took part in Operation 'Mascot' in July 1944, a second attack against *Tirpitz,* but her Barracudas were unable to successfully penetrate the defences.

The remaining half of the book covers *Formidable*'s deployment with the British Pacific Fleet, sailing from Scapa on 17 September 1944 with a full complement of 30 Corsair fighters and 22 Avenger torpedo bombers. However, a few days later she put into Gibraltar with a damaged gearwheel. Flint suspected sabotage, but a more likely explanation is that axial and torsional vibration on her centre shaft had damaged the teeth on her main reduction gear wheel. It took nearly three months to fit a new one, while the crew were able to enjoy what delights wartime Gibraltar could offer.

She finally reached Sydney in March 1945 to join four other fleet carriers in the BPF. Meanwhile the crew had been fully worked up, Flint's action station being part of the ammunition supply party to the starboard forward pair of twin 4.5in guns. *Formidable* then took part in the 'Iceberg' operations against airbases on Japanese-controlled territories. Crowdy gives a graphic account of the two *kamikaze* attacks on 6 and 9 May, where they slipped past the combat air patrols and AA gunfire to crash on the flight deck. Despite fires, her 3in armoured flight deck prevented major damage. More serious was a self-inflicted wound, when a careless aircraft fitter fired off a Corsair's guns inside the hangar, starting a fire from fuel and a risk of 500lb bombs cooking off. The ship was sent back to Sydney for repair, returning in time to take part in the July and August attacks on the Japanese mainland until surrender. Thereafter her aircraft were dispersed and she was used for repatriation of Allied prisoners from former Japanese-held territories to Australia, before finally departing for Portsmouth where she arrived on 6 February 1946. Little used and unmodernised thereafter, she was towed to Inverkeithing for scrap in May 1953.

The book provides a good flavour of carrier operations, although the naval terminology is not always correct; the routine descriptions of crossing the line ceremonies and shore exploits are of less interest. The photographs are of variable quality, with a large number of flight deck views and multiple shots of ceremonies. Overall, the author succeeds in his aim of producing a useful memoir of a short-lived ship's wartime service and her crew.

Ian Buxton

Jon Wise

The Royal Navy and Fishery Protection: From the Fourteenth Century to the Present

Seaforth Publishing, Barnsley 2023; hardback, 320 pages, 38 B&W and colour images and line drawings; price £25.00.
ISBN 978-1-39904-170-6

Drawing considerably on primary sources, Jon Wise's new book tells the story of the crucial role played by fishery protection in safeguarding the UK's maritime interests.

The book is set out in broadly chronological order, weaving operational activities into the geopolitical background. Beginning in 1379 with the first documented account of British ships protecting fishing vessels, the story proceeds with the Dutch wars and Anglo-French rivalry. From the outset there was recognition of the importance of fish as an economic asset, suggesting the philosophical question: 'Who owns the sea?' For several centuries, freedom of the seas (*mare liberum*) was a fundament principle of successive British governments; one over which they 'might be prepared to go to war with the strongest power in the world', and the principal driver of British actions right up to the latter part of the 20th century.

Following the Napoleonic Wars, Britain enjoyed unrivalled economic hegemony. The introduction of steam power revolutionised fishing; the wooden sailing boats using hook and line of the 1840s were superseded by the steel steam-powered trawlers of the early 1880s. Ashore, the railways revolutionised distribution, elevating fish from a food for coastal communities to an affordable staple in the diet of the burgeoning inland towns and cities. This spurred a boom in the fishing industry that eventually led to over-fishing and the need for international regulation, attempted by the North Sea Fishery Convention of 1882. By the 20th century there was growing opposition to the historic adherence to the three-mile territorial limit. The doggedly independent British fishermen stood their ground, forcing the Royal Navy to intervene in international disputes, particularly with the Soviet Union and Norway.

The two world wars only stalled the inevitable, the traditional roles being reversed when the fishermen and their trawlers and drifters took an active part in the conflicts. The usefulness and durability of the fishing boats, together with the experience and hardiness of their crews, proved invaluable, particularly in combating mines and enemy submarines. They suffered significant losses in the process.

A notable section of the book is devoted to the three Cod Wars with NATO ally Iceland, a series of escalating confrontations that encapsulate the broader themes of the narrative: national sovereignty, economic interests, and the evolving role of the Navy. Wise provides a detailed account of these conflicts, including the tactical and political manoeuvres employed by both sides. Throughout, the Royal Navy remained faithful to its mission of protecting the trawlers from interference and providing an environment in which fishing could continue uninterrupted. Perhaps inevitably, the Royal Navy lost the public relations battle and, with international opinion favouring Iceland, in June 1976 Britain accepted the 200-mile limit, effectively ending the deep-water fishing industry, with far-reaching consequences for the fishing communities.

The concluding chapters bring the history up to date, detailing the squadron's modern responsibilities and looking to the future. Duties now extend beyond traditional fishery protection to include protecting offshore installations and the monitoring of pollution. This shift reflects the increasing complexity of managing maritime security and national waters in a globalised world.

This is a well-written, thought-provoking and enjoyable read, making complex historical and legal issues accessible to a broad audience. It is richly illustrated with well chosen images and maps that enhance the reader's understanding of the narrative and depict the various vessels used for the task, many of which proved less than ideal for the arduous conditions they faced. Jon Wise is to be congratulated on producing a thoroughly researched record that must now be considered the definitive account of Royal Navy fishery protection, providing valuable insights into this exacting yet unsung role. This book is highly recommended for scholars, maritime professionals, and anyone interested in the rich tapestry of naval history.

Phil Russell

[Note: in October 2024, as this edition of Warship *was in the process of being prepared to go to press, it was announced that* The Royal Navy and Fishery Protection *had won the Maritime Foundation's award for the best book of the year.]*

Angus Konstam

Royal Navy Home Fleet 1939–41: The last line of defence at Scapa Flow

Osprey Publishing, Oxford 2024; softback, 80 pages, 36 illustrations (6 in colour), 6 maps; price £15.99.
ISBN 978-1-47286-148-1

Although it could scarcely be compared in numbers with the Grand Fleet of 1914, the Home Fleet of 1939, which included eight capital ships and two aircraft carriers, remained the most important unit of the Royal Navy. Its commanders, unlike their predecessors, probably did not expect to fight an all-embracing fleet action given the much-changed circumstances. However, its basic tasks, as Angus Konstam explains, were quite similar: to prevent the *Kriegsmarine* from breaking out of the North Sea, to guard Britain from invasion and to suppress German merchant shipping, while at the same time protecting the vital Atlantic convoys. These missions, which became much more difficult after the German occupation of France and Norway, were, in general, carried out successfully during the period in question, although the Home Fleet would scarcely ever operate as a combined unit.

This short book, copiously illustrated in Osprey's usual manner, provides accounts of the types of ships operated by the fleet; the recent advances in technology, especially in aircraft and radar, which often gave admirals a better knowledge of the enemy's dispositions than in the previous war; the organisation and subdivision of the fleet; a list of its principal commanders; the sources of intelligence and the means of communication available to them; the main features of its base at Scapa (itself the subject of an Osprey book); and several maps revealing the great extent of the areas where the fleet had to maintain a presence.

The concluding section recounts the actions fought by the fleet during 1939–41, the two major ones being the Norwegian campaign (when bad weather and inadequate intelligence prevented the interception of the invasion forces, although the *Kriegsmarine* still subsequently suffered serious losses), and the operation against the *Bismarck*, when the Home Fleet's strategy was successfully employed to sink Germany's greatest battleship, though at the cost of the tragic loss of the *Hood*. That action, shortly followed by the German invasion of the USSR, changed the strategic situation, as will no doubt be set out in a sequel to this useful book.

Kenneth Fraser

Mark Stille & Jim Laurier (illustrator)
Japan's Indian Ocean Raid 1942: The Allies' Lowest Ebb

Osprey Publishing, Oxford 2023; softback, 96 pages, illustrated with colour and B&W photographs, maps and illustrations; price £16.99.
ISBN 978-1-4728-5418-6

This book follows the standard Osprey format: a brief, informative text, interesting photographs and high-quality colour illustrations – although some of these are reproduced over two pages, with detail lost in the 'gutter'. Author Mark Stille has written a number of other books for Osprey on the IJN and USN; here he turns his attention to the IJN's ill-conceived Operation 'C', the incursion into the Indian Ocean, the attacks on Trincomalee and Colombo, and Admiral Somerville's responses.

Stille does a good job in covering the basics of the operation and is even-handed in describing the events. He criticises Admiral Nagumo for being too casual in his scouting, meaning that he missed a chance to spot and attack both elements of Somerville's inferior Eastern Fleet, thereby averting a naval disaster for the British – in Stille's view, Nagumo's flawed approach to this operation was a precursor of Midway. However, he is equally critical of Somerville, in command of a mixed bag of unsuitable ships with which to face the Japanese. In his opinion, Somerville failed to assess fully the intelligence on the composition of the Japanese forces, risking his fleet's annihilation with his proposal to attack the Japanese at night using Fairey Albacore torpedo bombers. He suggests that Somerville's attitude had been prompted by accusations earlier in the war about his lack of aggressiveness. As it was, the fleets barely evaded contact with each other.

There are several errors in the book: the author claims that the Dutch destroyer *Isaac Sweers* was a Royal Norwegian Navy vessel; and illustrator Jim Laurier seems to have used the same Blenheim bomber twice (if not three times) in his illustration of 11 Squadron's attack on the Japanese force. A detailed discussion can be found on-line, covering other errors and omissions. The reviewer felt that the book was useful for anyone wanting an overview of the operation, but that more detailed and accurate accounts are available in other sources.

Andy Field

Robert C Stern
Knife's Edge: South Pacific Carrier Battles from the Eastern Solomons to Santa Cruz

US Naval Institute Press, Annapolis 2023; hardback, 336 pages, 44 B&W photographs, 8 maps and 3 ship damage diagrams; price $41.95.
ISBN 978-1-68247-568-3

Robert Stern has written more than twenty books on a variety of naval and military subjects, the most recent of which, *Scratch One Flattop*, describes the Battle of the Coral Sea. An opening section explains the standardised format the author has used for nomenclature, dates, times, language, units, aircraft designation, geography and weather throughout the text, and he also explains that no attempt has been made to edit comments that would be considered today as racist from quotations in order to preserve their original context. Brief biographical descriptions of the major USN and Japanese *dramatis personae* give readers an understanding of their strengths and weaknesses. The C-in-C of the IJN Combined Fleet, Admiral Yamamoto Isoroku, is described for instance as 'easily the best known (and perhaps most overrated) Japanese naval commander of World War II'.

A historical work written by an author who has fully researched and understood both the strategic and tactical aspects of naval warfare is always a pleasure to read. Stern describes the carrier battles fought between the USN and IJN as both navies sought to reinforce their forces fighting on Guadalcanal Island and to influence outcomes. In the US case this was the 1st US Marine Division, which held onto the vital airstrip named 'Henderson Field' throughout the period in question. The decisions made by remote admirals and generals in theatre command and the reactions of those on the spot are well described. Stern fully understands the complexity of carrier operations and the difficulties inherent in navigating the aircraft of that era above a featureless sea over long distances. His description of the inherent weakness of USN contact reporting procedures and the impact these had on the battles described is an aspect of these operations I had not fully appreciated until reading the author's carefully-researched comments.

Stern's extended photograph captions enhance the text, and he is able to 'look deeply' into the images to describe everything that can be seen, even in the background, in impressive detail. The image on page 224 of the USS *Hornet* under attack has been used in earlier books, for instance, but he draws attention to the damaged 'Val' dive bomber and describes the result of its attack. A passing 'Kate' torpedo bomber provides clear evidence that a torpedo attack was carried out at the same time, and the author highlights the 'water curtain' thrown up by the ship's anti-aircraft fire to disrupt torpedo attacks, seen in the foreground. Stern's use of such images to inform and enhance the text is outstanding.

Overall this book uses a wealth of research to provide the best account yet published of these historic fleet actions, and the reasons why they evolved as they did; I am very pleased to add this book to my library and I thoroughly recommend it.

David Hobbs

Erminio Bagnasco (translation by Alessandro Nati Fornetti) Italian Assault Craft 1940–1945: Human Torpedoes and other Special Attack Weapons

Seaforth Publishing, Barnsley 2023; hardback, 232 pages, illustrated with 309 photographs in colour and B&W and 57 line drawings and diagrams; price £35.00.
ISBN 978-1-3990-5609-0

Erminio Bagnasco, who died in late 2022, was the editor of *Storia Militare*, Italy's leading military journal. He was widely regarded as the foremost authority on Italian warships and was the author of many books on the subject, including *Italian Battleships* and *The Littorio Class*, both published by Seaforth. His reputation is maintained with this impressive work which, following a general Introduction, is divided into two parts: Part 1 of the book describes the origins and development of the many types of craft, including accounts of their successes and failures, while Part 2 has a detailed description of all the craft and their operations, as well as the personnel and organisations involved. The book concludes with three appendices, a bibliography and an index.

There is a degree of repetition as a result of this layout, with mention of an operation in Part 1 then being recounted in detail in Part 2. Both sections are copiously illustrated with photos of craft, crew and targets as well as general arrangement drawings of the individual types. Both also cover the post-1943 armistice division of Italy and conclude with the postwar period. Acknowledged, if not in so much detail, is the Royal Navy's redesign of the 'slow running torpedo' as the famous 'chariot'.

One of the surprises of this volume is the large number of types that served with the *Regia Marina* under the overall heading of assault craft, ranging from explosive launches, via launches of various sizes employed both as assault craft in their own right or as mother craft to the human torpedoes, through to the submarines specially fitted for 'approaching' the target with the human torpedoes on board; Part 2 has no fewer than 14 headings covering the different types of craft.

The human story is also prominent, with many of the Italian Navy personnel involved pictured, often with a note in the caption that *x* was killed on 'such-and-such' a date, reflecting the hazardous nature of the role. As was the experience of the German *Kriegsmarine* in similar circumstances, losses were heavy, not just from enemy action, but often as a result of difficulties with the breathing apparatus or mechanical failure of the craft once sent into action. One can only marvel at the stoicism of men under such circumstances.

This is a book of value not only to the historian, but also to the casual reader interested in such operations or indeed a naval architect working with small craft today.

W B Davies

Angus Konstam, illustrated by Edouard A Groult German High Seas Fleet 1914–18: The Kaiser's challenge to the Royal Navy

Osprey Publishing, Oxford 2023; 80 pages, 28 photographs, 21 paintings & maps; price £15.99.
ISBN 978-1-4728-5647-0

This book is intended as a 'concisely detailed' account of the High Seas Fleet (HSF) and its operations, illustrated by a mix of photographs, paintings of ships in action, and coloured maps. It begins with a chapter on the background to the building of the fleet, including Admiral Tirpitz's 'Risk Theory', and the way that the realities of the First World War nullified the assumptions that underpinned the whole basis of the HSF. The second chapter covers the ships of the fleet and their underlying technologies, while the following chapter details the organisation of the Navy and the Fleet, including operational control, communications, intelligence, bases and logistics. An order of battle for August 1914 is provided here, although the changes to this during the war are to be found in the previous chapter, 15 pages earlier. The fourth and final chapter covers operations and their broader context, giving a useful overview of the principal activities involving the HSF from 1914 to 1916; it is, however, very sketchy regarding the period 1917–18, with no mention of the important role of some of the fleet's crews in the revolution that contributed to Germany's exit from the war.

A map of the North Sea is provided with the locations of the principal naval actions marked, together with a number of box features dotted throughout the book covering topics such as scouting strategy and specific parts of naval actions, supported by diagrams and maps. Their placement is somewhat odd, with one covering the HSF deployment of 18–19 August 1916 many pages before the chapter to which it should logically belong (but in which it is not mentioned at all), while one episode of Jutland is separated from the other parts of the action by 38 pages, and by box features covering the actions of 1914 and 1915.

There is a reasonably extensive bibliography, but while it lacks notable recent publications it does, however, include a number of dated sources, which may account for the statement on page 11 that the battleships of the *Nassau* class were 'laid down in secret'. Unfortunately this is by no means the only mistake in the book, with minor factual errors, incorrect spellings of ships' names and a number of the photographs wrongly or misleadingly captioned, as well as the repeated use of the definite article when referring to a ship as 'SMS'.

Thus, while an attractive overview of the German Navy in the North Sea during the First World War, this book has rather more significant errors than one might hope to find in such a work.

Aidan Dodson

John Grehan (ed)
Okinawa: The Last Naval Battle of WW2: the Official Admiralty Account of Operation Iceberg

Frontline Books, an imprint of Pen & Sword Books, Barnsley 2022; hardback, 237 pages, 16 photographs; price £25.00.
ISBN 9787-1-39909-193-0

The Publisher's Note on page ix describes the text as having been extracted from Admiralty Battle Summary Number 47. It aims to make the original text available to a wider potential readership that might not find it easy to gain access to the original summary, which is held at The National Archive (TNA) at Kew under the reference ADM 234/368. Unlike the original, it is illustrated by 16 well-chosen photographs. Battle Summary Number 47 was produced by the Admiralty with the title *Naval Operations OKINAWA – Operation 'Iceberg' March to June 1945* as a hardback volume within the complete Naval Staff History of the Second World War series, which was intended to give serving officers a better understanding of recent naval operations. This new book states inside its dust jacket that the original was a secret publication, but this was not the case; it was a confidential book identified as CB 3081 (32). It was later downgraded to restricted and re-identified as BR 1736 (43), when much of its source material was declassified. The text was drafted in 1948/49 based on both RN and USN reports of proceedings and it was issued by the Admiralty in 1950.

The reviewer has a copy of the original and was thus able to make a direct comparison. Surprisingly, the three pages of British and American sources and their references at TNA on pages x, xi and xii of the original have been deleted, a major detraction from the new book's value. They were identified with the 'M' numbers in use at the time and the compiler could have provided a useful service for modern readers and researchers by transposing them into the later ADM numbered identification system used to identify documents today. Disappointingly, the eleven fold-out maps in the original have also been deleted with no explanation or attempt to reproduce them at smaller scale within the pages. The original summary had 17 appendices A to Q. For some reason the compiler has decided to reduce these to seven, in which both A and G have been changed. This means that important information such as operating statistics for the British Pacific Fleet, bombardment data and Fifth Fleet ships sunk or damaged by enemy air attack are omitted. The reduced number of appendices leaves reference notes on the pages of the new book vulnerable to editorial error, and there are indeed references to appendices that existed in the original but not in the Frontline Books extract. The pagination has been completely changed and a new index had, therefore, to be compiled. This is smaller than the original, omitting everything but ship names.

In summary, Frontline Publishing's *Okinawa* has made the original text available, but the deletions are a disappointment, as is the failure to appreciate the value of source notes. No attempt has been made to balance the text with information that has become available since 1950, especially the role played by intelligence in making Allied plans and dispositions. I would recommend anyone with an interest in the Okinawa Campaign who can gain access to the original Admiralty summary to do so. This book is a second-best option that offers only an incomplete extract.

David Hobbs

George Hodgkinson
A History of Ship Launches and their Ceremonies

Pen & Sword, Barnsley 2023; hardback, 344 pages, illustrated; price £25.00.
ISBN 978-1-3990-4945-0

Alongside the evolving techniques of launching ships, the attendant rituals and procedures have been invested with a deep cultural significance since the dawn of recorded history. Homer described them in the *Iliad* as the precursor to a massive amphibious operation that would capture and destroy an enemy city, while every heroic journey in antiquity began with a ship launch. These superstitions persist; the psychological impact of an unsuccessful launch remains real to this day. The extensive catalogue of disastrous launches reflects the difficulty of controlling a process usually reliant on gravity and the absence of friction. The fatalities and failure to launch that attended Brunel's *Great Eastern* reflected a combination of inadequate means and the immensity of the challenge. Like so much of Brunel's work it prompted greater attention to detail. By contrast *Titanic*, like most White Star Company ships built at Harland & Wolff, did not have an official launch ceremony, but a shipyard worker was killed by flying timber. The ill-fated cruise liner *Costa Concordia* was marked for disaster when the ceremonial champagne bottle did not break.

Large warships built to embody the prestige of a nation

and its ruler often had costly launch ceremonies, complete with festivities and rewards for all involved. Warship launch traditions vary across the world, with the presence of the ruler being an important part of Russian launch events, even if they occurred in Denmark! Old superstitions meant, as Hodgkinson demonstrates, that the practice of having women launch ships is relatively new, the first high-profile launches being of Royal Navy warships, including Nelson's niece launching HMS *Trafalgar* in 1840. Religious ceremonies were introduced for British warship launches as late as 1875, in part because it had hitherto been impossible to guarantee when the ship would actually slide into the water. Even so the ceremony remained short, and could be hastened if the ship began to move.

This is an important book that delivers exactly what the title promises.

Andrew Lambert

Nicholas A Lambert
The Neptune Factor: Alfred Thayer Mahan and the concept of Sea Power

US Naval Institute Press, Annapolis 2023; hardback, 448 pages, 3 charts & tables; price $34.95.
ISBN 978-1-6125-1158-0

In 1890 Captain Alfred Thayer Mahan USN (1840–1914) published *The Influence of Sea Power upon History 1660–1783,* a work shaped by his upbringing at West Point and by his intellectual engagement with wider issues of American expansion, defence and strategic priorities. Commonly viewed as an apostle of the battlefleet, and an architect of the emergence of an American battlefleet Navy, his work also influenced thinking in Germany, France and Japan, and was used to validate British practice. Mahan opposed the 'Dreadnought', preferring larger numbers of 'standard' vessels. This important new book argues that Mahan's life and work have not been fully understood.

The key to Mahan's work can be found in economics, not history or strategy. He developed 'Sea Power' as a naval economic alternative to the 'decisive' continental warfare exemplified by Napoleon. His studies of the French Revolutionary and Napoleonic Wars, and the Anglo-American War of 1812, emphasised the ability of a dominant navy to crush the economic sinews of a land-bound foe. In 1890 the United States needed a powerful battlefleet to prevent the same outcome in a war with another great power – he did not expect that power would be Britain. Mahan condemned the current American strategy of *guerre de course*, a predatory cruising war against merchant shipping, as inconsequential, using statistics of merchant ship losses and economic growth to show how Britain not only survived but prospered between 1793 and 1815, despite heavy losses of ships, because it had command of the sea and could run an effective convoy system. By 1945 his thesis had been proven beyond doubt by the double defeat of German U-boat warfare.

At the same time Mahan saw American expansion as an economic necessity, to secure markets for industrial and agricultural surpluses in a world of rising import tariffs. Market access would be achieved by naval power and imperial expansion. Between 1890 and 1900 the United States acquired Puerto Rico, Guantanamo Bay, Samoa, Hawaii and the Philippines, and built a fleet of battleships. The 'new' Navy secured American hegemony in the Western hemisphere and collected debts from weaker powers. By 1914 the United States had the second largest battlefleet in the world. However, this could not prevent it from being drawn into the European conflict by the need to protect economic interests.

In this context Mahan worked hard to shape the legal context of war, ignoring his orders as a delegate to the 1899 Hague Peace Conference, where he worked closely with Admiral Sir 'Jacky' Fisher. He recognised that the ability to stop all commerce was the key to sea power strategy, and that America would be disarmed if the legal regime was altered. This may have been his greatest contribution to sea power theory and to American national security. Retiring from the Navy in 1896, Mahan became an internationally recognised defence analyst, with a suitably large income to support his comfortable lifestyle. This was possible because, as Lambert demonstrates, he lived and worked in the complex intellectual milieu of New York clubs in the 'Gilded Age' of plutocratic wealth, alongside railroad experts, economists, international lawyers, and statesmen, notably Theodore Roosevelt. His celebrity aligned with a dramatic increase in global naval construction, prompting British Prime Minister Arthur Balfour to urge that he be appointed Regius Professor of History at Cambridge, as the expert on the history and strategy of the British Empire as it faced fresh challenges. The offer was never made.

Andrew Lambert

Vice Admiral BB Schofield, CB, CBE, edited & expanded by Victoria Schofield
The Rescue Ships and the Convoys: Saving lives during the Second World War

Pen & Sword, Barnsley 2024; hardback, 201 pages, 34 B&W illustrations; price £25.00.
ISBN 978-1-03610-266-1

This new edition of a book first published in 1968 has lost none of the original's impact. It has been edited by Admiral Schofield's daughter, Victoria, who has corrected some of the errors and omissions of the 1st edition. In her Editor's Notes, she records that a fleet of only 30 small ships, averaging 1,500 gross registered tons, performed this valuable rescue work, and states that 'It truly is a record of "great deeds" in "little ships"'.

The origins of the book date from 1966 when Sir John McNee, temporary Surgeon Rear Admiral during the war, suggested to Lieutenant Commander Louis F Martyn, who had organised the Rescue Ship Service, that he contact Schofield, Director of the Trade Division between 1941 and 1943, to write their history. Martyn passed over 'a suitcase full' of the Masters' Reports while, to personalise the narrative, Schofield also contacted many former merchant seamen and naval doctors for their stories.

Schofield's account begins with a comprehensive description of the organisation and designations of the various slow and fast convoys, before discussing the origins of the Rescue Ships themselves. Losses of shipping and their valuable crews were rising by the end of 1940, and existing measures for saving survivors were proving inadequate. A dedicated Rescue Ship, equipped to pick up and care for the crews of torpedoed vessels, seemed a better option than relying on other merchant ships or escorts.

Small merchant ships were chosen for conversion. That they remained merchant ships, not naval vessels, was soon recognised to be important. Seamen on the ships in convoy not only felt safer with a Rescue Ship accompanying them, but felt that it also mattered that, if sunk, they would be picked up by 'one of their own'.

Schofield gives a full description of which ships were chosen and how they were modified. As they were not hospital ships they could be armed, and became invaluable additions to the convoy's defences. One captain described them as 'errand boy and char lady and God knows what', delivering messages, supplementing anti-aircraft defences, 'fixing' submarines with their HF/DF, and acting as plane guards for MAC ships. Peter Gretton, the famous convoy escort group leader, summed up their value, writing that 'It was a great thing ... to have a Rescue Ship with the convoy.'

Certainly, no one doubted their value. More were acquired in 1943, including five corvettes, converted while building. These, however, were never as successful as mercantile adaptations: their flared bows, overhanging sterns, cramped internal accommodation and lower endurance meant that they were just not as effective for the role in question. As long as convoys ran, rescue ships were crucial as morale boosters and supplements to the escorts. Their story was, and remains unique, and is well told in this book, which is highly recommended.

Andy Field

Przemysław Budzbon, Jan Radziemski, Marek Twardowski
Warships of the Soviet Fleets 1939–1945 Volume III: Naval Auxiliaries

Seaforth Publishing, Barnsley 2023; 320 pages, *c*300 photographs, *c*80 drawings; price £45.00.
ISBN 978-1-3990-2281-1

This book is the third and final volume of a magistral compilation of data on the warships that constituted the Soviet Navy during the Second World War (Volume I was reviewed in *Warship 2024*, 210). Its subtitle somewhat undersells its scope, as the volume also lists weapons and related equipment, ex-enemy vessels acquired at the end of the conflict, project numbers, and enemy losses attributed to Soviet war vessels.

The principal 14 sections begin with headquarters vessels and move through a wide swathe of types and support roles, before culminating in a 'miscellaneous' series of barges, launches and landing stages. Most units listed were converted merchantmen, with a leavening of naval construction, including the apparently-immortal submarine salvage ship *Kommuna* (ex-*Volkhov*), still in service a century on. Other Navy-built vessels are represented by hulks, while icebreakers represent major 'state-civilian' units, commissioned and armed during hostilities. Vessels are given such technical data as available, ranging from comprehensive to none, with summaries of career data, and ultimate fate where available.

Given the vast number of vessels involved and the obscurity of many of them, only a small proportion are illustrated, with a mix of photographs and good-quality line drawings. Other photographs accompany the two important sections that follow the main set of fourteen, which deal respectively with ex-enemy units in Soviet-controlled waters at the end of hostilities or allocated by the Inter-Allied Tripartite Commission in 1946, and with weapons and sensors in service with the Soviet Navy during the period in question.

The listing of the ex-enemy vessels is the most comprehensive ever published in English. Tabulations give Soviet and original names, dates added to the Soviet Navy list and commissioned, fleet allocation and fate. The latter notes are not necessarily complete or wholly accurate. For example, *Graf Zeppelin* and *Lützow* were sunk in explosives and target trials, not strictly speaking 'scuttled'. A good range of photographs is provided of ships in Soviet service, including some lesser-known ones.

Guns covered in the next section range from the 16in/50 weapons intended for the *Sovietskiy Soyuz* class down to 7.62mm machine guns, but exclude the UK and US guns acquired in vessels lent to the Soviet Union pending the transfer of Italian tonnage. They are followed by listings of torpedoes, torpedo tubes, depth charges, DC throwers, rockets, mines, minesweeping equipment and sensors (radar, sonar and hydrophones). A selection of photographs of key weapons and equipments is provided.

A full list of naval project numbers follows (covering 1926 to 1955), with two further lists applicable to certain civilian river vessels and to NKVD projects. Then comes the list of enemy losses attributed to Soviet naval action, including minefields laid by submarines; this amounts to no more than 54 successes in five years of war. Closing material begins with an Afterword, in which the authors highlight not only their own attempts to provide fully-balanced coverage, but also comment on the broader topic of Russian military capability in light of the attack on Ukraine in 2022. Second comes a set of addenda et corrigenda to the previous two volumes, third an appeal for corrections and additional data, and fourth a consolidated index for the full set of *Warships of the Soviet Fleets 1939–1945*.

One issue (also found in earlier volumes) is that transliterations from Cyrillic follow German/East European practice, writing 'j' for 'й', rather than 'y' or 'i' which is usual for works published in English (and is also used in Russia itself for transliterating into the Latin script). There are also a number of instances of technical terms being translated into English literally, rather than using the standard English terminology (*eg* 'sanitary', rather than 'hospital' ships), and of awkward/non-idiomatic phraseology. The series would have benefited from the attention of a native speaker familiar with the subject matter before going to press. However, this in no way takes away one's admiration for the splendid achievement represented by *Warships of the Soviet Fleets*.

Aidan Dodson

Alexander Hall
Soviet Cruisers 1917–45: From the October Revolution to World War II

Osprey Publishing, Oxford 2024; softback, 48 pages, 45 illustrations (11 in colour); price £12.99.
ISBN 978-1-47285-933-4

The cruisers of the prewar Soviet Navy comprised a small and miscellaneous group. The Bolsheviks had, in theory, inherited thirteen cruisers in 1917 but, after several years of civil war, one had been sunk, one taken abroad by the 'Whites', one returned to Turkey, and eight were in such poor condition that they had to be scrapped. There remained only the famous *Avrora*, which had fired the shot to signal the October Revolution, and the *Komintern* (ex-*Pamiat Merkuriia*). Eight new units of the *Svetlana* class had been building, but the Soviets could complete only three of them, one to a modernised design. Nothing is said here about why the non-standard 180mm calibre was selected for their main armament.

By the late 1930s, Soviet industry had expanded and the much superior *Kirov* class could be constructed, although only four were completed by 1941 followed by a further two later in the war. There should also have been the *Petropavlovsk*, purchased from Germany, but German procrastination ensured that she could never be finished. Also described is the former Imperial Yacht which after many years had been converted into the minelayer *Marti* (although the author omits her postwar renaming as *Oka*, when the French Communist whose name she had taken fell from favour).

For every class, tables are provided showing dates of construction, dimensions and armament, including alterations to the latter. As would be expected in an Osprey book, there are coloured illustrations including profiles, cutaway views and artist's impressions, with many photographs. However, the photo on page 7, captioned as *Askold*, appears to be a British cruiser of the *Cressy* class.

Having studied numerous Russian sources, the author presents accounts of the wartime careers of each ship. The Soviet Union, hard pressed on land, often had to employ the crews of large warships as ground troops and the ships themselves for coastal bombardment, anti-aircraft defence, or transport. The cruisers of the Baltic Fleet in particular were seldom at sea. His conclusion is that ultimately these ships did not have much opportunity to prove themselves in the traditional roles of cruisers.

Kenneth Fraser

Stan Fisher
Sustaining the Carrier War: The Deployment of US Naval Air Power to the Pacific

US Naval Institute Press, Annapolis 2023; hardback, 288 pages, B&W photo section, 20 tables, 7 figures; price $32.95.
ISBN 978-1-68247-847-9

This book is the latest in the US Naval Institute's thought-provoking series Studies in Naval History and Sea Power, edited by the late Christopher M Bell and James C Bradford. The author is a Commander in the US Navy and a former Navy pilot, and the work is based on a dissertation undertaken for his PhD.

On 15 June 1940 the US President Roosevelt authorised an expansion in the US naval air arm from a projected 3,000 aircraft to 4,500. Four days later this figure was increased to 15,000. This had huge implications for the training of pilots and technical aviation ratings. BuAer promptly raised its estimate of projected naval aviation personnel in 1945 from 33,110 to 120,533. The impending rise in the level of the administrative workload proved too much for the Personnel Division of BuAer, leading to a radical reorganisation of training.

Fisher gives a blow-by-blow account of the expansion of technical training in the trade schools and the many problems encountered, not least of which was finding the necessary instructors, many of whom were civilians and had no experience of classrom instruction or of naval aircraft/ordnance (and who were resented by their military counterparts because of their higher pay!). In March 1942 the primary school, NAS Jacksonville, transitioned from one shift per day to two. Even with 37,257 aviation technicians under training simultaneously – the course generally lasted four months – the Navy was still unable to meet the requirements of the fleet; the capacity of some schools was increased and a small factory training programme introduced. Overwhelmed, BuAer set up a new administrative structure reporting directly to the Secretary of the Navy, the NATTC (Naval Aviation Technical Training Command).

During the first half of the Pacific War, Carrier Task Forces (CTF) routinely returned to Pearl Harbor to refuel and to replenish stores, parts, aircraft and aircrew. However, all this changed in January 1944 in preparation for an assault on the Marshalls and a thrust into the Central Pacific, moves which coincided with the entry into service of the carriers of the *Essex* class and a new generation of naval aircraft. A variety of new organisations were established to support forward deployment, including the onboard CASD (Carrier Aircraft Service

Division), and the shore-based CASU (Carrier Aircraft Service Unit) and AROU (Aviation Repair and Overhaul Unit). The CASUs were forward-deployed and were capable of servicing and minor repairs; a typical CASU had 17 officers and 516 enlisted men. These were mobile units that could be quickly transferred to the next area of operations. The AROU was larger, but like the CASU was self-sufficient and mobile; the first deployed to the Pacific in April 1944. In addition to repairs they undertook the assembly and checking of new, crated aircraft delivered by the transport carriers. As the war neared the Japanese islands, large fleet anchorages were established and managed by floating mobile base units (Service Squadrons) with a mix of escort carriers, cargo ships, oilers and ammunition ships, and fleet tugs – for Okinawa SS-6 comprised no fewer than 73 ships.

With the delivery of large numbers of new aircraft by US industry – a total of 245 fighters was delivered in 1940, in 1945 the corresponding figure was 24,000 – there was a significant change in the Navy's approach to maintenance. With ever-larger air groups – the designed air group for *Essex* was 82, but these carriers were operating 102–03 aircraft by mid-1945 – hangar and deck space were at a premium, and the primary focus for the carrier maintenance crews was keeping the combat aircraft flying. Newly-delivered aircraft undertook a maximum of two combat tours before being retired from the front line, and heavily damaged aircraft were simply pushed over the side. Fisher accurately describes this as a 'throwaway culture'.

This is a detailed and engaging account of an under-reported subject, with some unusual insights into the hazards of maintaining aircraft on a moving, pitching ship with a crowded flight deck. It takes a chapter or two to get going, but it is certainly well worth the effort.

John Jordan

Mark Stille
Super-battleships of World War II: *Montana* class, *Lion* class, 'H' class, A-150 and *Sovetsky-Soyuz* class

Osprey Publishing, Oxford, 2022; softback, 11 illustrations in colour and 24 in B&W; price £11.99/$19.00.
ISBN 978-1-4728-4672-3 (pbk), 1-4728-4673-0 (e-book)

The ever-expanding Osprey series has now turned its attention to a group of battleships which, although none of them entered service, were remarkable for being the largest ever projected. After the event, commentators regarded the *Yamato* and her sisters as magnificent anachronisms that were already outdated by the threat of air attack. This had not been the view of naval administrators at the time: in 1939, about 30 capital ships were under construction, with a further 30 projected.

Their progress was little hindered by the restrictions of the interwar naval treaties, which by this stage had effectively lapsed. The US Navy reached the conclusion that a battleship with adequate protection against 16in guns required a displacement of 60,000 tons, and the Soviet Union commenced building ships of comparable tonnage (described in detail in *Warship 2021).* Japan, meanwhile, having kept it secret that *Yamato* mounted 18in guns, considered it likely that the Americans would find out and follow suit, so intended to adopt 20in in the next class. As the preliminary plans for these ships were destroyed in 1945, the author can deduce their appearance only by conjecture.

Much more is known about German intentions, which aimed to offset Britain's numerical superiority by ships of individually greater power. The designs up to H41 (77,000 tons) were intended to be built, whereas the larger ones were theoretical exercises characterised by the incorporation of more and more anti-torpedo protection. The British *Lion* class had more modest ambitions, at first observing the Treaty limits, but the author does not mention Norman Friedman's discovery that in 1945–46 designs exceeding 60,000 tons were considered before the cash-strapped Admiralty abandoned the super-battleship.

Kenneth Fraser

Thomas McKelvey Cleaver
The Tonkin Gulf Yacht Club: Naval Aviation in the Vietnam War

Osprey Publishing, Oxford 2021; softback, 378 pages, 24 photographs; price £15.99 (UK), $29.00 (Canada).
ISBN 978-1-4728-4594-8

A book by Rene J Francillon with an identical title was published by Conway Maritime Press in 1988. After receiving my review copy of McKelvey Cleaver's book, however, I found that it has no link to the earlier title. US Navy carrier air wings deployed off Vietnam referred to themselves unofficially as the 'Tonkin Gulf Yacht Club', and the author elected to repeat this name for his new book.

Francillon's book is not listed in the bibliography, and the new book has less specific detail and is less well illustrated. Whereas Francillon included a reproduction of the 'Tonkin Gulf Yacht Club' badge proudly worn by naval aviators who fought in the war on their flight suits, McKelvey Cleaver does not. Francillon's book had a section describing US carrier aviation in the 1960s followed by a detailed, chronological description of the naval air war. Appendices listed every carrier that operated off Vietnam together with complete lists of the squadrons that were embarked in them. Three-view drawings and technical details of aircraft and both plan and side-view drawings of the carriers added detail to the extensive selection of photographs. McKelvey Cleaver takes a different approach, with successive chapters describing the outbreak of the war, Rules of Engagement and different aspects of naval air warfare such as fighter combat and strike warfare. Operations 'Rolling Thunder' and 'Linebacker' have chapters dedicated to them, as does the creation of the 'Top Gun' school of fighter combat.

Having researched and written his book three decades

after Francillon, McKelvey Cleaver clearly had access to information released more recently than that available to his predecessor, and is able to place the failure of American political leadership into sharper perspective. This is especially so in the case of Secretary of Defense Robert S McNamara, who had no clear idea that he was placing lives at risk in a war that could not be won with the Rules of Engagement he imposed on American aircrew. Another point of difference is that McKelvey Cleaver has clearly studied the North Vietnamese Air Force, its fighter pilots and their units, and uses this knowledge effectively to add context to his description of American air activities.

McKelvey Cleaver's book is a good read that gives an interesting overview of US naval air operations over Vietnam; it is a positive addition to the historiography of the conflict and is recommended. It lacks the level of chronological and technical detail contained in the earlier book of the same name, although it can be said to complement it.

David Hobbs

Edward J Marolda
US Seventh Fleet, Vietnam 1964–75

Osprey, Oxford 2023; softback, 80 pages, illustrated in colour and B&W; price £15.99.
ISBN 978-1472-8568-14

The fourth instalment of Osprey's new 'Fleet' series combines a crisp text by a leading authority on the subject – Dr Marolda was an official historian of the war, which he experienced first-hand in the US Army – with well-chosen contemporary photographs, new maps, and excellent artwork.

The basic story will be familiar to most readers: the US Seventh Fleet supported South Vietnamese and American operations ashore, using naval gunfire, bombs, and latterly precision-guided munitions against North Vietnamese targets, and conducted an increasingly technological air campaign over Hanoi and other key locations, encountering Soviet-supplied fighters, air defence systems and missiles, along with the occasional Chinese jet.

This was all a far cry from the position the fleet had occupied between 1946 and 1949, when it had been based at Quingdao, the former German port of Tsingtao. In 1964 a North Vietnamese MTB attack on Seventh Fleet destroyers sparked American intervention in the war. A second attack, now known to be imaginary, prompted a retaliatory air strike on MTBs in ports.

While the Seventh Fleet had the firepower to achieve strategic effect that the Johnson administration, wedded to political science concepts that included 'graduated escalation', aimed for, the belief that American aims could be signalled with precision and understood by their opponents, thereby minimising the risk of Soviet or Chinese intervention, proved to be a delusion. The restricted use of naval and air force bombers in the 'Rolling Thunder' campaign signalled weakness and lack of resolve. It required a new President, Richard Nixon, to recognise reality, securing peace by deploying overwhelming force. The Seventh Fleet quickly wrapped up the conflict, air-dropped minefields closed key ports, and its support of the Air Force 'Linebacker' bombing campaigns compelled North Vietnam to make peace. The Fleet had the thankless task of evacuating Americans following the collapse of South Vietnam.

Seventh Fleet operations during the conflict hastened the introduction of new technology, from precision-guided munitions to improved operational safety on aircraft carriers following catastrophic deck/hanger fires. This is an excellent introduction to a complex conflict and the operational lessons learned. Modern carrier warfare came of age in this conflict.

Andrew Lambert

Stefan Dramiński
The Destroyer USS *Kidd*

Osprey Publishing, Oxford 2024; large format hardback, 336 pages, copiously illustrated with photographs, artwork, and line drawings in colour and B&W; price £45.00.
ISBN 978-1-4728-2741-8

This is the 30th 'Anatomy of the Ship' volume and is one of the new generation with significantly more illustrations built around the author's highly detailed, and historically correct 3D CGI models.

As with other books of the series, Section One has a brief but complete history of the vessel, together with a number of official US Navy photographs and some detailed line drawings. Section Two, some 50 pages long, consists of 'primary views' covering the overall appearance of the vessel in her main guises, with port and starboard profiles and an overhead for each stage of her career, together with perspective views. All these are in full colour, using CGI modelling. Finally, Section Three has some 260 pages of detailed artwork divided into eight separate sub-sections covering topics from hull structure to ship's boats. Again, each period is covered with a mix of line drawings and colour artwork.

The book is a dream for the ship modeller. The more casual reader may find the result a little overpowering in comparison with the original series format; however, this is a very welcome addition to anyone's reference library.

W B Davies

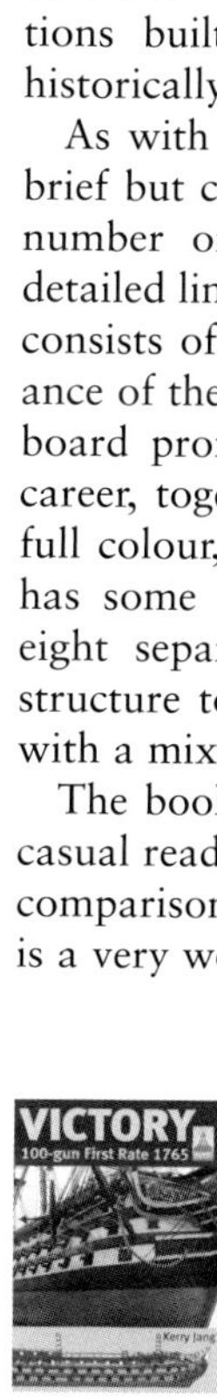

Kerry Jang
ShipCraft 29: *Victory*, 100-gun First Rate 1765

Seaforth Publishing, Barnsley 2022; softback, 64 pages, highly illustrated in colour and B&W; price £16.99.
ISBN 978-1-3990-8573-1

This book is the 29th in the 'ShipCraft' series, and follows the standard format: a brief history of the individual vessel, here broken down into design and construction and career highlights, followed by a comprehensive list of available model kits and off-the-shelf additions, pictures of completed models, colour plates of the *Victory* at appropriate times in her active life and selected references.

The first section has a detailed account of the ship's various actions under several admirals both before and after Nelson. It highlights the authority each admiral had over details of design and armament, particularly the number and type of gun embarked. The coloured profiles for the different periods described in this first section occupy the final section of the book. To be slightly critical, it is not known when her 'raised bulwarks' were fitted; otherwise the profiles are excellent.

The second and third sections are aimed directly at the ship modeller. The list of available model kits ranges from small (1/1200 scale) white metal kits for wargamers through to the largest plastic kit (1/100), then on to the various wooden kits, and finally the many accessory or 'aftermarket' sets that provide the small details not usually available in an off-the-shelf kit. The third section includes images of three completed models built to different scales, each ascribed to a named modeller. Two of these are scratch-built by Japanese modellers from a Tokyo modellers' group whilst the third illustrates what can be done with the Heller 1/100 scale plastic kit. All demonstrate an impressive level of skill on the part of the modellers in question.

The fourth section focuses on *Victory*'s appearance at the various stages of her career, including a note on armament changes and a collection of line drawings of the vessel at various stages in her life, bringing the story almost to the present day.

As with other books in this series, while the book is aimed squarely at the modeller, there is enough history to encourage the reader to delve deeper, and it is recommended.

W B Davies

Mark Stille
New Vanguard 322: US Navy Gun Destroyers 1945–88: *Fletcher* class to *Forrest Sherman* class

Osprey Publishing, Oxford 2023; softback, 48 pages, heavily illustrated in colour and B&W; price £12.99.
ISBN 978-1472855121

US Navy Gun Destroyers 1945–88 forms part of Osprey Publishing's 'New Vanguard' series. The publisher describes these books as being focused on the design, development, operation, and history of the machinery of warfare throughout the ages. Written by long-established Osprey author and former US Navy intelligence officer Mark Stille, this recent release broadly follows the pattern established by the series to date, squeezing an impressive amount of text and illustrative material into a compact 18cm x 24cm, 48-page format. A key feature of the series is its extensive use of specially commissioned colour artwork. This volume is no exception, incorporating a range of imagery that includes class profiles and a detailed cutaway drawing produced by digital illustrator Adam Tooby to supplement a wide range of black and white photographs.

As its title suggests, the book is focused on the US Navy's gun-armed destroyers of the Cold War era. This includes both the numerous vessels that were retained after the conclusion of the Second World War and the much smaller number of ships that were completed during the 1950s before the transition to the missile age took place. It encompasses the wartime *Fletcher* (DD-445), *Sumner* (DD-692) and *Gearing* (DD-710) classes, as well as the post-war *Mitscher* (DL-2) and *Forrest Sherman* (DD-931) types, together with the unique *Norfolk* (CLK-1/DL-1).

After a short introduction, Stille summarises these various designs against the backdrop of new fleet requirements that emphasised anti-air and anti-submarine warfare over the destroyer's traditional anti-surface role. This section includes helpful box-outs describing USN destroyer nomenclature and the Fleet Rehabilitation and Modernization (FRAM) programme that saw many of the war-built destroyers upgraded to meet these revised requirements. This is followed first by an overview of the weapons and sensors available at the time, then by an inevitably brief summary of service in the US Navy. These sections pave the way for more detailed descriptions of each of the various classes, which occupy approximately half the book. Tables providing key characteristics and ship lists are included. A short conclusion, together with a bibliography and index, bring up the rear.

There is clearly only so much information and analysis that can be accommodated within a 48-page book. Despite this constraint, Mark Stille provides an effective introduction to understanding the role and characteristics of a series of warships that played a major part in US Navy operations on the eve of a time of rapid technological transition. This interesting and informative overview left this reviewer keen to learn more.

Conrad Waters

Aidan Dodson
The Windfall Battleships: *Agincourt*, *Canada*, *Erin*, *Eagle* and the Balkan & Latin-American Arms Races

Seaforth Publishing, 2023; hardback, 288 pages, 200 B&W drawings and photographs; price £25.00.
ISBN 978-1-3990-6322-7

In August 1914 the battleships for export named in the title were appropriated by the Royal Navy. Aidan Dodson, finding that the published information on these four vessels was sketchy and sometimes inaccurate, has now given them comprehensive coverage while also placing them in the contexts of both the Latin-American and Balkan naval races and of the export of all warship types. This review focuses mainly on the dreadnoughts ordered from British yards.

The Latin-American dreadnought race began with Brazil ordering three battleships from Armstrongs. Their initial armament of twelve 10in guns was soon increased to twelve 12in, in two pairs of superfiring twin turrets fore and aft and two twin wing turrets without provision for cross-deck firing. Two ships were laid down at Armstrong and Vickers in April 1907. When the first, *Minas Geraes*, was completed in January 1910, she had, albeit briefly, the most powerful broadside worldwide, thanks to the Armstrong designers' early adoption of superfiring. The sighting hoods were quite different from those in contemporary British ships, but we are not told whether blast was excluded so that the upper turrets could fire when trained fore or aft.

In 1910, two Argentine battleships with 12in guns and similar turret layouts, but with cross-deck firing, were laid down in American yards. In response, the Brazilians wanted a more powerful broadside for their third battleship, and Armstrong duly submitted designs with guns up to 16in. However, the Brazilian naval minister had been persuaded by the Kaiser himself that 12in was sufficient. Armstrong devised a unique layout with seven centre-line turrets, and *Rio de Janeiro* was laid down in September 1911. Chile's response was to order two super dreadnoughts from Armstrong with ten 14in guns, their broadsides equalled only by the latest American battleships. *Almirante Latorre* was laid down in May 1912 and launched in November 1913 but the second, *Almirante Cochrane*, was not laid down until February 1913.

At this point Dodson switches to recounting the complex course of the Balkan naval race. In 1910, Armstrong and Vickers proposed battleships for Turkey with guns up to 14in, the final Armstrong design, the *Refladiyeh* with ten 13.5in guns, being laid down in August 1911. It compared well with the British *Iron Duke* class, first laid down in January 1912. *Refladiyeh*'s completion was expected in August 1914. By the time of the launch of *Rio de Janeiro* in January 1913, the Brazilians recognised that her guns were too small and, with an ailing economy, they decided to sell. Turkey outbid Greece, and the ship was renamed *Sultan Osman I*; she too was ready in August 1914.

Both Turkish ships were then commandeered by the Royal Navy and renamed *Agincourt* and *Erin*; they joined the Grand Fleet in September. Chile agreed that, when completed, its ships would serve in the Royal Navy but would be delivered once the war was over. *Almirante Latorre* was commissioned as *Canada* in August 1915; there had been time to equip her with a director and the first of the latest model of Dreyer Table, the Mark IV*. However, the hull of *Almirante Cochrane* remained on the stocks and was adapted for the pioneering aircraft carrier *Eagle*, completed after the war.

The wartime service of the three windfall battleships is fully described, including their participation in the Battle of Jutland. Dodson declares that *Agincourt* fired spectacular broadsides, but this is not mentioned in his cited source, Campbell's *Analysis*. *Agincourt* then had neither a director nor a fire control table, so was particularly dependent on spotting. Given that she could fire seven-gun salvos, these are more likely than broadsides. In 1920, following a brief refit, *Almirante Latorre* was returned to Chile, but we are not told whether she retained her almost up-to-date British fire control installation.

Dodson concludes that the battleships provide 'an interesting set of lenses for the study of naval history in the earlier twentieth century', and has written a thorough survey of a previously overlooked subject.

John Brooks

William Smith
Allied Convoys to North Russia 1941–1945: Policy, Strategy and Tactics

Pen & Sword Maritime, Barnsley 2023; hardback, 270 pages, 49 B&W photographs; price £25.00.
ISBN 978-1-39905-473-7

There have been many books which have dealt with the Arctic Convoys in the form of campaign histories and personal recollections, as well as detailed examinations of individual actions such as the PQ17 disaster, the Battle of the Barents Sea, and the attempts to sink the battleship *Tirpitz*. William Smith here offers an alternative approach to a much discussed topic. The sub-title, 'Policy, Strategy and Tactics', indicates that the book does not engage in the minutiae of events involving each convoy or associated action, instead assuming prior knowledge and providing just the briefest of synopses to orientate the reader. Rather, he focuses principally on the strategic narrative, using the relevant discussions, minutes, reports drawn from both British and German archival sources as well as the communiqués and private messages passed between the Americans, British and Russian leaders in the course of this four-year campaign.

Smith is refreshingly impartial with respect to each of Churchill, Stalin and Roosevelt. The usual cosy image of Winston, FDR and 'Uncle Joe' is dispensed with, as when Churchill tells Roosevelt politely but firmly, 'I beg you not to press us beyond our judgment in this operation'. Stalin comes across alternately as tetchy and demanding but at other times is praiseworthy and grateful for his allies' persistence in running these perilous convoys.

Nevertheless tensions ran high, especially in late 1942 after the PQ17 debacle and the crippling losses from U-boat and aircraft attacks. Other priorities intervened, such as the upcoming 'Torch' landings, which contributed to the prevarication on the part of the Western allies about resuming the convoys. Stalin's suspicions concerning British and American commitment came to the fore and there were some acerbic exchanges. This problem was exacerbated by a long-running dispute about the treatment of Western servicemen isolated in the inhospitable north Russian seaports, unable to return home because their hosts, amongst other things, were dragging their feet on the issuing of visas.

However, fortunes shifted during the winter/spring of 1943–44 due to several factors including the Battle of North Cape, the transfer of much of the *Luftwaffe* to the Mediterranean, the introduction of the escort carrier and the disabling of the *Tirpitz*. All these issues are thoroughly dealt with, enabling the reader to gain a comprehensive grasp of the overall strategic picture and to gauge the mood of the leaders involved, who shouldered the enormous burden of responsibility.

This book has a flaw, and an important one. The expected norm in a work of this kind is to use a balance of quotation and authorial analysis. In *Arctic Convoys*, Smith extensively employs verbatim quotations, sometimes stretching to over two pages of text, quite unnecessarily including opening and closing salutations, or perhaps copies of 'draft' memoranda, the gist of which is repeated later in a different format. As the book is comprehensively referenced and could easily be used as a basis for further academic study, there is simply no need for every word of these high-level exchanges to be printed. In other places, the author demonstrates his ability as a skilled analyst, so the resulting imbalance is regrettable and could have been avoided with judicious editing.

Despite this flaw, *Arctic Convoys* is a valuable addition to the literature relating to this campaign. Most importantly, by concentrating on the discussions at the highest political and service levels it goes beyond the normal 'what happened' narrative by addressing the vital 'why it happened' issues.

Jon Wise

Steven J Zaloga
Allied Warships vs the Atlantic Wall, Normandy 1944

Osprey Publishing, Oxford 2023; softback, 80 pages, illustrated with maps, photographs and paintings in colour and B&W; price £15.99.
ISBN 978-1-47285-415-5

The title suggests a book about the bombardment of the beaches on D-Day, an often overlooked or unduly simplified subject. In fact, rather than looking at warships (plural) at the Normandy landings, the book is primarily a case study of just one ship, USS *Texas*, in action against Batterie Hamburg at Cherbourg on 25 June 1944.

It starts well with a 27-page overview of the development of German defences and the Allied bombardment strategy, but the warship bombardment on D-Day itself warrants only a page and a half. The next chapter of 21 pages details the two combatants. The description of *Texas* and the evolution of her main armament and aircraft spotting is good, and there is a detailed history of Batterie Hamburg's design and growth.

After all this build-up, the combat section is quite short, with only 20 pages dedicated to the background to the attack on Cherbourg, the bombardment itself and the analysis (and this includes full-page maps and numerous illustrations). *Texas* is naturally the focus of the eleven pages detailing the battle, with the narrative primarily drawn from her after-action report. But this means that there is barely any reference to the other Allied warships involved, nor to the German batteries facing them.

Like all Osprey books this one is well illustrated with historic photos, maps, and illustrations of ships, bunkers and shells. It is an easy read and Zaloga knows his subject, so there is much about the wider Normandy situation here. There are a few easily spotted errors through the book: there are inconsistencies and conflicts in the author's account, and the unforgivable 'the HMS' occurs at least once. It is also a shame that the title does not accurately reflect the book's actual content.

Stephen Fisher

Les Brown
ShipCraft 33: British Light Cruisers: Town, Colony and Later Classes

Seaforth Publishing, Barnsley 2024; softback, 64 pages, heavily illustrated in colour and B&W; price £16.99.
ISBN 978-1-3990-3753-2

The latest in Seaforth Publishing's well-established ShipCraft series, this book is effectively a sequel to the same author's previous volume, on the *Leander-*, *Amphion-* and *Arethusa*-class cruisers, reviewed in last year's *Warship*. In common with most other volumes in the series, it adopts a standard 64-page, softback format and has a pleasing layout incorporating extensive use of both colour and black and white imagery. The primary intention is to provide warship modellers with the information needed to produce accurate representations of the ships covered in model form.

In general alignment with the series' approach, the book commences with chapters providing a design history of the 'Town' and 'Colony' classes and a description of wartime modifications. These are followed by a comprehensive, eleven-page review of the various model kits and accessories available. A further 17 pages are devoted to the series' traditional *pièce de résistance*: the 'Modelmakers' Showcase', illustrating a range of completed models in full colour. This is supplemented by a further four pages of colour camouflage profiles illustrating nine specific ships and a chapter describing the various paint schemes carried. Concluding sections encompass a review of post-war service (including the redesigned *Tiger* class), a selection of additional line drawings, and a short bibliography to support further research.

While providing valuable reference material at an attractive price, the 'ShipCraft' series suffers from a number of potential weaknesses. While 64 pages might be adequate to provide an overview of a distinct class or a small number of ships, this becomes a significant constraint when describing a total of 27 major warships built to several, distinct iterations. This limitation manifests itself in the complete omission of the review of operational wartime service contained in the previous volume. Equally, although the author has done a good job condensing a complex design history into a very few pages, many details are inevitably omitted. It would also have been beneficial if he had been given more space to utilise his clearly encyclopaedic knowledge of relevant modelling products to expand his invaluable reviews of the major kits available.

A further problem is the not-insignificant number of errors in a book targeted at helping model makers to build accurate representations. These are particularly apparent in the camouflage section, possibly reflecting limitations in some of the sources used. For example, MS4a came into use, alongside the other Royal Navy Standard Camouflage Colours, only in 1941, and was not employed for 'vessels which served with the Home Fleet pre-war'. This fundamental error sets the scene for numerous other mistakes and misunderstandings in the rest of the chapter.

The 'ShipCraft' books have become a well-used and popular series since the first editions were published under the then-Chatham Publishing imprint back in 2004. However, there are indications of a decline in quality in some recent volumes, and a thorough review of both overall design constraints and editorial standards is probably overdue if its well-earned reputation is to be maintained.

Conrad Waters

Angus Konstam
Cape Matapan 1941: Cunningham's Mediterranean Triumph

Osprey Publishing, Oxford 2023; softback, 96 pages, three double-page colour illustrations, many B&W photographs, colour maps and diagrams, bibliography, index; price £16.99.
ISBN 978-1-4728-5723-1

Enrico Cernuschi
Italian Battle Fleet 1940–43: 'La Squadra', The Pride of the Regia Marina

Osprey Publishing, Oxford 2024; softback, 80 pages, three double-page paintings, colour maps and diagrams, many B&W photographs, bibliography and index; price £15.99.
ISBN 978-1-4728-6059-0

Angus Konstam's book on the Battle of Cape Matapan forms part of a burgeoning number of other titles in Osprey's Campaign series in which key historical battles and campaigns are described in brief, lavishly illustrated and accessible accounts. *Cape Matapan* is attractively presented with easy-to-follow colour diagrams of key stages in what turned out to be four days of manoeuvring by the two navies to the south of Greece. Angus Konstam explains the operations of both fleets in meticulous detail.

Two key factors are emphasised from the start: first, the inhibiting influence on the Italian naval commanders of Mussolini's reluctance to allow his prized *Regia Marina* to get involved in a surface action with the British Fleet; second, the determination on the part of the celebrated Admiral Cunningham to catch the fleeing Italian battleship *Vittorio Veneto* and to bring her to action with his ageing capital ships.

The bibliography betrays the fact that the author has relied almost exclusively on existing British and American sources. Perhaps inevitably, this serves to distort the narrative. The hyperbolic strapline in the title, 'Cunningham's Mediterranean Triumph', suggests a carefully orchestrated tactical victory by the Admiral, whereas an impartial observer might also discern that chance and good fortune played a significant role. The faraway events of 1941 surely now deserve a more nuanced and less partial approach if a reader new to the subject is to acquire a well-rounded understanding of the naval history of the Second World War.

Enrico Cernuschi answers some of the above criticisms. The author claims that most British histories of the Mediterranean naval war rely heavily on misleading second-hand information based on outdated wartime propaganda rather than the British and Italian primary source information now available.

Italian Battle Fleet 1940–43 is part of Osprey's Fleet series and is similar in length and the quality of illustration to Konstam's *Cape Matapan*. In this book, Cernuschi provides a concise yet informative description of the Italian battle fleet's underlying *raison d'être*, of the ships that formed 'La Squadra' during the Second World War, and of its operational doctrine.

He does not shy away from detailing the shortcomings of the *Regia Marina*: the variable quality of its leadership, the limitations of the older, reconstructed battleships and the absence of carriers. On the other hand, Cernuschi stresses the quality of Italian warship design, its machinery, gunnery and its armoured protection, all of which, he claims, put the RM at least on a par with its opponents.

The second half of his 80-page narrative deals with all the naval engagements between the British and Italians during the three-year period 1940–43. While some actions might be classed as 'battles', others were fairly fleeting gunnery duels, some with debatable outcomes.

The author's principal aim is, with the support of more recently available evidence, to highlight the questionable nature of the accepted British version of events. One such example is the action off Calabria in July 1940, when the damage inflicted on Admiral Cunningham's flagship HMS *Warspite* was much more severe than has been recorded in English-language accounts.

Cernuschi is to be applauded for providing this revisionist history. However, he is not writing in his first language, and some assiduous editing would have resulted in a more cohesive, chronological narrative with fewer *non sequiturs*; these inhibit the reader's ability to gain a comprehensive grasp of this campaign.

Jon Wise

WARSHIP GALLERY

Herbert Charles Sparrow, Royal Marines, 1927–1930

The photograph albums of private individuals can be a rich source of interesting and unusual images. In the case of *Warship* these will very likely be those who served in their country's naval forces in some capacity or other. We published pictures from one such collection in 2009, and this year we feature another, comprising pictures taken by Royal Marine Herbert Sparrow during his time serving in China during 1927 and, subsequently, in the eastern Mediterranean. Introduced and captioned by the Assistant Editor, these provide a fascinating glimpse into service on two overseas stations at the beginning of a time of great upheaval and often violent change in both regions.

Herbert Charles Sparrow was born on 6 May 1900 in Trowbridge, Wiltshire, the son of Henry Sparrow, a factory worker. The day after his 18th birthday Herbert and a school friend went to Southampton to volunteer for service in the Royal Marine Artillery. While the friend was turned down (probably on health grounds), Herbert himself was accepted as a private, subsequently embarking on some seven months of initial training. In December, with hostilities over, he became a Gunner, 2nd class; his first ship, from October 1919 to January 1922, was the

Herbert Sparrow, RM, Shanghai Defence Force, February–October 1927, smart in his uniform, next to a sandbagged observation post overlooking British destroyers (that on the left probably being *Keppel*, leader of the 3rd Flotilla), a cargo liner and an unidentified smaller vessel moored out in the Whangpoo River.

Another European colonial power with major regional interests was the Netherlands, and an indication of the importance the Dutch attached to the area is the presence of one of their new cruisers, either *Java* or *Sumatra*, in this photo.

Sparrow and other signallers, with the tools of their trade, again depicted in an elevated position. Many of Sparrow's photographs were taken from such locations, which, as a signaller, were presumably where he spent much of his time on duty.

Keppel again, moored out in the river, with two more destroyers beyond her. The vessel to the right is the Italian cruiser *Libia*, originally ordered by Turkey but taken over while under construction during the Italo-Turkish War (1911–13). Fitted out for overseas service in 1924, she spent much of her career as Italy's Far East flagship.

While it has not been possible to identify definitively any of the vessels shown in these two photos, the two four-funnelled ships to the right of centre are probably the Japanese cruisers *Yahagi* and *Hirato*, both of which had been dispatched to Chinese waters at the time (see pages 108–109). Many of the ships present are dressed to celebrate a special occasion, though it is not known what this was.

Among the mass of local craft, visible to the extreme left is the Italian cruiser *Libia*, plus two large merchantmen and, on the right, what appears to be either *Yahagi* or *Hirato*.

The small unidentified vessel that is partly visible behind Sparrow in the first photograph, possibly a customs cruiser.

In one of the RAF's first major overseas deployments, six Fairey IIIDs of 441 Flight operated from a temporary base set up on Kiangnan Racecourse, Shanghai, where they were twice put ashore from *Argus* before eventually rejoining the ship in March 1928 for the voyage back to the UK. The dark stripe on the fuselage of each aircraft was a recognition feature.

The aircraft carrier *Argus*, surrounded by local small craft. In contrast to other RN ships depicted, she sports a dark grey colour scheme, as she was actually part of the Atlantic Fleet, from where she was temporarily detached in February 1927 for service in the Far East, after aircraft from *Hermes* and *Vindictive* had proven their worth in the increasingly complex operating environment of the time. *Argus* embarked Fairey IIIDs of Flight 441 at Malta on her way out to the Far East, where she remained until the following March. Note the two 4in guns mounted at the extreme stern, and the pair of large cranes for recovering seaplanes from the water – the Fairey III could be converted from wheel- to float-based configuration in about two hours.

battleship *Barham*, flagship of the 2nd Battle Squadron, Atlantic Fleet. Herbert's records show that his character was consistently rated as 'VG' and his ability generally as 'Sat'[-isfactory] throughout his service. After *Barham*, and with the Royal Marine Artillery and Royal Marine Light Infantry being amalgamated into a single force – the Royal Marines – he spent the next few years mostly in the Portsmouth Division before departing in January 1927 for China as part of the 12th Royal Marine Battalion. Now a signalling specialist, Herbert was one of 1,000 Marines who comprised the specially-raised force. While the official line was that they had been specially picked, the reality was that the force was made up of whoever was available. While the primary purpose of the Marines' deployment was in one of their traditional roles, to support the land defence of naval bases and associated establishments – in this case that related to the British presence in Shanghai – on a wider scale they were part of an unprecedented increase in British military and naval presence as international relations in the area became increasingly strained.

China was becoming a more and more troubled and divided country at that time, torn between different factions and political entities, with the nationalist leader

A Japanese destroyer, probably *Urakaze*, with two sampans alongside, an unidentified vessel astern and then, beyond, *Argus* (see above). The destroyer is not only in an unusually pale colour scheme for a Japanese naval vessel, but has also had her identification number painted out.

A Japanese hospital ship with, ahead, what appears to be either *Yahagi* or *Hirato* and a destroyer.

Japanese sailors enjoying some shore-based recreation.

The Japanese cruiser *Sendai* (distinguishable from her sisters by the shape of her bow), part of the 1st Dispatched Squadron, sent to China in March along with *Isuzu* to reinforce Japan's influence as local tensions rose. A fast, modern vessel, like the British *Hawkins* she represented a very different type of presence to the big but dated *Yakumo* (see overleaf). She remained in the area until September.

Sun Yat Sen only recently deceased, the Communists on the rise, and local warlords and brigands threatening one another and the increasingly resented foreign presence. Ferocious attacks on members of the latter in a number of cities prompted the strong naval presence from not only Britain but also the USA, France and Japan, as well as lesser powers still with an imperial or trade-based interest in the region, such as the Netherlands and Spain. Shanghai, as a major economic and trading centre, was responsible for around half China's overseas trade and a similar proportion of its manufacturing industry. The city was home to international 'settlements' enforced on the population by the four above-mentioned major powers, and was a regular port of call for their warships, mainly for 'showing the flag', though also for recreational purposes. Major vessels generally moored in The Bund, a waterfront and protected historical area of the city. While servicemen of different nationalities would generally mix happily with one another (albeit often in various states of intoxication), relations with the local inhabitants were mixed, and increasingly so as the wider situation worsened. In addition, by 1927 relations between the various foreign powers, though officially still cordial, were also becoming more strained, prompting the deployment of the Marines, as well as assorted ships from the Atlantic and Mediterranean Fleets, to help strengthen Britain's position vis-à-vis countries who, while nominally current (or at least recent) allies, were also either historical or more contemporary rivals – a cause of growing concern in London.

The period that Herbert spent in the city included the infamous 'Shanghai Massacre' of April 1927, when Chiang Kai-shek's Nationalists and gangs of local thugs murdered some 12,000 Communists and suspected sympathisers, setting off a horrific chain of similar killings across the country and sparking what would become more than 20 years of civil war. It is not known whether Herbert witnessed any of these atrocities (and if he did he never mentioned it to his family), but given the scale of what happened he can hardly have been unaware of it. However, the pictures he took with a newly-acquired camera (and, judging from his later photo-

The British flagship *Hawkins* illuminated at night.

It wasn't always just overseas warships that were present. This is one of China's three small German-built cruisers, either *Hai Yung*, *Hai Chou* or *Hai Chen*. All three were eventually scuttled on the Yangtze when Japan launched its full-scale invasion in 1937, the wrecks being broken up for scrap in the late 1950s and early 1960s.

Partially obscured by warehouses, the cruiser *Dragon*, like *Argus* a temporary detachment, this time from the Mediterranean Fleet. Of the two merchantmen midstream, the one on the right is one of the Blue Funnel Line quartet of *Patroclus*, *Hector*, *Antenor* or *Sarpedon*.

A British sentry on duty in the port area.

A series of large cruisers lined up in 'Man o'war row' on the Whangpoo river. From left to right: *Yakumo* (IJN, at this time serving as a training ship), USS *Pittsburgh* (ex-*Pennsylvania*), *Jules Michelet* (France) and HMS *Hawkins* (flagship, RN China Station). The first three had all been powerful units when first completed around the turn of the century, but were now obsolescent; in contrast the British vessel was only a few years old, and indicative of a rather different approach to the issue of 'presence' in an increasingly strife-torn region, though the Americans, it seems, regarded her with some puzzlement. The smaller crew of the more modern ship compared to her foreign neighbours meant less available manpower for the numerous potential missions that might need to be carried out at short notice, hence the need for other forces to be available, such as the Marines.

As well as cruisers, the various overseas powers also sent destroyers and other smaller warships to the region. This is the American 'flush-decker' *MacLeish* (DD-220).

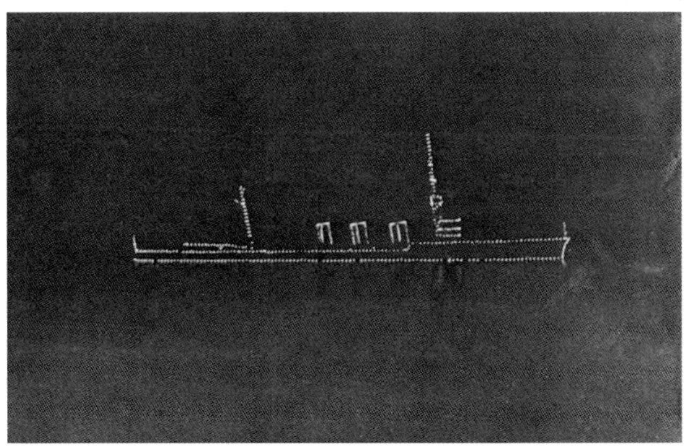

Another new cruiser belonging to a European navy, the Spanish *Blas de Lezo*, depicted during the day and lit up at night. Completed in 1924, she had a short career, especially by the standards of many Spanish warships of the time, being wrecked in 1932.

Between February 1928 and August 1930 Sparrow was aboard the newly-commissioned aircraft carrier *Courageous*, serving in the Mediterranean Fleet, during which time he witnessed the dramatic incident depicted here. The carrier, having spent the first eighteen months of her commission engaged in the usual round of cruises, manoeuvres and port visits, embarked the 2nd Battalion, South Staffordshire Regiment in August 1929, bound for Jaffa (Palestine) for peace-keeping operations after violence between the local Jewish and Arab communities. (Some of her aircraft briefly operated from Gaza for the same purpose.) The following month, while off Jaffa, Fairey III F S1220 (Pilot Officer PJH Halahan, RAF) of 445 Fleet Spotter Reconnaissance Flight ran off the port side of the flight deck into the deck-edge palisades. The latter performed exactly as intended and, despite the hair-raising angle at which S1220 came to rest, both aircraft and crew were successfully recovered onto the deck of the carrier.

graphs, possibly a second) record both his service and the wider life of the still vibrant, cosmopolitan Shanghai of the time. They depict events such as concerts in the Bund Gardens, sporting activities, local people of all ages going about their daily lives, the cityscape and surrounding countryside, and the Marines and their foreign equivalents engaged in their duties. He also took many photographs of the warships and merchant vessels that were a constant presence on the Whangpoo River, and it is these that form the basis of this year's Gallery. While the quality of the images is very variable, their value as an historical record is undeniable.

The Marines did not stay long: manpower problems meant they were soon returning to Britain, in Herbert's case being home in October 1927. After a short spell back in Portsmouth he found himself departing overseas again, this time as part of the RM contingent on board the newly recommissioned aircraft carrier *Courageous* in the Mediterranean. Again he took his camera, over the following two-and-a-half years recording visits to Athens, Egypt and the south of France, but also the RN in the Suez Canal area and its engagement in peace-

Photographed from below decks, the aircraft carrier *Eagle*, which *Courageous* joined in the Mediterranean Fleet during 1929–30. With two carriers operating together, operational concepts still in their relative infancy, such as combined air operations and night flying, could be worked on.

Two photos showing various ships, both naval and merchant, at a crowded Port Said. Beyond the bustle of local craft can be seen the aircraft carrier *Hermes* (presumably on her way to or from the Far East) and the battleship *Royal Oak*, part of the Mediterranean Fleet, as well as a number of cargo liners and other vessels.

Coaling ship, Port Said.

Port Said: the Suez Canal Company building with, in the background, a British 'V&W'-class destroyer of the Second Flotilla.

keeping operations off the coast of another troubled land, the British protectorate of Palestine.

His subsequent service was divided between further periods in Portsmouth, where he had married and settled, and a further Mediterranean posting, this time to Malta, before leaving the Marines at the start of 1939 and becoming an electricity inspector. Recalled to service in September of that year, he does not seem to have had a particularly action-packed war – his age and signalling specialism may be behind this – other than a brief posting to Ramsgate to man coastal defence guns at the height of the invasion threat in 1940. Discharged in September 1945, Herbert had by now returned to Wiltshire, settling in Chippenham with his young family and working as a travelling salesman, and later as a gateman at the Westinghouse Brake and Signal Co. Sadly he was stricken by cancer and died aged just 66. His photograph album from China and the Mediterranean remains a much-treasured family possession.

Perhaps finishing off a film after the end of his travels, Herbert took this atmospheric shot of the Forth Bridge on a gloomy day, with *Argus* (like him, home again) and what appear to be a battleship and a cruiser dimly visible in the murk beyond the bridge and the island of Inchgarvie.

Acknowledgement:

All photographs are reproduced by permission of Penny Ives, Herbert Sparrow's grand-daughter and current custodian of his photograph album. Introduction and captions are by Assistant Editor Stephen Dent, using information from Herbert Sparrow's service record, ADM-159-92-216357, family notes, and various published and online secondary sources. Additional thanks go to John Jordan, David Hobbs, Aidan Dodson, Matthew Heaslip and Ian Sturton for help with background information, and to Simon Castens of Wild Swan Books (www.wildswanbooks.co.uk) for bringing the photograph album to *Warship*'s attention in the first place.

Warship would be pleased to hear from any readers with access to similar collections of images which could form the basis of a future Warship Gallery.